The Institution of
Engineering and Technology

Guide to
Earthing and Bonding for AC Electrified Railways

Publication information

Published by The Institution of Engineering and Technology, London, United Kingdom

The Institution of Engineering and Technology is registered as a Charity in England & Wales (no. 211014) and Scotland (no. SC038698).

© The Institution of Engineering and Technology 2022

First published 2022

The Institution of Engineering and Technology
Futures Place
Kings Way, Stevenage
SG1 2UA, UK

Copies of this publication may be obtained from:
The Institution of Engineering and Technology
PO Box 96, Stevenage, SG1 2SD, UK
Tel: +44 (0)1438 767328
Email: sales@theiet.org
www.electrical.theiet.org/books

ISBN 978-1-83953-645-8 (paperback)
ISBN 978-1-83953-647-2 (electronic)

Typeset in India by MPS Limited
Printed in the UK by Hobbs the Printers Ltd, Brunel Road, Totton, Hampshire, SO40 3WX

Contents

Contents

Contents

Acknowledgements

The IET gratefully acknowledges the advice and assistance provided by the following people and organizations in the development of this Guide.

Technical Authors:
Dr R. White BSc FIEE CEng Cert Ed
A. McDonald MSc MIET CEng

Special thanks to:
P. Dearman
J. Morris
R. Catlow
G. Keenor
R.J. Calder
M. Anderson
G. Brindle
R. Wilson
Dr S. Hanrott
A. Gavrilakis
E. White
P. McDonald

Additional Contributors:
Terradat UK Ltd

Section 1

Introduction

The distribution and use of electricity within modern societies is accepted as an essential requirement, although it is sometimes not appreciated that the dependency on electricity has developed over approximately 150 years. Pioneers such as Thomas Edison, Charles Parsons, Sebastian Ferranti and Nikola Tesla, and companies such as Siemens, Westinghouse, Brush, and Crompton drove the widespread adoption of electricity for lighting and power and, for which, railways and tramway systems were early adopters. Although DC electric systems were the first to be developed at the end of the nineteenth century, AC electric traction systems soon followed.

The first applications of AC electrification were constrained by the technology available at the time and, in particular, by the use of DC series commutator traction motors used on railway vehicles. It was understood that by increasing the electrical frequency the lag between the field flux and the armature current increased due to eddy currents in the iron. These early twentieth-century systems in Switzerland and Germany were therefore developed using a frequency of $16^2/_3$ Hz at 15 kV, and they remain in use today. In the United Kingdom, AC electrification at 25 Hz and 6.7 kV was provided on a number of south London lines from 1909, although these were later replaced by a DC electric traction system following the railway companies amalgamation that formed the Southern Railway. From the mid-twentieth century, the limitation imposed by DC traction motor and control technology disappeared through the development of mechanically robust and compact mercury arc rectifiers, and then semiconductor-based technology that enabled AC to DC rectification to be installed on trains. This enabled the AC electric traction system voltage to be increased and standardized at 25 kV. More significantly, the frequency could be raised to 50 or 60 Hz, which is the same as used in most national power distribution networks, therefore enabling traction supplies to be taken directly from these networks. Subsequently, in the 1990s the technology for three-phase traction drives and the use of AC traction motors became common.

It was understood from the earliest experiments on electrical phenomena that electricity had the ability to kill, shock, burn or injure humans. But while these experiments were confined to laboratories, those involved appreciated the severity of the electrical hazards. Promoters of the widespread use of electricity understood that electrical systems, equipment, and products within that system, or those powered from an electrical source of energy, had to be designed, constructed, maintained and operated to protect the public, who had no understanding of electricity, from electric shock. Electric railways at 25 kV AC use the rails as the traction return circuit and have to position the live conductors of the overhead catenary system to suit the railway gauge. The rails are not insulated and can be touched by railway staff and by the public at accessible locations, such as road crossings, maintenance depots and stations. Live overhead catenary system conductors have to be placed clear of persons standing on platforms and the lineside, but there will be bridges that cross the railway that potentially allow persons to come in close contact with these live conductors. Earth faults on the overhead catenary can result in a large current that creates potentially hazardous voltages flowing in the lineside infrastructure before making its way back to the source via the rails. The public and railway maintenance staff therefore need to be protected from electric shock under normal conditions and under fault conditions by protective measures that include clearance, insulation, barriers and obstacles to prevent direct contact and by the earthing and bonding arrangements of the electric traction system.

The complexity of integrating the earthing of a 25 kV electrified railway with various electrical distribution systems and exposed-conductive-parts, means that it is impossible to prescribe one earthing and bonding design solution that addresses the needs of every railway. Many public standards, codes of practice and company documents exist to guide the design of the earthing and bonding of AC electric traction systems, but in practice there remains gaps, overlaps and conflicts between them. In many instances the underpinning rationale in these is not defined or is opaque. It is not uncommon for earthing and bonding designs to follow a discipline-specific scope, rather than being holistic in their approach, to produce designs and installations appropriate to an individual railway even though the type of electric traction system is the same.

Section 1 – Introduction

This Guide aims to assist students and electric traction system practitioners to understand the earthing (grounding) and bonding of AC electric traction systems by describing the basic concepts around the mass of Earth in the traction return circuit, and the nature of electric shock, the derivation of safety limits and protective measures. Typical protective measures that are applied to the electric traction system and lineside infrastructure are explained as well as the often-neglected consideration of lightning protection for the alignment of the railway. A section on the safety of staff during maintenance, renewal or decommissioning is included as often this is only considered towards the end of an electrification project. During these activities, many of the designed protective provisions provided for an intact electric traction system may be removed or degraded to allow the works to take place. An understanding of how the electric traction system is to be maintained safely in degraded states is essential when new systems or extensions to existing systems are planned and implemented. Much of the content used in this Guide can be found in various international standards and railway codes of practice, and it is not intended that this Guide will replace these. Rather it is hoped that this Guide provides an overarching framework that places these standards in context and assists in the overall understanding of the earthing and bonding of AC electric railways.

 **Section 2**

AC electrification distribution system

2.1 AC railway electrification

AC electrification systems have been developed specifically for railway traction purposes and to utilize the availability of universal 50 and 60 Hz HV public electricity supplies. The main feature of this Section is to describe the principle operation of the railway AC traction system as a single-phase overhead electrification system with one pole intentionally earthed.

2.2 AC railway electric traction system voltages

The technical strength of all railway electrification is that the primary source of energy is removed from the train. By taking power from the public AC 50/60 Hz electricity supply, there is almost unlimited power available, which has been efficiently produced. However, the capital cost of the railway electrification infrastructure needed to transmit power to the train has become its economic weakness. The cost of providing the 25 kV electrification scheme, as well as the civil works and protective measures for telecommunication and signalling systems, can only be justified where the operational requirements include services such as high speed intercity, mixed passenger, suburban metro or heavy freight traction loads. Areas where traffic density does not necessitate full electrification schemes would be better served by hybrid trains that are battery and diesel-powered. The operational and maintenance costs of using diesel-powered hybrids (diesel and 25 kV) can negate the long-term benefits of implementing full electrification schemes.

Electric traction systems are designed to meet the needs of different types of train service that run on particular railway track layouts (multi-track network with many junctions and crossovers, or simple long two- or four-track networks with few junctions and crossovers) using one or more types of traction characteristic. This requires an electric traction system that is reliable and provides a high degree of security of electrification supply for the train operators. This security is necessary to ensure that the electric traction system can provide the required power levels to fulfil the performance required by train timetables. If the service or loads are increased, then the performance of the electric traction system should also be reviewed.

Modern traction drives now take the 25 kV AC power through a pantograph. This is rectified on the train and a three-phase inverter drive provides the variable frequency drive for a three-phase induction motor. Electric trains are more efficient than non-electrified and also have the advantage of being able to regenerate, during braking, into the 25 kV supplying other trains operating in the feeding section. This improves operational efficiency and reduces wear on the mechanical brakes. When compared to diesel trains, these electric 'vehicles' are cheaper to build, easier to maintain, and are more reliable. Other key factors driving the move to electrification include the requirement to reduce emissions of greenhouse gases and the necessity to improve local air quality, particularly in towns and cities.

2.2.1 AC low frequency system voltage (15 kV 16.7 Hz and 11 kV 25 Hz)

A train powered from a high voltage (HV) AC supply requires a transformer to reduce suitably the voltage to be rectified and fed to DC traction motors. It was found that mercury arc rectifiers (used for DC electrification) were too fragile for use on the traction unit; but it was possible to design AC traction motors with characteristics similar to DC traction motors, which could operate at a low frequency supply.

Section 2 – AC electrification distribution system

Choosing 15 kV HV enabled high power transmission with low losses. The 15 kV AC system has been powering traction motors since the beginning of the twentieth century in Germany, Austria, Switzerland, Sweden and Norway. HV electrification began at $16^2/_3$ Hz, exactly one-third of the public electricity supply frequency of 50 Hz. This choice facilitated the operation of rotary converters supplied from the public electricity supply and allowed dedicated railway power generators to operate at the same shaft speed as a standard 50 Hz generator by reducing the number of poles by a factor of three. Therefore, a generator turning at 1,000 rpm would have two poles rather than six. Since rotary converters work more efficiently with lower frequency supplies,16.7 Hz and 25 Hz were common for this application.

The first generators were synchronous AC generators. Subsequently, with the introduction of modern double-fed induction generators, the control current induced an undesired DC component, leading to problems with poles overheating. This was resolved by changing the frequency slightly, away from one-third ($16^2/_3$ Hz) of the grid frequency to 16.7 Hz which was then chosen to remain within the tolerance of existing traction motors. Austria, Switzerland and southern Germany switched their power plants to 16.7 Hz on 16th October 1995.

Electrical distribution for the overhead contact line is fed by rotary or static converters that are located at typically 30–40 km for medium power (refer to EN 50388 Table D.1). As the 15 kV 16.7 Hz traction frequency is electrically unrelated to the frequency of the three-phase public electricity supply, the traction substations can be 'mesh' connected with no requirement for neutral sections. This feeding arrangement has the benefit of improved voltage regulation and extended lengths of electrically connected railway that is good for energy recovery through regeneration. This distribution arrangement is operationally resilient as the adjacent substations share the same load during equipment failures or planned maintenance

Rotary converters have been steadily replaced over the past 70 years, initially with mercury arc rectifiers, and more recently with solid-state rectifiers. More recently, static frequency converters (SFCs) have been adopted, which provide a more cost-effective and reliable supply when compared to the rotary converter.

Amtrak operates a 25 Hz traction power system along the southern portion of its Northeast Corridor; the route between Washington, D.C. and New York City is 225 miles (362 km), and between Philadelphia and Harrisburg, Pennsylvania, the route is 104 miles (167 km).

One major disadvantage of low frequency ($16^2/_3$ Hz and 25 Hz) traction units, when compared to 50/60 Hz locomotives, is that it requires a heavier transformer. Low frequency transformers have heavier magnetic cores and larger windings for the same level of power conversion. The heavier the transformer, the higher the axle load, which would lead to increased track wear and the need for more frequent track maintenance. Modern traction drives now take the 15 kV AC power, which is rectified on the train, and a three-phase inverter drive provides the variable frequency supply for a three-phase induction motor.

2.2.2 AC 25 kV 50/60 Hz system voltage

The electric traction distribution system requires a single-phase supply from the public electricity HV system, with traction loads between 2 MW and 20 MW (EN 50388 Table D.1 'Characterisation of AC electrified lines'). The 25 kV AC 50/60 Hz electrification systems have been developed specifically for railway traction purposes and utilize the availability of universal 50/60 Hz HV public electricity supplies. The main characteristic of the electrification system includes a single-phase system with one pole intentionally earthed. Even though railway administrations have their own individual electrification configurations, the basic design of all these systems remains similar with minor variations.

The single-phase 25 kV traction loads will cause an unbalance within the three-phase HV system and introduces negative phase sequence (NPS), harmonic currents and voltage distortion to the public HV supply. To minimize the effect on public HV supplies, railway connections are generally made to the supply networks at 132 kV or 275 kV, and more recently for autotransformer feeding arrangements at 400 kV.

To address unbalance of the three-phase HV public supply system, it is necessary to connect the single-phase traction supply transformers to different phase pairs of the public supply at successive feeder points at HV grid sites, providing some balancing of the HV supply system. This electrical feeding arrangement does require neutral sections within the overhead line to ensure the segregation of each 25 kV supply.

Where access to the national grid requires a remote operation, distribution of the railway network at 50 kV has been used and has been implemented in the Deseret Railway, USA, and the Sishen–Saldanha railway in South Africa. This voltage is only implemented where the clearances to fixed overline structures would have the necessary clearance for basic insulation.

2.3   AC feeder switching station

The feeder switching station is usually a railway-owned asset, and its prime function is the distribution and protection of the electrical supply to the overhead contact lines. The grid electrical supply to the railway is usually a double circuit. The feeding arrangement of the overhead line is single end-fed, with the two circuits separated via a neutral section, as shown in Figure 2.10.

The feeder switching station performs several operational functions, and these are controlled from the railway electrical control room (ECR) through the supervisory control and data acquisition (SCADA) system. Operations and functions that are managed include:

(a) operational functions:

    (i)   control the feeding arrangement of the electric traction system: normal and degraded states;
    (ii)   operation of circuit-breakers and other switching devices; and
    (iii) provide isolation of the electric traction system for planned maintenance and in an emergency;

(b) monitoring the HV AC system:

    (i)   monitoring of the protection equipment and circuit-breakers;
    (ii)   monitoring of system voltage: undervoltage and overvoltage;
    (iii) metering of electrical energy (import and export);
    (iv) maximum demand; and
    (v)   monitoring of the traction return current;

(c) monitoring and operation of the LV systems:

    (i)   status of the security system;
    (ii)   LV supplies; and
    (iii) uninterruptible power supplies and standby generator.

2.3.1   Circuit-breakers

Circuit-breakers provide various key functions. They ensure the unimpeded flow of current in the overhead contact line and return circuit under normal operating conditions, and are necessary to interrupt the flow of excessive current in a fault condition when detected by the protection relays. They provide switching so that feeding sections can be fed differently under normal and degraded states of the electric traction systems and to support the operation of train services. Circuit-breakers also provide points of isolation and earthing for the overhead contact line during planned maintenance and repair.

The design of the circuit-breaker relies on the reliable mechanical operation of the HV components and good electrical design to ensure that the circuit-breaker can satisfy the electrical voltage stress. During the opening sequence, an electric arc occurs between the breaker contacts. This arc occurs in either a vacuum or gas and is tolerated in a controlled manner until there is a natural current zero when the arc is quenched.

Figure 2.1   25 kV Structure-mounted outdoor switchgear Gowkthrapple Feeder Switching Station (image reproduced with permission by R. Catlow)

2.3.2   Protection against electric shock

The protection system detects if there is an electrical fault on the 25 kV system and automatically disconnects the 25 kV supply to the feeding section or equipment in which the fault has occurred.

Electrical protection for overhead contact lines

Electrical protection of the overhead contact line limits damage caused by earth faults, lightning, or excessive overloads. The protection, usually, cannot prevent faults from occurring but can limit their effects.

Electrical protection systems also prevent electrical overload, which could exceed the design rating of cables, conductors, damage traction loads, and prevent interference or damage to other utilities by conducted or induced currents corrupting communications.

Section 2 – AC electrification distribution system

Protection against electric shock

Protection from 'indirect contact' (during fault protection) is required for public, operational and maintenance staff working on or near the electrification system. The protection system is required to ensure electrical safety from an unacceptable risk of harm caused by electric shock, which causes pathophysiological effects resulting from a current passing through a human or animal's body. The protective measures involve protective equipotential bonding and automatic disconnection of the supply.

Protection of the public from 'direct contact' (or basic protection) with the 25 kV electric traction system mainly occurs on station platforms, maintenance platforms, near overbridges and at level crossings. Railway maintenance and operational staff may be exposed in other locations as they do have access to the entire electrification system. The protective measures aim to keep people away from exposed live parts and therefore include the provision of obstacles such as fences and separation by placing live parts out of reach, particularly on station platforms.

This subject is addressed in detail in Section 5.

2.3.3   Failure modes of the AC electric traction system

The transformer, overhead lines and traction return system must be capable of sustaining an HV AC earth fault until the protection and circuit-breakers clear the fault.

Most earth faults of the overhead contact line system (OCLS) primarily occur under bridges and where electrical clearances may not be optimal, and where there is regular movement through the mechanical forces from train pantographs, ambient temperature variation and thermal expansion through electric current. Most faults are caused by a 'live to earth' fault, and the following are the most likely categories:

(a) insulation failure of the overhead contact line:

   (i) bird roosting, flying debris that provides a conductive path through the air gap insulation to live parts;
   (ii) lightning strike to the AC overhead lines;
   (iii) failure of overhead line insulator (flashover) or switchgear;
   (iv) flashover to the underside of a bridge, for example, by ice formation or water ingress from the bridge soffit which provides a conductive path through the air gap insulation;
   (v) vandalism and trespassing; or
   (vi) inadvertent shorting by maintenance equipment;

(b) mechanical damage to the overhead contact line:

   (i) failure of the overhead line registration;
   (ii) failure of the train pantograph; or
   (iii) flying debris or falling objects;

(c) failure of equipment mounted on the overhead contact line system (OCLS):

   (i) electrical insulation failure of a pole-mounted 25 kV transformer;
   (ii) failure of isolating and earthing switch; or
   (iii) failure of the 25 kV automatic switching;

(d) failure on the train:

   (i) failure of the train pantograph includes a broken or partially broken pantograph head allowing the contact wire to become detached from the pantograph contact strip;

(ii) fault between 25 kV supply and train vehicle body;

(iii) protection failure; or

(iv) failure of the train to detect neutral section 'power off';

(e) insulation failure of electrical equipment:

(i) switchgear, power transformers, measurement transformers, insulators, cabling and cable sealing ends;

(ii) AC switching causes transient overvoltage and breakdown of the overhead line insulators;

(iii) electrical damage due to excessive electrical traction load in an overhead line section;

(iv) protection relay that has failed or been incorrectly set; or

(v) harmonic frequencies causing an overvoltage and insulator flashover;

(f) operational failures:

(i) the fault is created when a pantograph runs across a section insulator into an earthed section in that category (unplanned and inadvertent scenario).

2.4 Traction return circuit

The traction return circuit has a combined neutral and rail (earth) return. This return circuit requires the rails of each track to be bonded together to share the traction return current and requires that the rails are bonded to the adjacent overhead line structures to control the touch potentials.

To provide this common traction return system it is, therefore, necessary to connect the main earthing components of the railway, including:

(a) the 25 kV transformer neutral earthed at the grid substation;

(b) the return current busbar at the feeder and switching stations;

(c) the running rails as an earth mesh; and

(d) the structural foundations of the overhead line masts.

One or two of the running rails of each track carries the traction return current, which is determined by the type of signalling train detection used. These rails are then bonded all together creating a meshed traction return system, and then bonded to the earth conductors, thereby forming a distributed earthing system and return circuit. Electrically, the rails are required to provide a continuous traction return circuit with a level of redundancy.

Aerial earth conductors (AECs) are mounted on the overhead line masts and are then bonded to the rails at the track-to-track cross-bonds. The AEC acts as a circuit protective conductor, providing an earth fault return path and also as a conductive path for traction return current. The benefit of using AECs is that there are considerably fewer connections required to the rails. The AEC may be used additionally to bond signal gantries, conductive bridges and flashover strips on partially conducting bridges.

The traction return circuit is mostly uninsulated; and consequently, voltages are present and accessible to both maintainers and the public during normal operation and fault conditions of the 25 kV system. Therefore, this is required to be controlled strictly to manage the risk of electric shock. The return circuit is also used by some signalling systems, and it is necessary to avoid interference with the signalling detection circuits and any other track-connected systems.

This is addressed in more detail in Section 3.

Section 2 – AC electrification distribution system

2.5 Overhead contact line sectioning and phase separation

Electrical sectioning of the overhead line is required where there is a need to control the energization, isolation and earthing of the overhead line.

2.5.1 AC overhead line section breaks

Section breaks enable control of the energization (and de-energization) where there is the need to maintain the overhead line, to separate circuits at crossovers, turnouts, maintenance pits, or deal with operational incidents, without having to isolate the entire electrical supply system. The section break is normally arranged to correspond with tension lengths of the overhead conductors (i.e. at the location of construction overlaps).

The section break or insulated overlap enables the pantograph to cross the electrical break with continuous contact with the 25 kV contact wire and therefore uninterrupted power to the train; hence it is necessary that the overhead contact line either side of the section break is fed from the same traction supply.

A section break can also be implemented with a discrete device called a section insulator, which involves an in-line insulator equipped with overlapping skids at each end, but is used in slower speed applications (see Figure 2.2).

Figure 2.2   Section insulator at Dollands Moor (image reproduced with permission by R. Catlow)

2.5.2 The neutral section (or phase break)

On single-phase 25 kV 50 Hz electrified systems, it is normal to power different HV grid substations from different phases of the public electricity supply to reduce unbalance of the three-phase supply network. The voltage difference between consecutive substations where this voltage is taken (from alternate line-to-line phases of the national grid) will then be $25 \times \sqrt{3} = 43.3$ kV.

Section 2 – AC electrification distribution system

To connect two unsynchronized electrical phases of the electrical supply is unacceptable; therefore, a means of electrical isolation within the overhead contact line is required. This is called the neutral section (or phase break), which prevents the two different electrical phases from becoming electrically connected even when the train traverses the section with the pantograph raised.

A simple 'section break' is not sufficient to provide this point of isolation, as the pantograph would briefly connect the overlap of the section break. A phase separation section or 'neutral section' provides a distinct separation of the electrical feeding sections, even during the passage of the pantograph.

Long and short neutral sections are defined, where the long variety is usually (but not exclusively) made up of consecutive overlaps, and may be 400 m in length (carrier wire neutral section), and the short is made up of discrete insulators spliced into the contact wire, and is less than 8 m long. Carrier wire neutral sections are the only option at speeds greater than 200 km/h.

Neutral sections are used at the boundary between different railway authorities, or differing voltages, or for billing purposes. Neutral sections are described regarding electrical functionality in EN 50388, and regarding spatial geometry and mechanical performance in EN 50367.

2.5.3   Trains traversing conventional neutral sections

The train should not draw power through the neutral section, as this would cause an arc to occur between the contact wire of the outgoing section and the earthed centre of the neutral section, creating a short circuit on the 25 kV system (see Figure 2.3). As this may cause damage to the neutral section and

Figure 2.3   Power control of the traction rolling stock

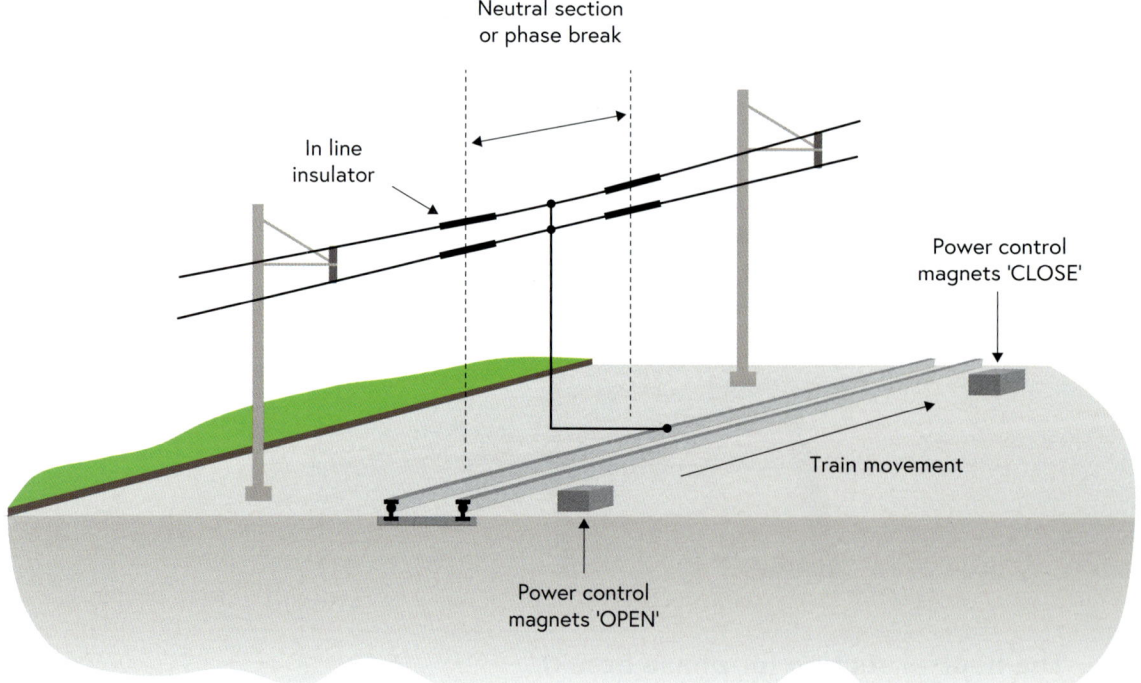

Section 2 – AC electrification distribution system

train, many administrations use trackside signs (or other operational controls) to warn the driver to open the train's 25 kV circuit-breaker and coast through the section.

Some railways, particularly in the UK, have installed trackside automatic power control (APC) magnets. The train has a detector coil that is activated by the trackside magnets and controls the train main circuit-breaker. When the pantograph has cleared the neutral section, a second magnetic device closes the 25 kV circuit-breaker and power is restored to the train (see Figure 2.3).

This 25 kV power control requires the train to coast for a certain distance through the neutral section, and hence the positioning of the neutral section in relation to slow line speeds, signal locations, station, and on rising gradients. This is a matter for consideration at the design stage of a project.

In the future, there may be alternative methods available for operating the traction power control, including using European Rail Traffic Management System ERTMS.

The development of polytetrafluoroethylene (PTFE) and glass fibre as an insulating material enabled the design of the in-line insulator. The Arthur Flury neutral section (Figure 2.4) is just within the 8 m length limit of the UK Railway Group Standard GLRT1210.

Figure 2.4   Arthur Flury neutral section Auckland, New Zealand (image reproduced with permission by R. Catlow.)

2.5.4   Carrier wire neutral section

The carrier wire neutral section (CWNS) is typically used at speeds greater than 200 km/h but is also required where a suitable short neutral section cannot operate at the required line speed. Varying configurations of CWNS are possible but typically consist of three or four overlap spans in series, depending on whether an earthed centre section is employed or not. The number of overlaps required in a CWNS is a function of the number of pantographs that a single train is allowed to raise.

The CWNS works with the train's pantograph passing from a live wire run to a second floating section and then back to a second live wire run. The train's 25 kV power is switched off and on by track detection devices in the normal way.

2.6 Single-phase AC overhead electrification system

Single-phase electrification is a two-conductor electric traction system. The two conductors are derived from the HV traction supply transformer secondary winding. One terminal of the transformer winding is earthed and is also connected to the running rails. The live terminal of the HV winding is connected to the overhead lines via the feeder switching station. The current draw is typically lagging the transformer electromotive force (EMF) due to the inductive nature of the overhead lines. A typical phasor diagram for a 25 kV system is shown in Figure 2.5.

Figure 2.5 AC single-phase phasor diagram

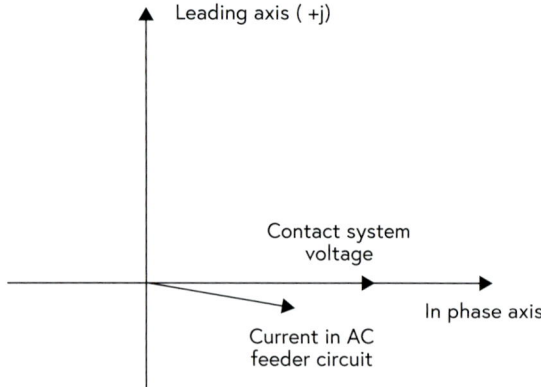

2.6.1 High voltage public electricity supply connection

The public electricity supply usually provides incoming HV substations at typically 132 kV (in the UK). However, if this is not available or not suitable, it is necessary to consider 275 kV. Two incoming circuits are usually made available at the public electricity supply HV substation. Both of these circuits are capable of individually carrying the total traction load under normal traffic conditions and provide a 25 kV traction power supply with a high degree of security.

With the connection shown in Figure 2.6, the 25 kV supply to the railway can be fed from independent parts of the public electricity supply HV network. If there is a failure in one of the HV feeds, this does not interrupt the supply to the second railway feed. The two HV railway supplies can be independent or be banked with 33 kV or 11 kV transformers feeding local industry or distribution networks.

With the electrical supply for traction being single-phase, the public electricity supply may need to balance its load by supplying feeds from a different phase at adjacent railway substations. This does not

Section 2 – AC electrification distribution system

Figure 2.6   Typical HV feeding arrangement for a single-phase 25 kV electrified railway

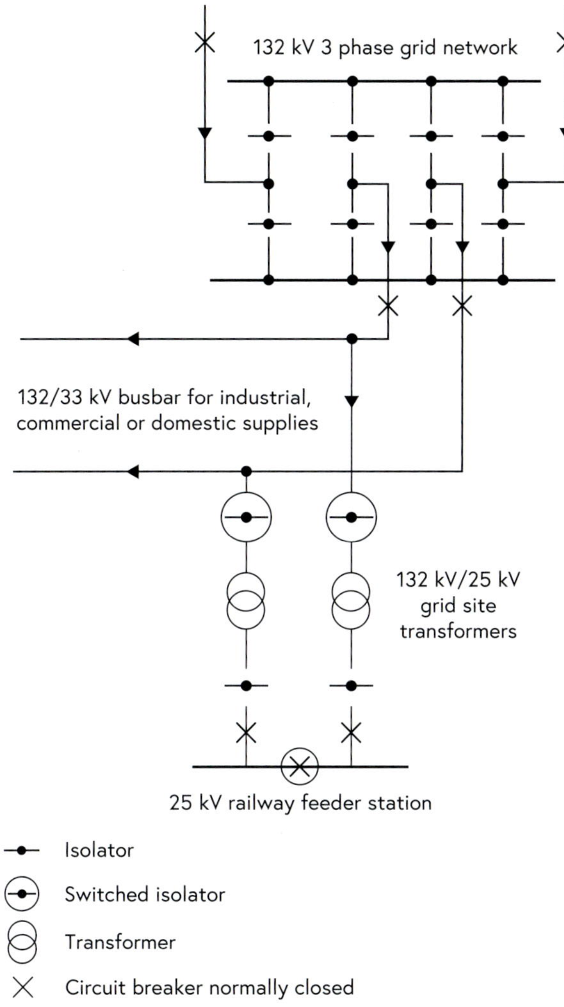

eradicate all the problems associated with unbalanced loads, but it does help to reduce the adverse effect on the supply system. The unbalance in the three-phase system will cause negative phase sequence currents and the possibility of overheating of electrical motors.

If there is a total loss of supply to the traction feeder switching station, both HV circuits are isolated, the 25 kV bus-section coupler at the feeder switching station is opened, and the adjacent feeder switching stations provide the supply to the overhead contact lines up to the opened bus-section coupler. This new feeding arrangement gives rise to some loss of train performance due to the increased voltage drop in the extended feeding area. Any loss in time to the traction unit due to the outage of a feeder switching station should be recoverable in the next normally fed 25 kV electrification section.

2.6.2   Public electricity supply grid transformer

At each HV grid site, the public electricity supply provides HV switchgear, single-phase grid transformer (132/275 kV to 25 kV), 25 kV protection equipment, 25 kV overhead line or cabling between the grid site and the public electricity 25 kV disconnection compound located next to the railway feeder switching station.

Section 2 – AC electrification distribution system

Substations have transformers that are typically rated for natural cooling at 5, 7.5, 10, 15 and 18 MVA. The nominal impedance is typically between 10 % and 15 %, and this is in an agreement between the railway and the public electricity supply:

(a) **typical transformer rating 10 MVA**: 132 kV/25 kV fixed ratio: This transformer is normally used on suburban electrified lines. The transformer is usually oil-immersed and naturally cooled; if an oil-circulating pump and forced air cooling are installed, the rating can be increased to 14 MVA.

(b) **typical transformer rating 18 MVA**: 132 kV/25 kV variable-ratio: (from 100 % to 112.5 % in steps of 2.5 %). It is necessary to adjust the transformer tap setting, to give the most suitable no-load voltage; this is determined by the characteristics of the incoming supply feeder and requires not exceeding 27.5 kV permanent while seeking to provide a full load voltage of 25 kV. Once set it is only necessary to adjust the tap setting if the incoming power system characteristics are significantly altered. The transformer is usually oil-immersed and naturally cooled; if an oil-circulating pump and forced air are installed, the rating can be increased to 26.5 MVA.

The public electricity supply may also fit balancers to control any unbalance at the point of common coupling and static VAR compensators to regulate the transmission voltage and improve power quality. This includes a set of electrical devices for providing fast-acting reactive power on HV electricity transmission networks. HV side harmonic filters may also be found in HV sites for compliance with the national Grid Code harmonic limits.

2.6.3   AC traction return circuit

The nature of the existing public telephone network has had a pronounced bearing on the solutions of electrification schemes that have been adopted. To reduce inductive interference in communication lines, different railway authorities have developed and adopted alternative return systems. In the 1960s Japan introduced booster transformers (BTs) with a winding ratio of 1:1, which were installed every 3–4 km along the trackside. Sweden and Norway used booster transformers; the British Railways electrification installed booster transformers at intervals of 2 miles (3.2 km).

The 'single-phase 25 kV' traction return circuit (running rails) has a combined neutral and earth return. The return path may include several parallel conduction paths, including rails, booster transformers, return conductors, ground (auxiliary) conductors and AECs.

Different railway authorities have developed and adopted alternative return systems, and the following arrangements have been commonly used:

(a) rail return (Network Rail);
(b) rail return with earth conductors (SNCF, Network Rail);
(c) rail return with aerial and auxiliary earth conductor (Channel Tunnel, Hong Kong West Rail);
(d) rail return with return conductors (Network Rail, Hong Kong East Rail); and
(e) rail return with return conductors and booster transformers (Japan, Network Rail, SNCF, Malaysia).

The traction return system is addressed in more detail in Section 3.

2.6.4   AC overhead line and return circuit conductors

The overhead contact line system (OCLS) and return circuit typically use the following bare and insulated conductors.

25 kV overhead line conductors: these conductors are supported by catenary clamps, links, pulleys and droppers. The registration arms that are mounted on the overhead line masts ensure the correct

Section 2 – AC electrification distribution system

Figure 2.7 Typical single-phase 25 kV conductor positions – open route

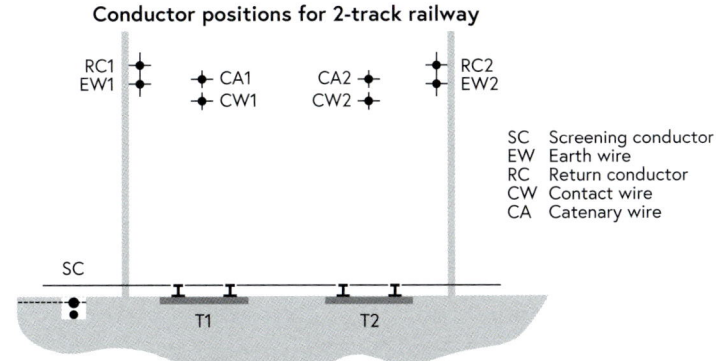

Conductor positions for 2-track railway

SC Screening conductor
EW Earth wire
RC Return conductor
CW Contact wire
CA Catenary wire

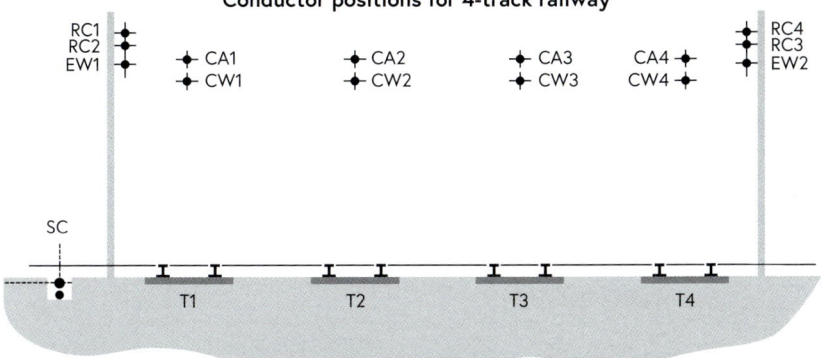

Conductor positions for 4-track railway

horizontal constraint. The contact and catenary are rated to supply the full traction load during the peak hours of traffic. Copper and copper alloys are normally specified for their reduced conductivity, tensile strength and hardness. The following are typical conductors used:

(a) the contact wire – typically 107/120 mm^2 copper alloy (Cu–Ag or Cu–Mg); and
(b) the catenary wire – typically 60 mm^2 copper or bronze 19/2.1.

Return conductors: these conductors can also be used with or without booster transformers. The return conductors are bare conductors that are mounted with insulators on the overhead line masts. These conductors form part of the return circuit and are bonded to the traction return system at a specified location. Where booster transformers are used, this is located midway between booster transformers. Return conductors with booster transformers rely on the Ampere-Turn Law and are the most efficient single-phase system at retaining return current in the return conductor.

Return conductors without booster transformers rely on the mutual coupling between conductors to retain current in the rails and return conductors. Return conductors are more efficient at retaining return current in the return conductors than rail return-only systems.

The rail return-only system relies on the mutual coupling between conductors to retain return current in the rails.

2.6.5 Principle of booster transformer and return conductor systems

The booster transformer is a type of current transformer designed to produce an AC current in the secondary winding which is proportional to the current flowing in the primary winding. The booster

Section 2 – AC electrification distribution system

transformer is a 1:1 current transformer and is part of the traction return system. It has the primary connected in series with the 25 kV contact wire and the secondary connected in series with the return conductors. The return current is therefore constrained to return in the return conductors due to the ampere-turn balance of the booster transformer.

2.6.6 Booster transformer electric traction system configuration

Booster transformers are positioned at typically 3.2 km intervals, with one booster transformer provided for each 25 kV track. The return conductors are mounted on overhead line masts and connected to the rails at the mid-point between booster transformers. The return conductors are positioned to provide magnetic flux balance with the current flowing in the overhead contact/catenary conductors. This arrangement is shown in Figure 2.7. The return current thereby counterbalances the current in the contact wire and reduces the level of induction into railway and third-party lineside conductors.

The train returns the traction current through the distributed traction return circuit to the return current busbar of the 25 kV feeder switching station and then to the neutral of the public electricity supply grid site substation. The operation of the booster traction return circuit is shown in Figures 2.8 and 2.9.

Figure 2.8 Booster transformer train located between the booster and mid-point

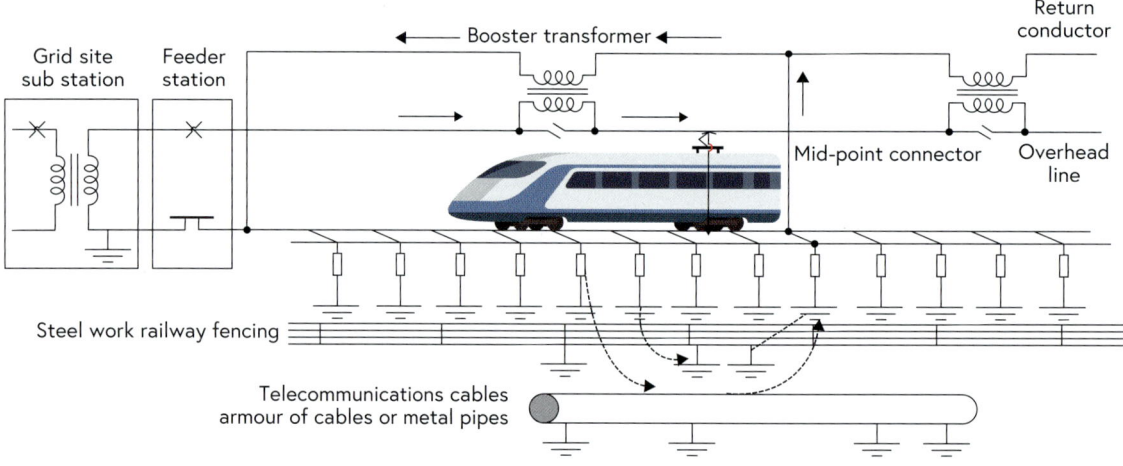

Figure 2.9 Booster transformer train located between mid-point and booster

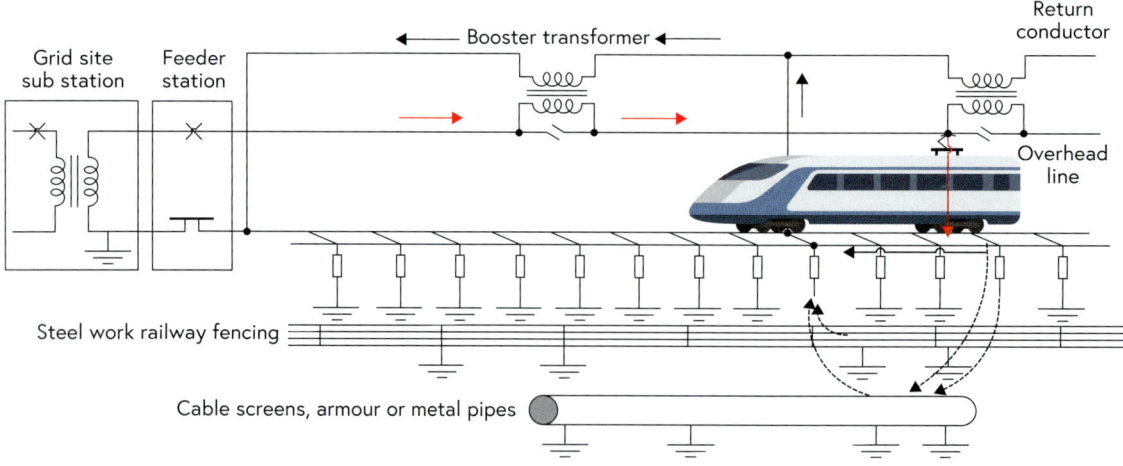

Section 2 – AC electrification distribution system

2.6.7 25 kV distribution and electrical sectioning

The traction substations where a supply is taken from the public supply authority are typically provided at intervals of between 40 km and 60 km. This spacing is specified so that the voltage regulation meets the specification as detailed in EN 50163 Table 1 'Nominal voltage and their permissible limits in values and duration'.

At the feeder switching station, each incoming 25 kV feed has its circuit-breaker on the 25 kV busbar. The two feeds can then be isolated using the bus-section circuit-breaker, therefore allowing both feeds to be independent.

The 25 kV feed to the overhead contact line is through a separate track circuit-breaker supplying the feed to each overhead line in each direction from the feeder switching station. It is necessary, therefore, to provide four-track feeder circuit-breakers for a two-track railway system, and eight-track feeder circuit-breakers for a four-track railway system. This somewhat complicated switching arrangement does provide a versatile system under planned maintenance or outage conditions.

Single-phase AC normal distribution

The 25 kV feeder switching station and midpoint switching stations (MPSSs) have a neutral section (or phase break). This ensures phase separation of adjacent HV supply points when operating in normal and emergency feeding.

An MPSS is located at the mid-point between feeder switching stations. The function of the MPSS is to provide electrical separation between feeds from adjacent public electricity supply points and also provide sectioning and track paralleling of the 25 kV system. This arrangement provides a high level of security and the best feeding arrangements without the loss of supply. Intermediate track switching stations are positioned between the feeder switching station and the MPSS, and their function provides paralleling of the overhead line and sectioning similar to that of an MPSS. Intermediate switching stations (ISSs) are not able to terminate a feeding section.

The overhead contact lines between traction substations are electrically split into sections and subsections to allow for alternative and emergency feeding and maintenance of the overhead lines. In this arrangement, each contact line can be independently fed. The overhead contact line circuit-breakers control the electrical feeding and provide isolation and earthing during maintenance of the overhead lines. Sectioning of the overhead lines is provided by 25 kV circuit-breakers at feeder switching stations, ISSs, and the MPSSs and section breaks. A typical feeding arrangement for single-phase 25 kV railways is shown in Figure 2.10.

Figure 2.10 Typical 25 kV feeding and distribution – normal feeding condition

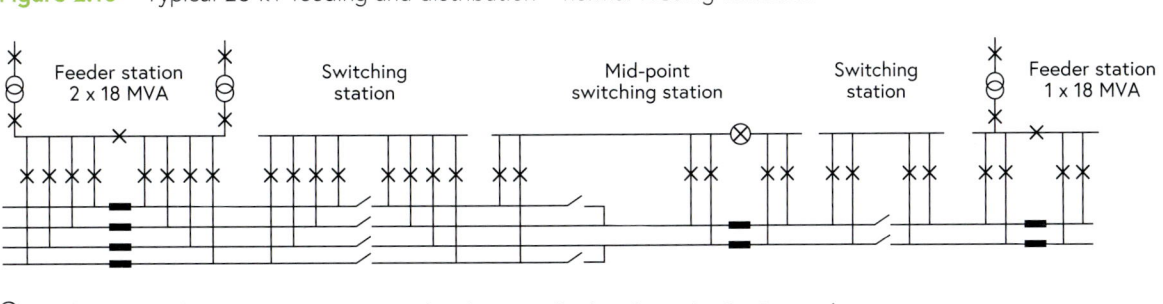

Section 2 – AC electrification distribution system

Under normal feeding arrangements, each traction feeder switching station supplies the feed to an MPSS in each direction. Further sectioning and paralleling of the overhead line is provided by the switching stations situated between the feeder switching station and the MPSS. Under normal feeding arrangements, it is necessary to have the bus-section coupler in the feeder switching station and MPSS open, ensuring that the feed from the feeder switching station to the MPSS is single-ended.

The single-phase 25 kV AC feeding configuration is normally designed with a typical fault level of 6 kA. This fault level was probably chosen due to the switchgear rating (12 kA) but also needed to be restricted due to interference between the 25 kV (go and return current) with the lineside signalling and telecommunications circuits.

It has been possible in some applications to operate transformers in parallel, assuming that the phasing is common, giving a fault level of 12 kA.

2.6.8   Short-circuit characteristics (typically 6 kA)

The booster transformers are located at 1.6 km, 4.8 km and 8 km (at intervals of 3.2 km). The graph in Figure 2.11 shows the short-circuit characteristic, as if the short circuit is applied in steps between the feeder switching station (0 km) and the ISS (9.6 km). The impedance of the overhead line and traction return system increases with distance from the feeder switching station. The short-circuit current reduces with distance from the feeder station, and where the current is required to go through the booster transformer, the impedance of the booster transformers create a step in the short-circuit characteristic.

The public electricity supply transformer impedance in the UK and in Hong Kong is set for a transformer impedance combined with the source impedance of the public electricity supply to give a maximum 6 kA at 27.5 kV, giving a transformer impedance of approximately 4.58 Ω.

The electrical performance of the 25 kV electric traction system is determined by the impedance of the 25 kV supply, overhead lines and traction return system. This impedance is essential as it determines significant operational characteristics, including:

(a) the effective touch potentials (EN 50122-1 (IEC 62138-1)):

    (i)   short-term $t < 200$ ms: effective touch potentials – these limits are imposed during an earth fault; and

    (ii) long-term $t < 300$ s: effective touch potentials – these limits are imposed during train operation;

(b) 25 kV voltage regulation of the overhead line:

    (i)   the voltage regulation determines the voltage of the overhead contact line and therefore, the maximum operational train timetable, and still maintaining a system voltage within limits specified in EN 50163 Table 1 'Nominal voltage and their permissible limits in values and duration'.

The short-circuit characteristics are determined by the electrical system impedances of the 25 kV system, which are given below:

(a) substation and transformer parameters:

    (i)   transformer source impedance;
    (ii) grid supply impedance; and
    (iii) resistance of the substation earth mat;

(b) parameters of electrical conductors as shown in Figure 2.7:

    (i) the self-impedance of the overhead line conductors;
    (ii) the self-impedance of the rails;
    (iii) the mutual impedance between conductors and rails; as determined by their height and spacing; and
    (iv) the impedance of the booster transformers;

(c) other electrical system parameters:

    (i) the resistivity of the ground;
    (ii) the resistance of mast foundations; and
    (iii) leakage to earth of the rails (Ω.km).

The graph below shows the short circuit as if it is applied in steps between the feeder switching station (0 km) and the intermediate track switching station (9.6 km). Where the current is required to go through the booster transformer, the impedance of the transformers creates a step in the short-circuit characteristic.

Figure 2.11 Typical short-circuit characteristic for a 6 kA booster transformer electrification system

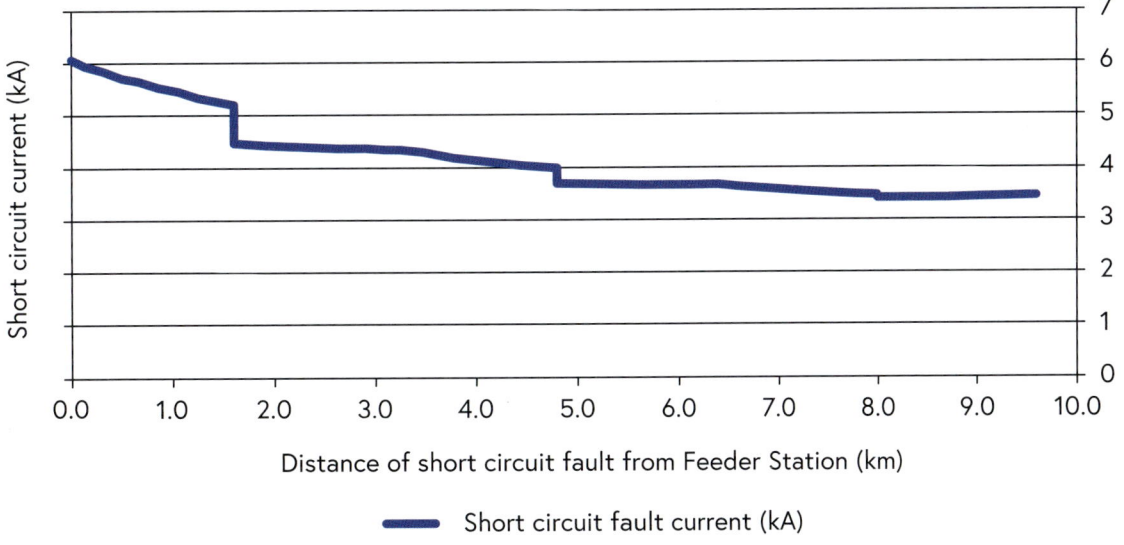

Distance of short circuit fault from Feeder Station (km)

Short circuit fault current (kA)

2.7 Autotransformer feeding and distribution system

The autotransformer electrification system is a two-phase, three-conductor electric traction system. The three conductors have been derived from the HV traction supply transformer secondary windings. The centre tap of the transformer winding is earthed and is also connected to the running rails. The positive voltage live terminal of the winding is connected to the overhead lines via the feeder switching station and the negative voltage terminal is connected to the autotransformer feeder conductor as depicted in Figure 2.14. The current drawn is typically lagging the transformer EMF due to the inductive nature of the overhead lines. Figure 2.12 shows the autotransformer phasor diagram depicting the +25 kV and -25 kV

Section 2 – AC electrification distribution system

Figure 2.12   Autotransformer two-phase phasor diagram

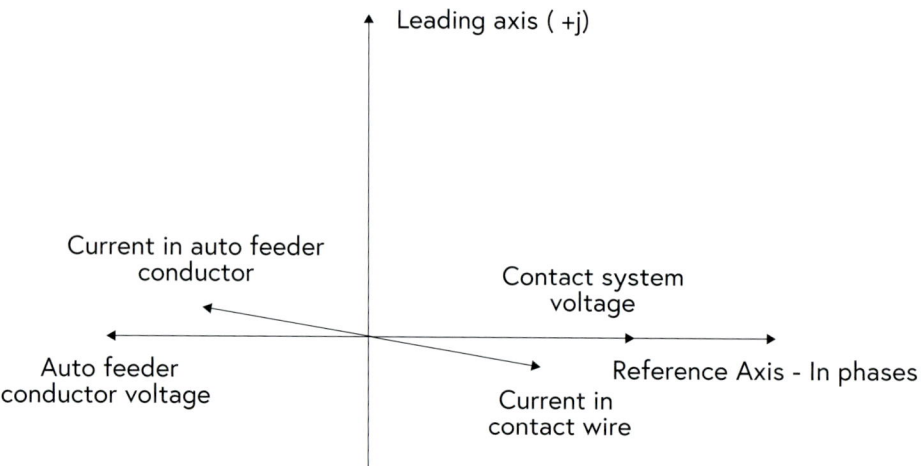

configuration. For simplicity, in the text this is stated as 25 kV-0-25 kV. The contact wire (+25 kV) is 180° out of phase with the AT feeder (-25 kV) and the current drawn is typically lagging the supply transformer EMF (25 kV) due to the inductive nature of the overhead lines.

Autotransformers (ATs) were first used for railway electrification design in 1913 when the New York, New Haven and Hartford railway electrification was extended to New Haven. The AT design reduced the line loss and the inductive interference on the 11 kV 25 Hz electrification system. The idea was then adopted by the Japanese in 1962 for the extension of one of their railway lines, a 25 kV-0-25 kV system. Subsequently, in 1981 SNCF adopted the 25 kV-0-25 kV 50 Hz autotransformer for use on a section of the Paris-Lyon TGV line. This new arrangement proved so successful that SNCF adopted it as the standard feeding arrangement for each of their new TGV lines.

In recent years the railway industry has been under significant commercial pressure from the aviation industry. To compete with airlines over 500–1,000 km, it has become necessary for railways to increase their maximum line speed to 300 km/h.

This necessity to increase train load, operational speed, and the frequency of trains required the introduction of a stronger and more rugged electrification distribution system. Autotransformers have been introduced in several countries including Australia (Blackwater and Gregory Coal line), Chinese Railways (Datong to Qinhuangdao), Russia (Vjaz'ma to Orsha), Japan (Shinkansen bullet train), France (TGV Lines), Belgium (TGV Lines), Hungarian State Railway (Lake Balaton), New Zealand (North Island line), USA Amtrack and Spanish High Speed Lines. Currently, the UK has AT systems on the High Speed 1 (HS1) route (Channel Tunnel Rail Link), West Coast Main Line, Midland Main Line, the Great Western Main Line, East Coast Main Line, and Crossrail (TFL). The UK High Speed 2 will employ an autotransformer system.

2.7.1   High voltage public electricity supply connection

The introduction of 50 kV distribution, with or without autotransformers enables the distance between feeder station and switching station to be increased significantly, alleviating the problem of access to the electricity supply network. The autotransformer distribution system is a two-phase three-wire system and is derived from a transformer with the two phases at 180 degrees apart.

The ideal connection for an autotransformer system is to the existing public electricity supply grid site with a high level of security of supply and high fault level. Figure 2.13 shows a typical double-circuit input from a

Section 2 – AC electrification distribution system

Figure 2.13 Typical HV feeding arrangement for a 25 kV-0-25 kV autotransformer electrified railway

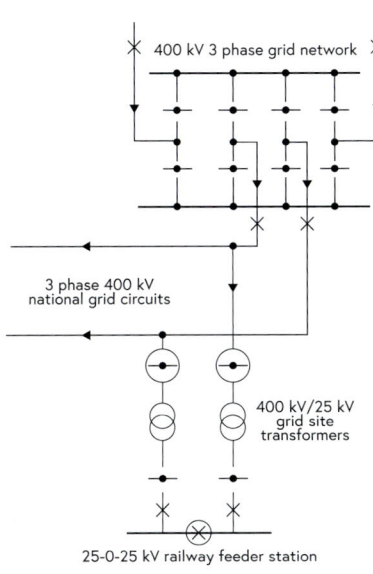

- ━●━ Isolator
- ⊙ Switched isolator
- ⊖ Transformer
- ✕ Circuit breaker normally closed

Table 2.1 Railway voltage limits and national grid system voltages (UK) The following are typical 400 kV and the anticipated 25 kV voltage limits

EN 50163:2004+A2:2020 Railway 25 kV voltage limits	UK National Grid Technical Specification TS1 (RES) Issue 1, May 2018
25.0 kV nominal voltage	NGC 400 kV nominal system voltage
26.25 kV	NGC 420 kV max continuous system voltage
27.5 kV highest permanent voltage	NGC 440 kV max system voltage < 15 min
29 kV highest non-permanent voltage	

400 kV supergrid supply point. The primary is connected across two phases of the supply; the secondary winding is centre tapped with one pole connected to the contact/catenary wire and the other to the autotransformer feeder wire supply with the centre tap connected to rail earth. The secondary is centre tapped with one pole connected to the contact/catenary wire supply and the other to the autotransformer feeder conductor with the centre tap connected to rail earth. The two poles are therefore configured in anti-phase.

2.7.2 Public electricity supply grid transformer

On new high speed lines, the grid substation transformers are typically rated between 80 MVA and 120 MVA, 400 kV to a 25 kV-0-25 kV fixed ratio. The grid voltage will fluctuate so the operational 25 kV system voltage is based on the operational voltage detailed below from EN 50163 and NGTS 1 (UK National Grid Technical Specification).

Grid supply transformer (400/50 kV): The grid transformer is usually dual-wound, oil-immersed and naturally cooled (ONAN, which stands for oil natural air natural), with centre tapped secondary winding.

Section 2 – AC electrification distribution system

The nominal impedance is between 10 % and 16 % and this limits the short-circuit current and contact line voltage regulation. The fault current is typically limited to 12–15 kA, and the percentage impedance agreed between the railway and the public electricity supply. The supply transformer will have a typical secondary complex impedance of $0.04175 + j2.751\ \Omega$ (based on one 25 kV transformer winding rated at 40 MVA at 16 %) with distribution autotransformers this will limit the short-circuit current at the feeder switching station to the 12 kA limit.

Where the national grid connection point is remote from the railway, cables are buried in the ground in a trefoil arrangement. A typical autotransformer connection arrangement is shown in detail in Figure 2.14 and Figure 2.15, including the public electricity supply substation, 25 kV disconnection compound, and 25 kV feeder switching station.

Figure 2.14   Typical 400 kV grid substation to 25 kV feeder switching station

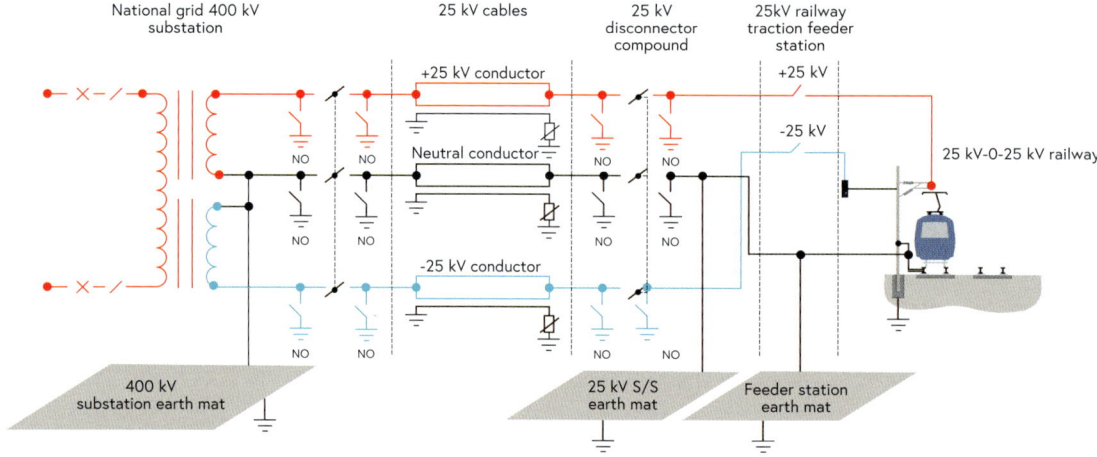

Figure 2.15   400 kV public electricity supply substation Patford Bridge (image reproduced with permission by R. Catlow)

Section 2 – AC electrification distribution system

2.7.3 Autotransformer electric traction system

Autotransformer supply schemes are increasingly being used for high speed AC electrification. The supply takes advantage of 2 x 25 kV (50 kV) power transmission to the trackside autotransformers. This has the benefit of reducing the level of the line current by 50 % and also acts to suppress the railway magnetic fields, and subsequently the 50/60 Hz induction into lineside conductors.

The autotransformers are located at intermediate sites spaced at intervals of 5–10 km. They are connected across the 50 kV supply and rated significantly less than the grid transformer. The rating is based on the traction load and the distance between autotransformer sites. The autotransformer system works on the principle that the train is supplied between the contact wire (+25 kV) and rail return (0 V) with the autotransformer feeder (-25 kV) acting as the return path between the autotransformer and the feeder switching station. This is shown electrically in Figure 2.12 and diagrammatically in Figure 2.18.

The operation of the booster transformers is different from the autotransformer arrangements, such that the booster transformer is only energized when a train is in section, while the autotransformers are energized whenever the supply is available in section and is independent of train position. In practice, however, most of the train current is supplied from the two adjacent autotransformers (see Figure 2.18 and 2.19).

The physical arrangement of the overhead line conductors of the autotransformer 25 kV-0-25 kV allows for the same electrical clearance equivalent (25 kV) when compared with the existing single-phase 25 kV overhead electrification schemes. The main difference is that the autotransformer arrangement has two conductors that require clearance from structures and overbridges. When upgrading a 'single-phase 25 kV' electrification scheme to an autotransformer, the autotransformer arrangement does add significant cost. However, the improved system performance is cost-effective.

The main advantage of the autotransformer system over the booster transformer system is the reduced impedance of the overhead line. This can be seen by comparing Figure 2.11 and Figure 2.19. The voltage drop in the supply system (nominally 50 kV) is less, with the autotransformer capable of supplying more power with fewer system losses, providing there is a train in section.

The International Telecommunication Union-Telecommunication (ITU-T) have specified the following requirements for the autotransformer:

(a) 52.25 kV, 10 MVA, oil-immersed and naturally cooled;
(b) leakage impedance, $0.17 + j0.92$ (ITU-T); and
(c) magnetization current, 1 A at 26,250 V (assumption).

2.7.4 Autotransformer conductors (open route)

Overhead line conductors: typical overhead line conductors used on autotransformer electrification schemes include:

(a) the contact wire – typically 120 mm^2 copper alloy (Cu–Ag or Cu–Mg); and
(b) the catenary wire – typically 60 mm^2 copper or bronze 19/2.1.

Running rails: one or two of the running rails of each track carries the traction return current and this is determined by the type of track detection system used. These rails are then bonded all together and in turn bonded to each overhead line structure, thereby forming a distributed earthing system and typically have an overall resistance to earth of less than 1 Ω.

Section 2 – AC electrification distribution system

The return rails are also connected to the return current busbar of the feeder switching station and all traction switching stations and then to the earthed neutral of the 25 kV grid site substation transformer.

Earth conductors: this can include AECs and ground earth conductors. The AEC acts as a circuit protective conductor providing an earth fault return path and a conductive path for traction return current. The benefit of using AECs is that they act as a protective conductor, and there are considerably fewer connections required to the rails. The AEC may additionally be used to bond signal gantries, conductive bridges and flashover strips on partially conducting bridges. Typical arrangements of the earth conductors are shown in Figure 2.16.

Figure 2.16   Autotransformer overhead line conductors – open route

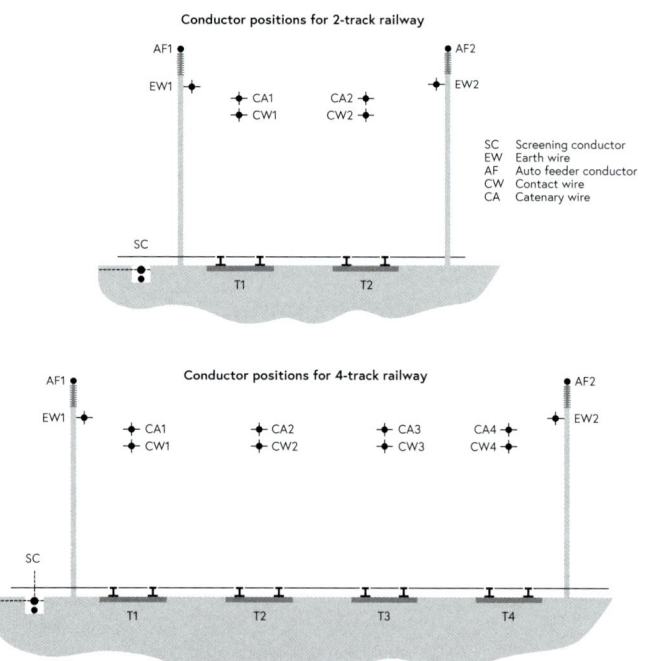

Autotransformer feeder conductor: this is usually a bare conductor suspended or supported on an insulator from either the portal or cantilever overhead line structure. At bridges, the bare conductor may not have adequate clearance to the structure, and in this case, it may be necessary to terminate and run as an aerial or ground-level insulated cable through those locations. At stations, the bare conductor is not electrically safe vertically above platforms, and in this case, it may be necessary to terminate and run as an insulated cable through the station.

2.7.5   Principle operation of the autotransformer

Where the two windings of a transformer are interconnected in anti-phase as shown in Figure 2.17, it is called an autotransformer. An autotransformer may have a single continuous winding with a centre tap for the secondary winding or two separate windings. The secondary winding is common to both primary (N_1) and secondary (N_2) circuits. The windings are normally arranged as a double winding on separate legs of the magnetic core, as this creates less leakage flux. Half of the windings are placed concentrically over each leg of the magnetic core to increase magnetic coupling, enabling all of the magnetic lines to go through both the primary and secondary windings. This winding arrangement on the magnetic core ensures that energy is transferred from the primary circuit to the secondary circuit through magnetic coupling.

Section 2 – AC electrification distribution system

Figure 2.17   Principle of the autotransformer

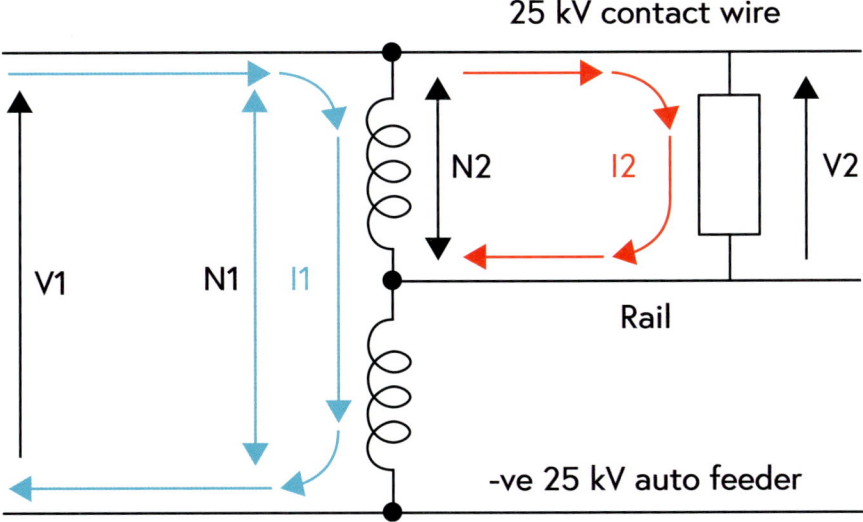

All transformers work on the Ampere-Turn Law detailed below. The magnetomotive force (MMF) can be calculated using Ampère's Law. The equation 2.1 for MMF is:

$$MMF = N \times I \tag{2.1}$$

The SI unit for the magnetic field strength or MMF of a coil or electromagnet is called the ampere-turn or amp-turn

where:

MMF is magnetomotive force in ampere-turns
N is the number of turns
I is current in amperes (A).

Kirchhoff's voltage and current equations 2.2a–c governing the 25-0–25 kV autotransformer are:

$$N_1(\text{primary 50 kV}) = 2 \times N_2(\text{secondary 25 kV}) \tag{2.2a}$$

$$I_{\text{primary}}(I_1) \times N_1 = I_{\text{secondary}}(I_2) \times N_2(\text{Ampere} - \text{Turn Law}) \tag{2.2b}$$

$$\text{Current in the upper winding is } I_{N2} = I_2 - I_1 \tag{2.2c}$$

The principle operation of the transformer is based on the electromagnetic flux within the magnetic core. The transformer is energized (magnetic flux) due to the magnetization current flowing from the 50 kV supply. This creates a magnetic flux in the core and induces 25 kV in both windings (N_1 and N_2).

When a load is in the section between autotransformers, a current is drawn from the secondary 25 kV winding (N_2) which reduces the magnetic flux in the core (Lenz's Law) and consequently reduces the EMF generated on the primary winding (N_1) causing a current to flow into the primary windings. The ampere-turn balance is thereby maintained.

Section 2 – AC electrification distribution system

2.7.6   Current distribution in the overhead line

The autotransformer distribution arrangement is radial with 25 kV-0-25 kV autotransformers connected in parallel and supplied from a grid transformer. The autotransformers are energized by being connected to the 25 kV-0-25 kV grid transformer secondary winding, hence the autotransformer magnetization current is drawn whether there is a load in the electrical section or not.

The autotransformer primary current circuit is supplied through the contact/catenary and the autotransformer feeder conductor. In the idealized arrangement it is assumed that there is no voltage drop along the contact/catenary or the autotransformer feeder conductor or the rails, which means that all the parallel connected autotransformers have the same primary voltage and the currents are in phase. The secondary current is drawn from the top windings of the autotransformer, as depicted in Figure 2.17.

Figures 2.18 and 2.19 show that the 50 kV distribution current is half that in the 25 kV traction load. The current returns to the grid transformer via the autotransformer feeder conductor, with little (or none) in the rail. Where a train is located between the two autotransformers the electrical load is supplied mainly from adjacent autotransformers.

Figure 2.18 shows the loop currents with the autotransformer section occupied by the traction load:

(a) This is a simplified analysis based on the magnitude of the currents, assuming adjacent autotransformer currents are in phase.
(b) The current in the contact/catenary is the phasor addition of the primary (50 kV) currents of all the autotransformers in parallel, plus the 25 kV current required by the train.
(c) The current flowing through the top winding of the autotransformer is the phasor addition of the current from the 50 kV in the primary winding ($I_1 = 100$ A), the current from the 25 kV secondary winding feeding the load ($I_2 = 200$ A). These are 180° in phase opposition; the phasor addition and magnitude of the secondary circuit is given below:

$$I(N_2) = I_1 \angle (\theta + \pi) + I_2 \angle (\theta) \tag{2.3}$$

The magnitude of the current secondary winding $|I(N_2)| = |I_1 \angle (\theta + \pi) + I_2 \angle (\theta)| \; |I(N_2)| = |I_1 \angle (\theta + \pi) + I_2 \angle (\theta)|$

Figure 2.19 shows the branch currents and simplified analysis based on the magnitude, and that the adjacent autotransformer currents are in phase. The split of currents to and from the train is broadly shown (as explanatory) to be the same because of the way the autotransformer system effectively supplies both ends of the section. In real life, the currents from each autotransformer will be dependent on train position.

2.7.7   Autotransformer distribution and electrical sectioning

The most commonly used autotransformer system is a two-track high speed railway, using a two-track mesh feeding arrangement with autotransformers typically spaced at 5–10 km. In this arrangement, it is common practice on French and UK HS1 (two-track railways) for the contact wire and corresponding autotransformer feeder conductor to be double-pole switched.

The overhead line is sectioned with neutral sections located (also referred to as phase break) as shown in Figure 2.20 at the feeder switching stations and mid-point switching stations. The grid transformer arrangement ensures the phase separation of adjacent grid supply phases when operating in normal and emergency feeding. The sectioning of the overhead lines is provided by 25 kV circuit-breakers at feeder switching stations, and autotransformer switching stations. The overhead contact lines between the switching stations are electrically split into sections and subsections to allow for alternative and emergency feeding during maintenance of the overhead lines.

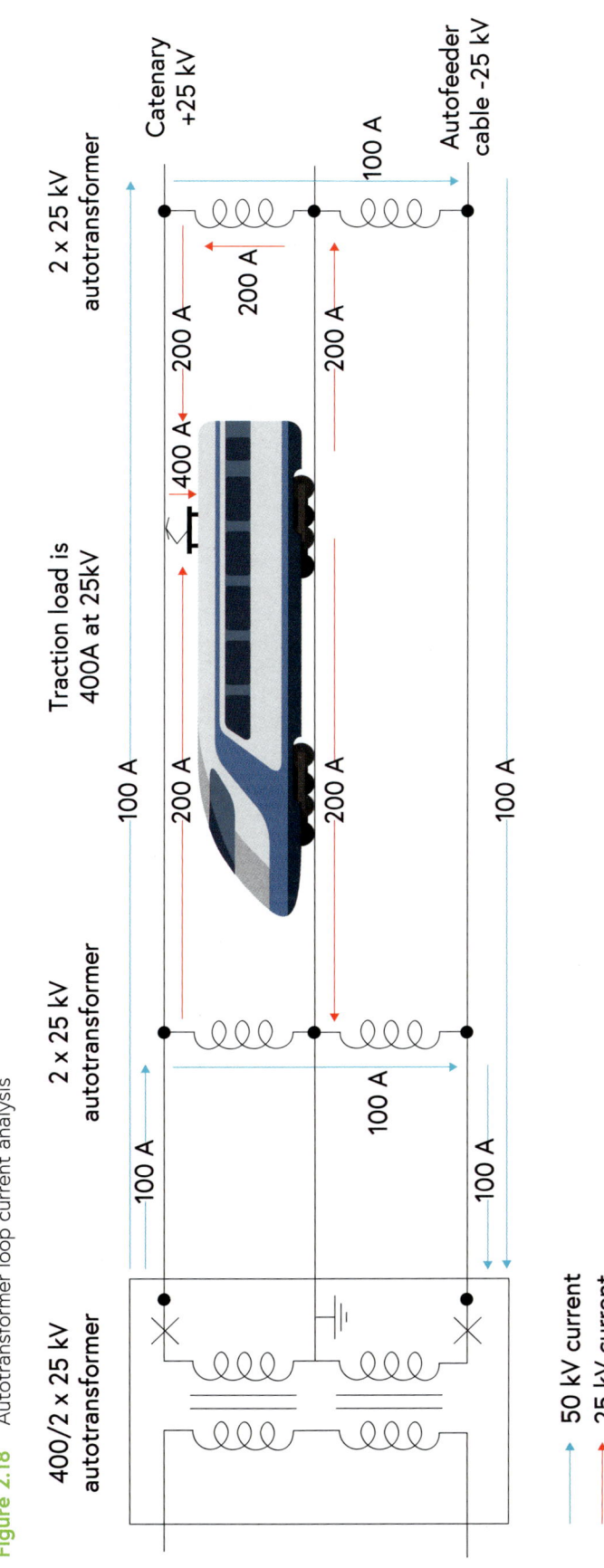

Figure 2.18 Autotransformer loop current analysis

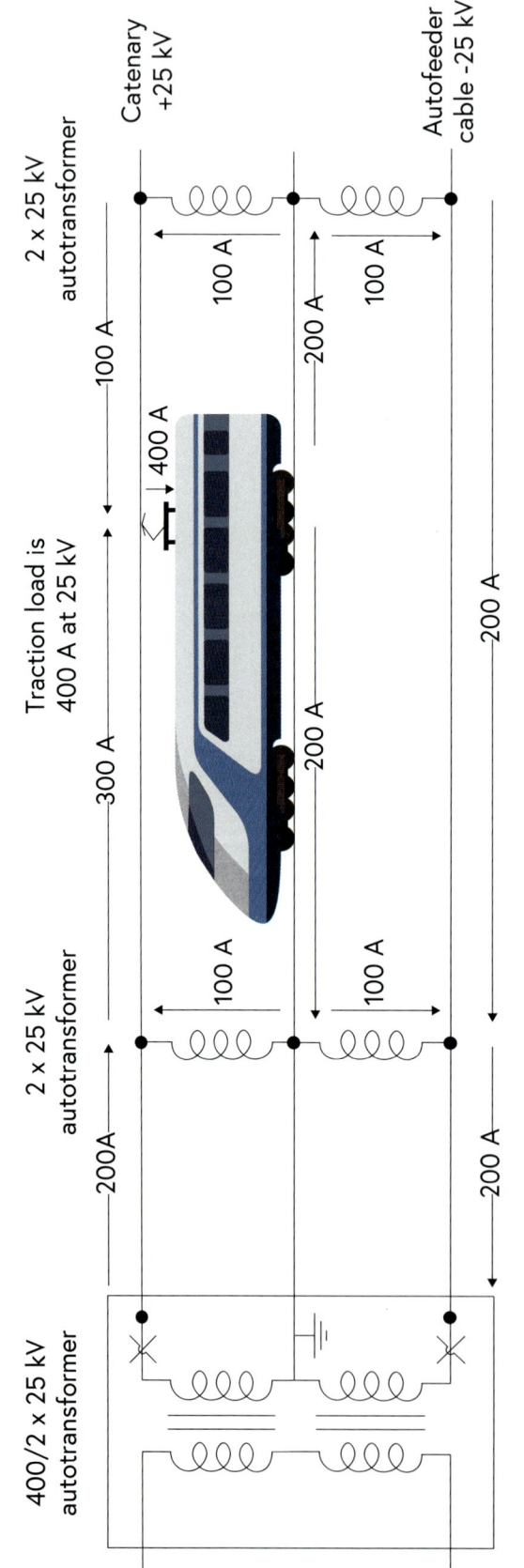

Figure 2.19 Branch current calculated by superposition of the loop currents

Section 2 – AC electrification distribution system

Figure 2.20 Autotransformer distribution and sectioning

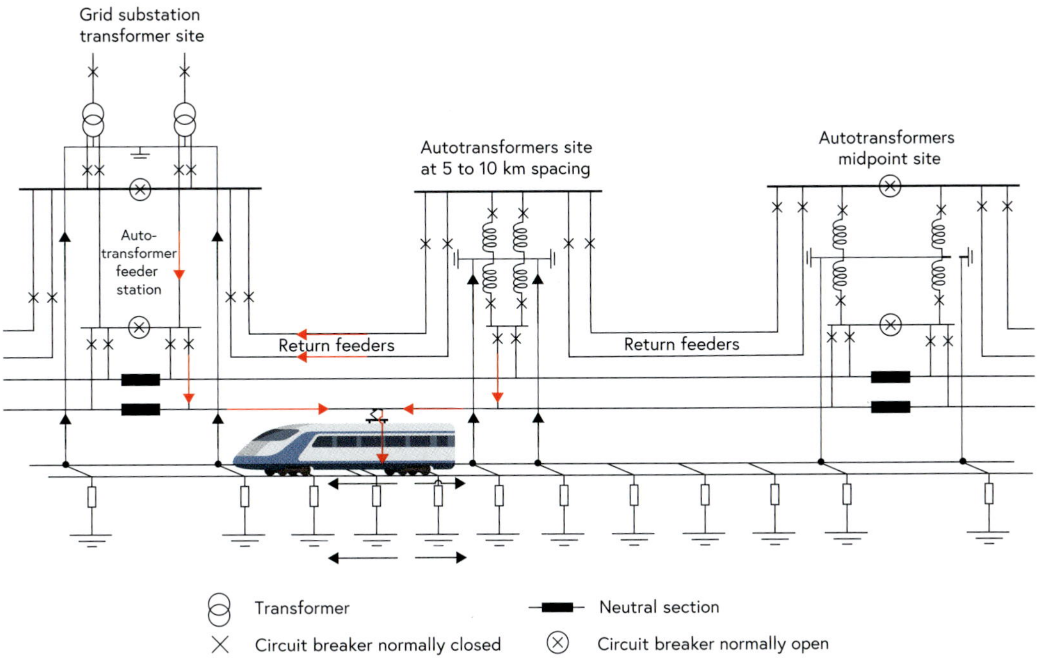

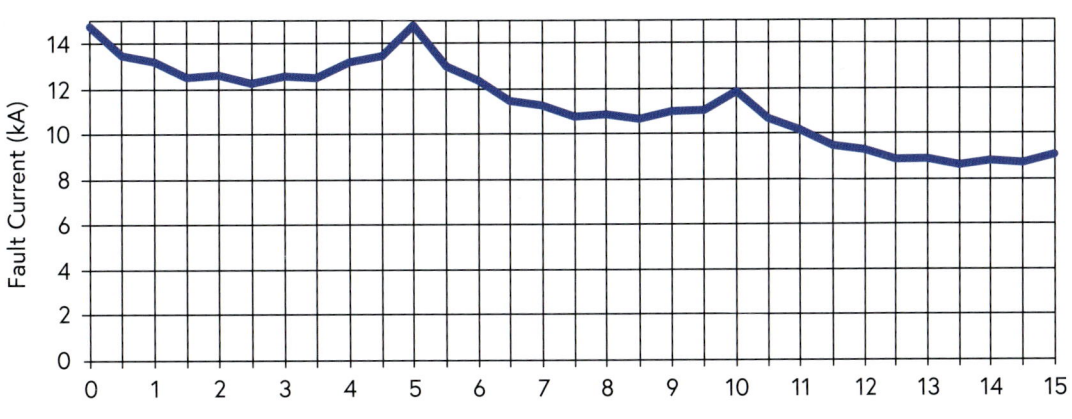

Where there is an end of the line, this often does not have a terminating autotransformer; therefore this section will be single-phase and rail return. The control of the return current and immunization of lineside copper circuits in this section may require special mitigation.

2.7.8   Short-circuit characteristics, typically 12–15 kA

The impedance of the overhead contact line and traction return circuit increases with distance from the traction autotransformer substation. As the impedance increases with the distance, the short-circuit current reduces as depicted in Figure 2.21. The grid transformer impedance and the autotransformer leakage impedances are defined to give a short circuit of 12–15 kA at 27.5 kV.

Figure 2.21 Typical 12–15 kA short-circuit characteristics

Section 2 – AC electrification distribution system

For example, a 12 kA fault level gives a magnitude of the transformer impedance of approximately 4.58 Ω per 50 kV winding. When the component provided by the autotransformers is added, the overall short circuit is between 12 kA and 15 kA. The leakage reactance of the transformer dominates this short-circuit characteristic; however, the characteristic of the short circuit as given in Figure 2.21 is determined by the electrical system parameters given below:

(a) transformer characteristic impedances:

    (i) transformer leakage impedance; and
    (ii) resistance of the substation earth mat;

(b) the local autotransformers:

    (i) transformer leakage impedance: and
    (ii) distance between autotransformer sites;

(c) conductor parameters:

    (i) the self-impedance of the overhead line conductors;
    (ii) the self-impedance of the rails; and
    (iii) The mutual impedance between each conductor and each rail; this impedance is determined by their height and spacing, as shown in Figure 2.20:

(d) system parameters:

    (i) the resistivity of the ground;
    (ii) the resistance of mast foundations; and
    (iii) leakage to the earth of the rails (ohm.km).

The autotransformers are typically located at 5, 10 and 15 km. Figure 2.21 shows the short circuit as if it is applied in steps between the feeder switching station (0 km) and autotransformer (5, 10 and 15 km).

 **Section 3**

Mass of Earth as part of a traction return circuit

3.1 Mass of Earth

The earth is a reference point in an electrical circuit from which voltages are measured, a common return path for electric current, or a direct physical connection to the earth. An earthing system, or grounding system, which is US terminology, connects specific parts of an electric power system with the ground, typically the earth's conductive surface, for safety and functional purposes. The choice of earthing system can affect the safety and electromagnetic compatibility of the installation.

The mass of Earth is an integral part of the AC traction return circuit with the running rails inextricably linked to Earth as a result of multiple leakage paths across rail chair insulators, rail chairs, sleepers, and ballast and the foundations of the overhead line masts; the masts are usually electrically bonded to the running rails. Therefore, this Section looks at the electrical characteristics of earthing elements, including the rails, overhead line masts, and earth mats to remote Earth.

The complexity of integrating the earthing of an AC electrified railway with various electrical distribution systems and exposed-conductive-parts means that it is impossible to prescribe one earthing and bonding design solution that addresses the needs of every railway. This creates different earthing designs for individual railways and also when there are changes in an existing design.

The AC traction return circuit has a combined neutral/earth return path and, therefore, the rail voltages fluctuate with the flow of both traction load current and fault current. The current flows in the earth are responsible for conductive and inductive effects on the traction return circuit and auxiliary systems. Computational analysis has been provided in this Section to provide a deeper understanding of the characteristic behaviour of both 'single-phase' and 'autotransformer' distribution and return circuits.

The definition of earth is given in EN 50122-1 and IEC 62128-1 'earth is the conductive mass of the earth, whose electric potential at any point is conventionally taken as equal to zero' [IEC 60050-826-04-01].

3.2 Statutory legislation

Electrified railways need to incorporate specific earthing and bonding arrangements to meet national statutory and railway administration requirements on direct and indirect contact and interfaces with other trackside electrical equipment, often fed from various other electrical sources.

Earthing and bonding principles on AC electrified railways are relatively straightforward, and the measures available are almost universally accepted. Statutory legislation, and key national and international standards normally include the following requirements:

(a) AC railways are required to be 'earthy', and therefore the whole of the railway is connected to the mass of Earth;

(b) the traction return circuit (earthing) is principally to provide protective provision for humans from direct and indirect contact;

(c) the traction return circuit is required to be a robust protective conductor and provide a robust electrically continuous circuit back to the substation;

(d) the earthing of the railway, its components and bonding conductors should be capable of distributing and discharging normal load and fault current without exceeding thermal and mechanical design limits based on backup protection operating times;

(e) the earthing system should maintain its integrity for the expected lifetime of the installation with due allowance for corrosion and mechanical constraints; and

(f) earthing performance should avoid damage to equipment due to excessive potential rise, potential differences within the earthing system and excessive currents flowing in auxiliary paths not intended for carrying parts of the fault current.

3.3 Elements of the earthed AC earthed railway

The return current path of the AC electrified railway flows in both the rails, earth conductors, earth mats, structure earth and additionally remote earth (current in the ground). The return current flowing in the rails can pass into deep earth and this is primarily determined by the resistance of structure foundations and the leakage resistance of rail fasteners and sleepers. The ground under the rails is made up of layers, with varying resistivity. The mutual coupling between the overhead conductors and the rails determines the percentage of the current remaining in the rails and the resistivity of the ground determines the depth of the return current in the ground.

3.3.1 Traction return system

A high voltage (HV) AC railway is required to be earthed to maintain safe voltages on the traction circuit. This earth must be provided at the supply point and throughout its length. Earthing the return circuit is necessary to control the rail voltages and the touch voltages between exposed-conductive-parts. If the return circuit were not earthed, then the traction return current in the rail would create excessive voltages due to the relatively high self-impedance of the rail $0.62 \angle 69° \Omega/\text{km}$ (including both internal and external reactance).

The traction return circuit is primarily connected to the earth (or the ground), to control voltages on exposed-conductive-parts of the railway. The connection to earth is primarily dependent on four elements:

1. the running rails which are mounted on anti-vibration pads and sleepers;
2. the overhead line structure foundations;
3. the earth mats (grids) at traction switching stations, passenger stations, tunnels and portals; and
4. the connection to the transformer secondary neutral which is earthed at the HV grid substation following standard national grid practice.

The rails and overhead line structure foundations create a type of earthed grid network that extends over the entire length of the railway, which can be hundreds of kilometres. The negative return of the 25 kV is required to produce an uninterrupted return circuit back to the HV electricity grid substation. Whenever this neutral return is broken, due to maintenance or a broken rail, high touch and accessible potentials on either side of the break in the return system can be created. The use of insulated rail joints (IRJs) in the AC traction return system should therefore be strictly limited to where they are required for the signalling system, and there is a mechanism to pass traction current around them.

Figure 4.3 and Figure 4.4 show how the interconnection of the rails and masts produces a distributed earthed return system such that if one part (or bond) becomes disconnected the remainder of the system shall still maintain safe voltages across the whole interconnected infrastructure.

A typical 2×25 kV electrification system in Figure 4.1, 4.2, 4.3 and 4.4 shows how the interconnection of the rails and masts produces a distributed earthed return system such that if one part (or bond) becomes disconnected the remainder of the system shall still maintain safe voltages across the whole interconnected infrastructure.

3.3.2   Structure earth

The structure earth consists of several earth electrodes that are inter-connected to each other by conductors. This interconnected conductive structure is mostly constructed from the reinforcements of concrete structure. Example of structures include passenger stations, maintenance sheds, viaducts, tunnel, overline and underline bridges piers and foundations.

3.3.3   Earth electrodes

Earth electrodes are one or more conductive parts that are in direct contact with the ground, and which provides an electrical connection to the earth (USA terminology - ground). This is usually the metallic or steel reinforcements of concrete structures and copper grids or rods used as the electrical reference for an electrical supply. To achieve adequate earthing does require early planning within the project programme to ensure provision of electrical cross bonding and terminals.

3.4   Electrical characteristics of conductors and structures

The railway return system is dependent on this ground resistivity to control the voltages of the running rails, the overhead mast foundations and other interconnected structures including bonded civil structures.

3.4.1   Ground resistivity

The ground can be composed of different soils with varying resistivity from location to location with some areas made up of five layers of different soil types, for example, marshy soil, clay, gravel etc. This can make the control of touch and step voltages challenging to manage, particularly in extreme locations like mountainous or sandy regions. The variation of soil resistivity varies widely and is detailed in Table 3.1

Table 3.1   Soil resistivity for frequencies of alternating currents (source BS 7430:2011+A1:2015)

Type of soil	Soil resistivity (Ωm)
Marshy soil	5 - 40
Loam, clay humus	20 - 200
Sand	200 - 2,500
Gravel	2,000 - 3,000
Weathered rock	Mostly below 1,000
Sandstone	2,000 - 3,000
Granite	Up to 50,000
Moraine	Up to 30,000

The traction return circuit is an earth return system and the main elements comprise the overhead line structure foundations (Ω ohms), the running rails (Ω.km called rail leakages), the earth of AC grid substation, AC switching stations and LV substations (portal, shafts and passenger stations).

Section 3 – Mass of Earth as part of a traction return circuit

3.4.2   Overhead line mast foundations

Overhead line masts foundations are primarily designed to structural requirements rather than to meet a target earth resistance. However when the masts are interconnected by earth conductors or rails the overall average earth resistance per foundation is low even though the earth resistance at some individual OCLS foundations could be high.

The foundations of overhead line masts are generally made of concrete, which is hygroscopic and therefore attracts moisture. When buried in the ground, a concrete structure behaves as a semiconducting medium with a typical resistivity of between 30 Ω.m and 90 Ω.m.

The structure foundations of the overhead line masts include reinforced steel bars with a depth of 2–3 m, which creates a typical resistance of between 10 Ω and 20 Ω where the ground resistivity is 30 Ω.m.

Where there are 30 foundations per kilometre, the overall resistance of structure foundations for a kilometre of the railway is typically 0.6 Ω.km (foundation 20 Ω). If this resistance is of a low enough value it is not usually necessary to drive earth rods. This resistance can significantly change with the ground condition. Therefore, a dry climate with poor ground conditions can increase each foundation's resistance to 100 Ω or more. It is worth noting that the mast structure foundations do provide a significantly better connection to the earth than the running rails.

Calculations for the structure foundations can be undertaken using equations 3.1, 3.2 and 3.3 which can be found in BS 7430:2011+A1:2015 *Code of practice for protective earthing of electrical installation:*

1. Reinforced concrete structure foundations, resistance of an electrode encased in low resistivity material, e.g. conducting concrete

The resistance of a backfilled electrode R_b in ohms (Ω) may be calculated from:

$$R_b = \frac{1}{2\pi L}\left\{ (\rho-\rho_c)\left[\log_e\left(\frac{8L}{D}\right)-1\right] + \rho_c\left[\log_e\left(\frac{8L}{d}\right)-1\right]\right\} \tag{3.1}$$

where:

    p is the resistivity of soil, in ohm metres (Ωm);
    p_c is the resistivity of the conducting material used for the backfill, in ohm metres (Ωm);
    L is the length of rod, in metres (m);
    d is the diameter of the rod, in metres (m)
    D is the diameter of the in-fill, in metres (m).

2. Reinforced concrete structure foundations.

Fagan and Lee [B65] use the following equation for obtaining the ground resistance, R_{CE-rod} of a vertical rod encased in concrete:

$$R_{CE-rod} = \frac{1}{2\pi L_r}\left(\rho_c[In(D_C/d)] + \rho[In(8L_r/D_C) - 1]\right) \tag{3.2}$$

where

ρ_c is the restivity of the concrete in Ω m
ρ is the restivity of the soil in Ω m
L_r is the lenth of the ground rod in m
d is the diameter of the ground rod in m
D_C is the diameter of the concrete shell in m

3. Steel pile foundations.

The resistance R_{ta} in ohms (Ω) of a strip or round conductor may be calculated from:

$$R_{ta} = \frac{\rho}{2\pi L} \log_e \left(\frac{L^2}{\kappa h d} \right)$$

(3.3)

where:

p is the resistivity of soil, in ohm metres (Ωm);
L is the length of the strip or conductor, in metres (m);
h is the depth of the electrode, in metres (m);
d is the width of the strip or the diameter of the round conductor, in metres (m);
κ has the value 1.36 for strip or 1.83 for round conductor.

3.4.3 Protective provision of earth electrodes and earth mats

Earth electrodes and mats are needed at sites where there are electrical supplies or switching of HV systems. This is necessary to control the touch and step potentials that can occur during normal operation and following short circuit of the transformer primary or secondary system. Earth mats are provided at HV grid sites, 25 kV switching stations and LV substations (portal, shafts and passenger stations).

This is specifically addressed in Section 6.

3.4.4 Conductors forming the earthed traction return system

The traction return circuit is made up of the following circuit elements and conductors:

(a) Running rails

In single-rail traction return areas, only one running rail of each track is designated as a traction return rail. This rail is usually located closest to the cess or wayside, to allow ease of bonding to the traction return rail.

In double-rail traction return areas, both running rails of each track are designated as traction return rails. Earthing and bonding connections may be made to either rail, subject to the requirements of the function of the track circuit or the use of impedance bonds in track circuit areas.

Where there are multiple tracks, it is necessary to provide diverse routing for the traction return current, which is achieved through track-to-track cross-bonding. By providing parallel paths, the track-to-track bonding can reduce the voltage drop within the traction return circuit.

The ballast resistance is a critical requirement for the correct operation of the signalling track circuit. The signal engineers need a minimum of 2 Ω.km between rails for the track circuit to work. There are no international standards that specifically define the leakage characteristic, and therefore signal engineers must specify this in the 'design specification'. The value of rail leakage is dependent on several variables, including the rail insulation pad and clip, the construction of the sleeper (usually reinforced concrete) and the type of ballast. The ballast (preferably granite) has a good resistivity. Other factors that affect the rails' leakage resistance are the weather – with low resistance on wet days and high resistance on dry days. The lack of maintenance of the ballast and cleanliness over time reduces the leakage value.

There is no EN or IEC standard that specifies typical rail insulation but, as a guideline, ballast rails have leakage to the earth of typically 50 Ω.km when 'installed' and 8 Ω.km at their 'end of life'. Therefore, the resistance to Earth of a two-track railway with all four rails bonded together is 12.5 Ω.km (as installed) and 2 Ω.km (end of life).

In most new railways, axle counters and transmission-based signalling systems are used and, in this case, the minimum value of rail insulation is not a requirement.

(b) Sheathed bonding conductors

Bonding conductors are required to bond the rails, and to bond electrification equipment and structures to the rails. These conductors are required to provide three fundamental functions:

1. a return path for traction load current;
2. a return path for 25 kV fault current; and
3. equipotential bonding of structures to traction return circuit.

A sheathed bonding conductor is used for bonding, typically an all-aluminium alloy conductor (AAAC) (19/3.25 – 157.6 mm^2).

(c) Aerial earth conductors

The aerial earth conductor (AEC) is usually a bare conductor that electrically bonds the overhead line masts, civil structures (overline and underline bridges) and signal gantries collectively to Earth and also to the traction return circuit. The AEC is normally common-bonded to the traction rails and so electrically forms part of the traction return circuit. Where a sheathed conductor is connected to the traction return circuit at both ends, it is necessary that the traction load and fault current be assessed in the calculation for the rating of conductors.

The AEC is normally a bare conductor and is typically an aluminium alloy conductor (AAC) or all-aluminium alloy conductor (AAAC) 19/3.25 – 157.6 mm^2, or a copper stranded conductor 150 mm^2. The conductor is not usually reinforced with steel as they are not highly (mechanically) tensioned.

(d) Buried earth conductors

The buried earth conductor is usually a bare copper conductor that is buried underneath the troughing or cable containment, mainly in open route areas. The buried earth conductor is common-bonded to the traction rails at track-to-track bonds and therefore electrically forms part of the traction return circuit. This conductor provides a reference earth for LV non-traction supplies and exposed-conductive-parts of signalling and telecommunications assets.

The buried earth conductor on SNCF TGV lines is a non-insulated 35 mm^2 aluminium/lead and on UK-HS1 is a non-insulated 35 mm^2 copper (UK-HS1) conductor. The size of the conductor chosen is dependent on several factors including the root mean square (rms) traction load current flowing in the conductor and the 25 kV fault level.

Buried earth conductors vary in size and are dependent on several factors including the rms traction load current flowing in the conductor, the 25 kV fault level, and touch potentials. The conductor's size varies due to the temperature rise of the conductor.

Section 3 – Mass of Earth as part of a traction return circuit

Buried earth conductor leakage can be calculated from the following standards:

(a) BS 7430:2011+A1:2015 Figure 14 'Impedance to the earth of horizontal earth electrodes buried in homogeneous soil';

(b) ENA Engineering Recommendation EREC S34 Issue 2, November 2018 A guide for assessing the rise of earth potential at electrical installations 'Appendix F: Typical values of earth resistance of long horizontal electrode, Figure F.1'; and

(c) EN 50522:2010 Annex J.1.

We can see from Figure 3.1 that for a 1 km earth bare conductor this gives a leakage of 0.5 Ω for resistivity of 50 Ω.m – this is commonly quoted as 0.5 Ω.km.

Figure 3.1   Typical values of earth resistance of the long horizontal electrode (image reproduced with permission by BSI)

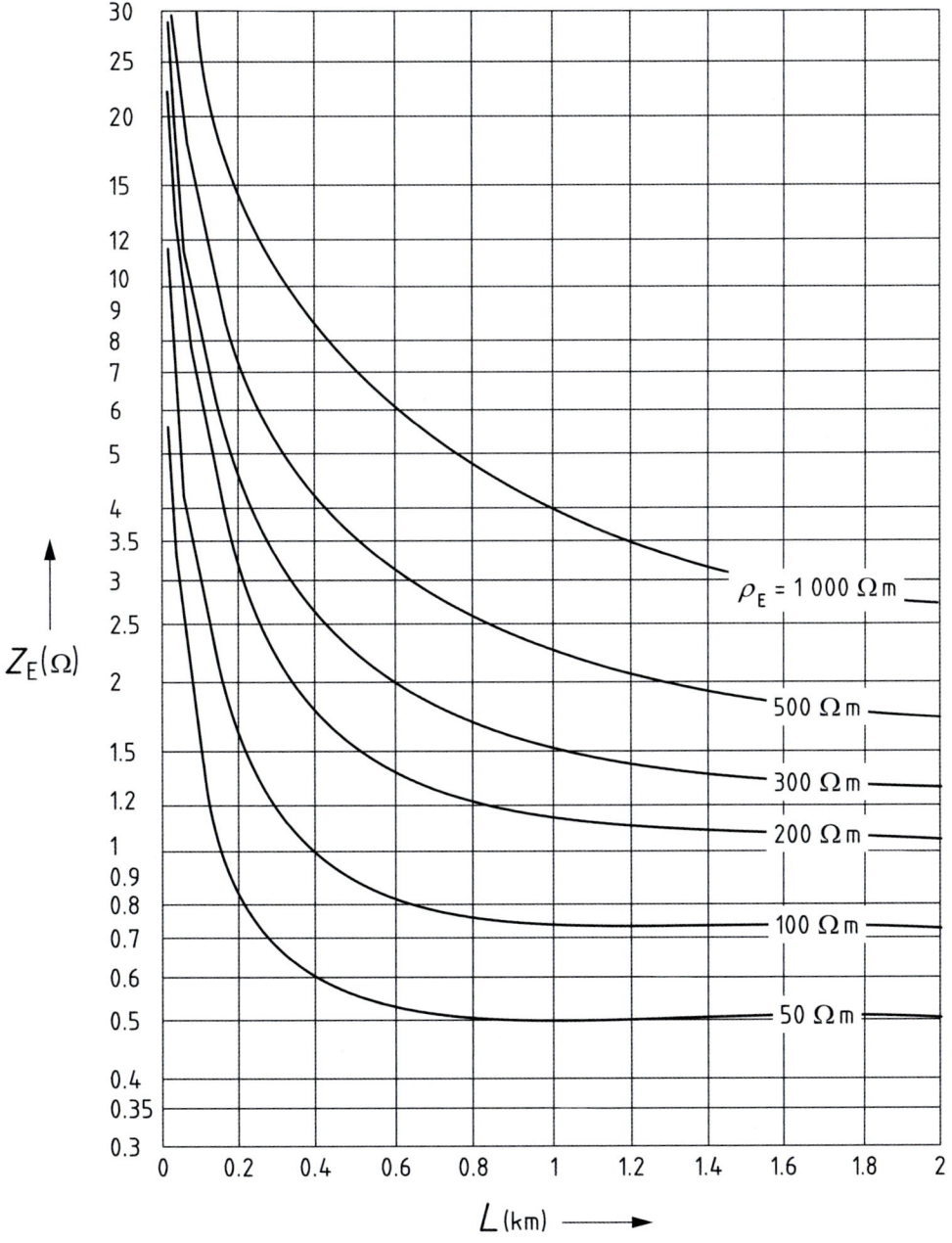

(e) Earth conductors at ground level (tunnels and viaduct areas)

Where the open route railway runs into a tunnel or onto a viaduct, the buried earth conductor becomes a ground conductor (or auxiliary earth conductor). The ground conductor is either a copper or aluminium conductor located in the cable containment system or mounted on the civil tunnel lining or viaduct parapet. The conductor is common-bonded to the traction rails at track-to-track cross-bonds and therefore electrically forms part of the traction return circuit. This conductor provides an earth reference for LV non-traction supplies and exposed-conductive-parts of signalling and telecommunications assets.

Examples of ground-level earth conductors include Hong Kong West Rail (300 mm^2 bare copper earth tape), where it is mounted on a viaduct parapet or tunnel lining, and UK High Speed 1 (an auxiliary earth conductor 35 mm^2. The size of the conductor is dependent on the overall system design and the current flowing in the conductor.

Ground conductors vary in size and are dependent on several factors including the rms traction load current flowing in the conductor, the 25 kV fault level, and touch potentials. The conductor's size varies due to the temperature rise of the conductor.

In calculating the cable size, the worst-case assumption is that the conductor is operating during traction load at a maximum operational temperature of 70 °C (thermoplastic) or 90 °C (thermosetting). This is followed by a 25 kV fault current flowing in the conductor. Additionally, in some rating calculations, allowances may be made for the traction return system's degraded modes.

(f) Return conductors (booster transformer)

The AC single-phase booster transformer system has return conductors mounted with a low voltage insulator on the overhead line masts. The conductor is then connected to the rails, typically every 3.2 km. The return conductor may be used on its own but is more commonly used with a booster transformer that acts as a forced return system.

The return conductor will typically be a bare AAC or AAAC conductor (19/4.22 – 265.7 mm^2 or 19/3.25 – 157.6 mm^2). Under a bridge or a station canopy, and where the clearances are tight, it is usual that the conductor is sheathed.

(g) Screening conductors (50/60 Hz)

A 50/60 Hz screening conductor is commonly used in 25 kV electrified railways to reduce the 50/60 Hz induction into non-traction systems. The screening conductor is located in the lineside trough, or on cable containment in tunnels or viaducts. The screening conductor in the troughing is required to have close magnetic coupling to communications-insulated conductors. It is ordinarily common-bonded to the traction rails and therefore electrically forms part of the traction return circuit. The screening conductor applies the principle of Lenz's Law (see Section 3.8.5) and provides electromagnetic screening for data and communications copper circuits.

The choice of conductor size will determine the magnitude of the current induced into the screening conductor. Therefore, to achieve a significant current and screening factor, a PVC insulated cable is required: typically, AAC 19/4.22 – 265.7 mm^2 or 19/3.25 – 157.6 mm^2. Self-screened cables with a screen of 10 mm^2 when earthed at both ends would only provide very poor screening factor at 50/60 Hz.

3.4.5 Substation earth mats, grids and fences

Substations require an independent earth mat for operation and electrical protection. The primary purpose of an earthing system in a substation is to provide a surface that has a uniform potential that is as near as possible to zero or absolute earth potential. The earth mat provides this common earth and provides an earth reference for structures and other non-current-carrying metallic objects in the substation.

3.4.5.1 Design requirements for the earth conductors

The substation earthing grid consists of a buried and cross-bonded earthing mesh. So that the earthing conductors function as intended, it is normal practice to bury copper earthing tape/conductors in a horizontal grid at a depth of between 0.5 m and 1 m. To ensure that the earth grid is adequately protected from mechanical damage, it is situated below the frost line and in surrounding earth that will not dry out. The earthing conductors provide connections to all equipment and structures and must have an adequate thermal capacity to pass the highest substation fault current for the substation's protection settings. The earthing conductors should also be designed with sufficient mechanical strength and be resistant to corrosion. The earth grid is usually constructed with 3 mm × 25 mm copper tape.

This subject is explored in more detail in Section 6.

3.4.5.2 Earth rods are driven vertically

Additional earth rods can be used where the ground is made up with a low-resistivity strata beneath the ground surface layer, it can be advantageous to drive vertical earth rods down into this, as shown in Figure 3.2 which has been reproduced from BS EN 50522:2010.

3.5 Traction earth return systems

In combination with appropriate measures, the earthing of the railway is required to maintain step, touch, and transferred potentials within the voltage limits based on the normal operating time of the protection relays and breakers. Therefore, the connection to the mass of Earth is required to limit permissible effective touch voltages and earth potential rise to acceptable levels as detailed in Section 5.

Controlling voltages on the rails and exposed conductive parts is achieved through the creation of a common traction return circuit with the interconnection of main earthing components, including the AC transformer neutral earthed at the grid substation, the running rails (called rail leakages) and the earth of the overhead line structure foundations.

The traction return current is conducted into the rails through the wheel set axle brush of the traction unit. The return path for the traction load current is via the rails, earth conductors and the earth itself (typically 30 %). The depth of the earth return path will be determined by the resistivity of the ground and can be 100 m or more.

The rails and AECs are then all cross-bonded every 300 m to 500 m. The current in the rails will disperse based on Ohm's Law. Where there are parallel tracks and cross-bonding, the current disperses between all the parallel rails. Any differences in the rail current are due to the different mutual coupling of the contact wire. The voltage on the earthed rails will vary quite significantly over the full length of the railway, and is dependent on both the traction load and the forced return system, and cross-bonding.

Section 3 – Mass of Earth as part of a traction return circuit

Figure 3.2 Resistance to earth of earth rods, vertically buried in a homogeneous soil (image reproduced with permission by BSI)

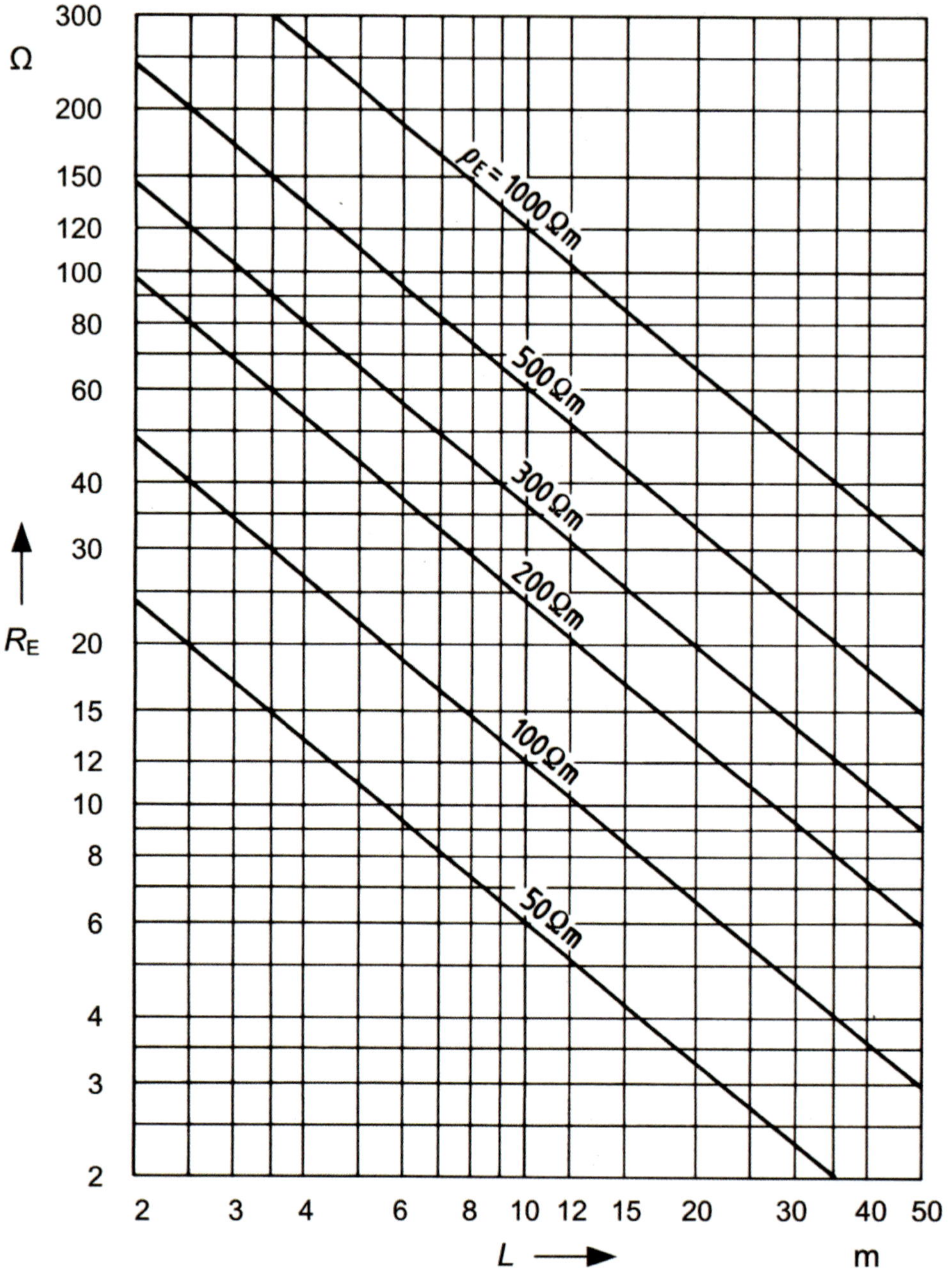

3.5.1 Traction units

The traction unit position determines those locations where the return current enters the rail and earth return system. Wherever there is a train injecting current into the rails, the voltage on the rails is typically between 10 V and 40 V and is dependent on the magnitude of the traction load current and impedance of the traction return circuit. The voltage at the grid substations is dependent on the current returning through the rails or the ground and the substation grid earth mat (typically 0.2 Ω). If 10 A returns through the earth and the substation earth mat, then the grid earth mat's voltage will be 2 V. Invariably the voltage on the grid earth mat is less than 10 V.

The voltage on the earthed rails will vary quite significantly over the full length of the railway and is dependent on both the traction load and the forced return system, and cross-bonding.

3.5.2 AC overhead earth faults

AC earth faults are most likely to occur where there are reduced clearances between the overhead contact wire or pantograph and civil structures (for example, bridges) or following a dewirement or a dewired pantograph. When an earth fault occurs, the short-circuit current flows into the traction return circuit through the rails and into the remote earth via the rail leakages and the structure foundations. The fault current and the resistance to earth produce a potential rise with respect to the AC grid transformer earth and remote earth. The resulting potential can create hazardous voltages between exposed-conductive-parts. It is required to verify the rail potentials in the design and, if necessary, adjust the spacing of the track-to-track cross-bonds.

In an integrated railway (see Section 4) the railway traction return circuit creates a type of Faraday cage. This includes the rails, the overhead line masts and other common-bonded structures (for example, stations). Due to the Faraday cage effect, the voltage between two exposed-conductive-parts within this area is significantly less than the earth potential rise (EPR). Therefore, passengers and maintainers are protected from hazardous voltages.

3.5.3 Potential gradient measured from the overhead contact system mast

The EPR of the rails is created by either the traction load or an earth fault following a dewirement or flashover. This EPR on the rail creates a potential gradient that is at a right angle to the railway and that falls away with distance.

Any system or earthed structure having an independent earth experiences a percentage of this rail potential compared to remote earth. Where the public, railway maintainers or third parties can access this voltage, it must be assessed for touch and step potential.

Guiding values for the rail potential gradient measured at a right angle away from the railway, with a homogeneous soil resistivity, are given in Figure 3.3 which has been reproduced from BS EN 50122-1 *Railway applications. Fixed installations. Electrical safety, earthing and the return circuit – Part 1. Protective provisions against electric shock.*

3.5.4 Single-phase rail return circuit

The current supplied to the traction load is from the 25 kV grid transformer. The feeding lengths are dependent on traction loads and can be between 15 km and 40 km, the feeding lengths for 15 kV can be between 20 km and 60 km (EN 50388 Table D.1).

The traction current that is conducted into the rail typically leaves the rail within 1 to 2 km (see Trace 1 in Figure 3.4). The position of the traction unit determines those locations where traction current can conduct through the rail leakages into the earth. The current that remains in the rail is held there by the mutual coupling between the overhead contact and catenary wires and the return conductors – in this case the running rails (Trace 2, Figure 3.4).

The current in the rails disperses based on Ohm's Law. Where there are parallel tracks and cross-bonding, the current disperses between all the parallel rails. Any differences in the rail current

Section 3 – Mass of Earth as part of a traction return circuit

Figure 3.3 Rail potential gradient measured from the mast (image reproduced with permission by BSI)

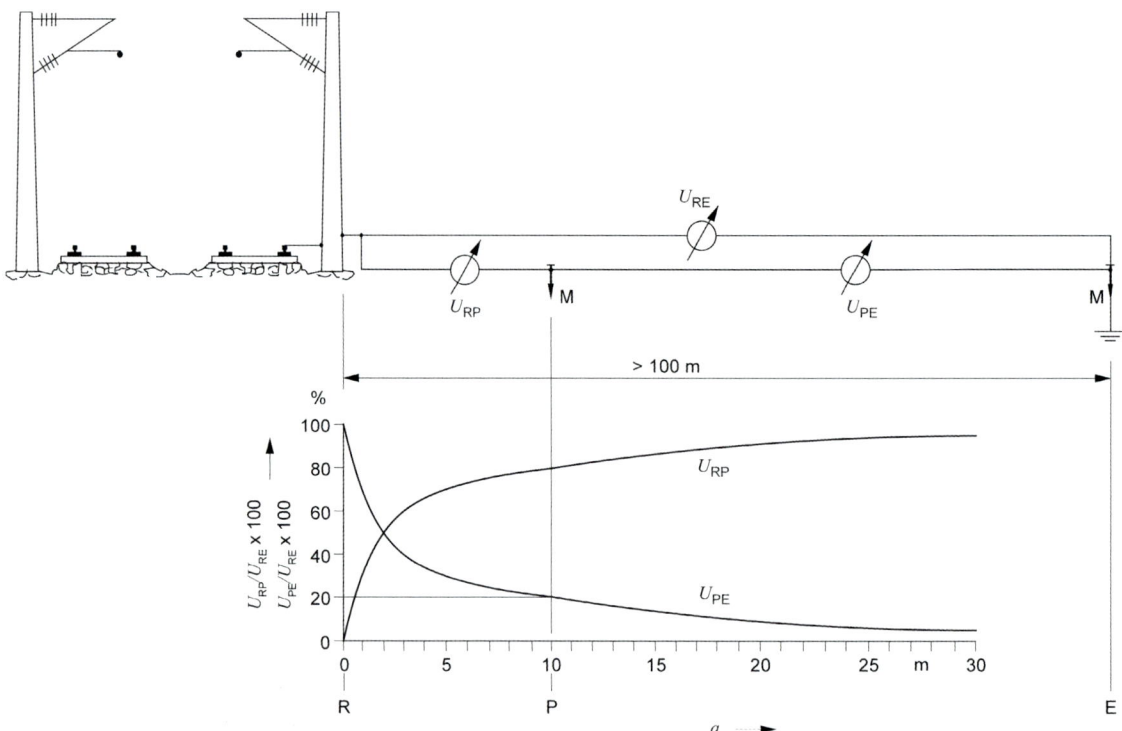

Where
a	is distance between running rail (mast) and measuring point
E	is Earth
M	is measuring electrode
P	is measuring point
R	is rail
U_{RP}	is rail potential
U_{RE}	is voltage between running rail (mast) and measuring point
U_{PE}	is voltage between measuring point and earth

are due to the different mutual coupling of the contact wire and the rails this is expressed mathematically in equations 3.4 and 3.5.

Faraday's Law of electromagnetic induction determines the mutual coupling current, and Ohm's Law is related by the following equations 3.4 and 3.5:

Ohm's Voltage Law (mutually induced voltage)

$$E = -(I.j.\omega.M)(V) \tag{3.4}$$

Section 3 – Mass of Earth as part of a traction return circuit

Ohm's Current Law (expression branch current in a matrix operation)

$$[I] = [V] \times [Z^{-1}] \quad (\mathrm{A}) \qquad\qquad (3.5)$$

where:

> E is electromotive force (EMF) (V)
> I is current (A)
> M is mutual inductance (Ω)
> j is complex operator
> Z is impedance (Ω)
> V is voltage drop. (V)

The total current within the rail is made up of two components, by the principle of superposition: the conducted current (Trace 1) and the mutually induced current (Trace 2). The calculation of these currents requires complex multi-conductor modelling and matrix inversion. The currents cannot be determined by hand calculation as the analysis is too complex. Typically, 30 % of the current returning to the substation is in the ground and occurs throughout the feeding section's full length up to the traction load. This simplification is shown diagrammatically in Figure 3.4.

Figure 3.4   Rail return system

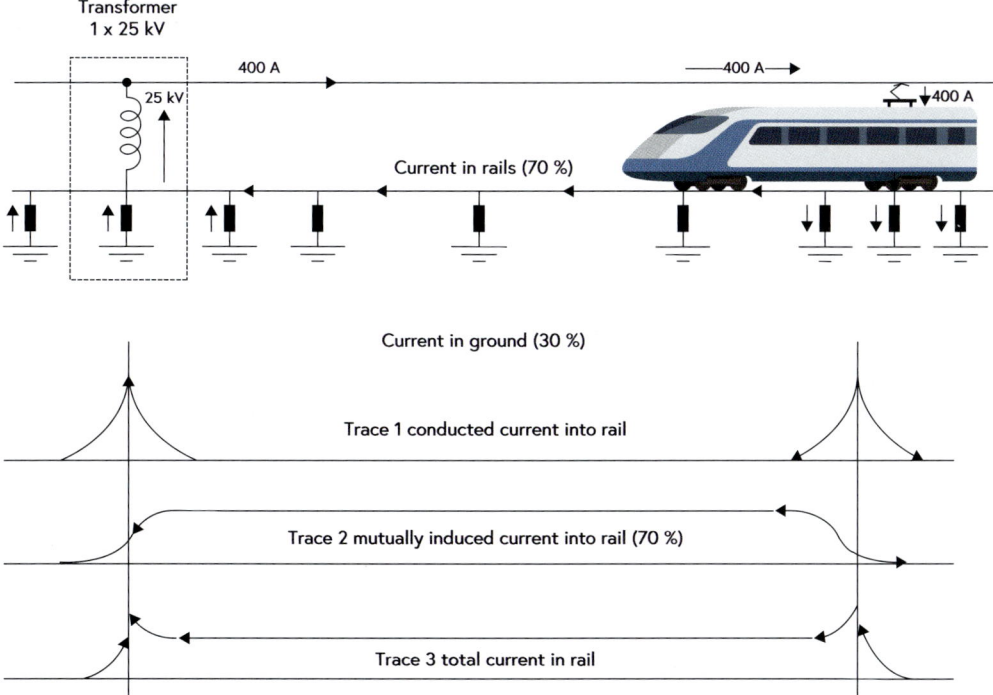

3.5.5   Autotransformer return current

The current supplied to the traction load is from the national grid transformer. The feeding length is dependent on traction loads and can be between 20 km and 50 km, the feeding lengths for 15 kV can be between 30 km and 60 km (EN 50388 Table D.1).

The position of the traction unit determines those locations where traction current can conduct through the rail leakages into the earth. The current supplied to the traction load and the return current in the earth is predominantly between the autotransformers on either side of the traction load. This simplification is shown diagrammatically in Figure 3.5 for a train at the mid-point between autotransformers.

Section 3 – Mass of Earth as part of a traction return circuit

Between the two autotransformers, the return path for the traction load current includes the rails, the earth conductors and the earth itself. The return current in the ground is predominantly between the two adjacent autotransformers, and the resistivity of the ground determines the depth of the current in the earth return path.

Section A – the current between the local autotransformers

The return current in the ground occurs within this autotransformer section that includes the active traction load. This simplification is shown diagrammatically in Figure 3.5.

The traction current that is conducted into the rail conductively typically leaves the rail within 1 to 2 km (Trace 1 in Figure 3.5). The current that remains in the rail is held there by mutual coupling between the overhead line 25 kV conductor and the rails (Trace 2). The total current in the rail is approximately 70 % of the traction load current (Trace 3), and comprises both a conductive element (Trace 1) and a mutually induced element (Trace 2). The current in the ground is approximately 30 % of the load current and is responsible for electromagnetic induction into lineside cables.

Section B – the current upstream of the first autotransformer

Figure 3.5 shows that the current feeding the first autotransformer is 200 A in the contact wire and 200 A in the autotransformer feeder conductor (50 kV). The currents at 50 kV are equal and opposite and 50 % of the 400 A (25 kV) traction load current. The current in the ground in this section of the railway is therefore minimal.

Figure 3.5   Autotransformer return system

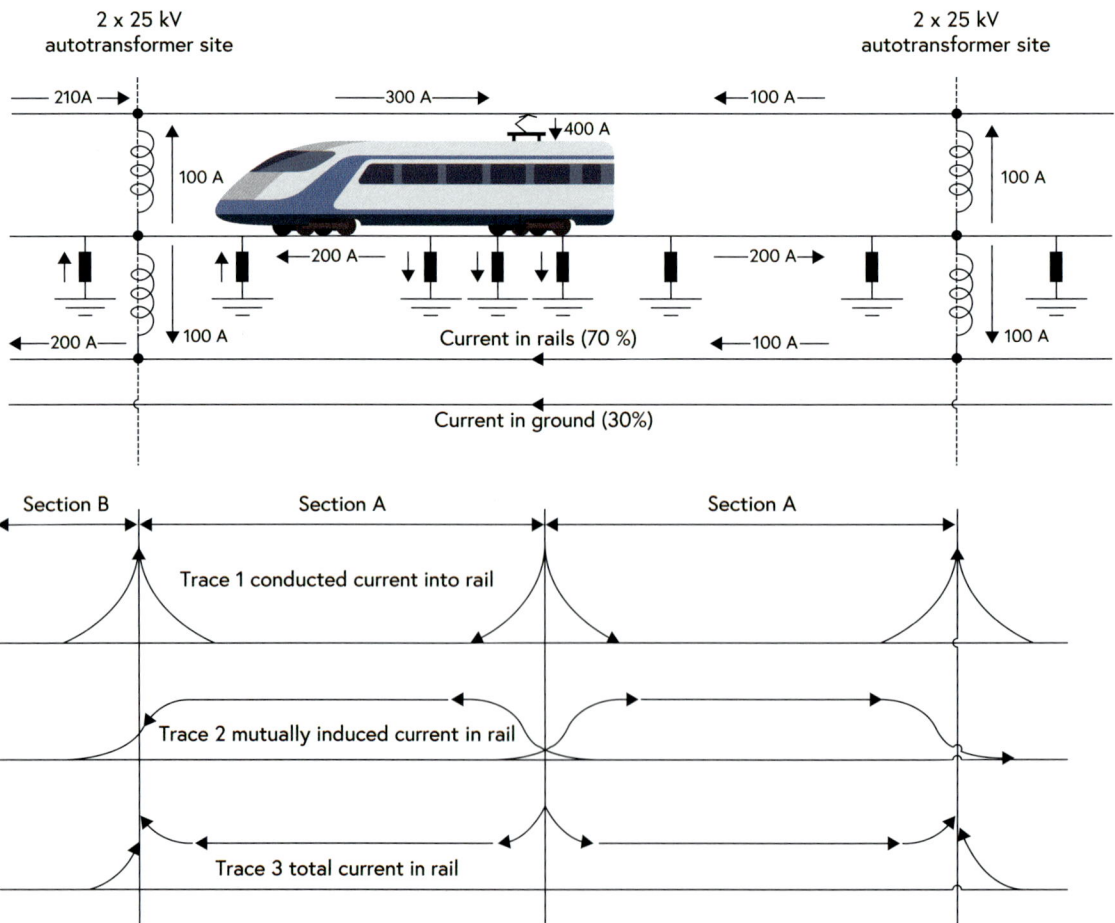

3.5.6 Single-phase booster transformer forced return current

The current supplied to the traction load is from the national grid transformer. The feeding length can be up to 15–40 km (EN 50388 Table D.1).

The return current leaking from the rail into the ground occurs within the booster transformer section that includes the active traction load. This simplification is shown diagrammatically in Figure 3.6 and Figure 3.7. The current into the rails disperses based on Ohm's Law. Where there are parallel tracks and cross-bonding, the current disperses between all the parallel rails. Any differences in the rail current are due to the different mutual coupling of the contact wire and the rails this is expressed mathematically in equations 3.4 and 3.5.

The return path for the traction load current is back to the 'mid-point connector' via the rails and the earth. At the mid-point connector, the current returns via the return conductor and the booster transformers (secondary) to the return current busbar at the feeder station.

In Figure 3.6 the train load is located prior to the mid-point connector, and in Figure 3.7 the train load is located beyond the mid-point connector.

Figure 3.6   Booster transformer with forced return system

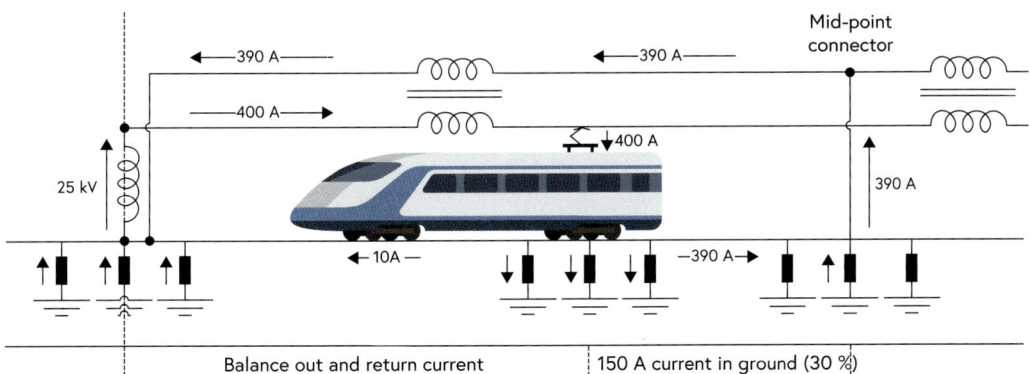

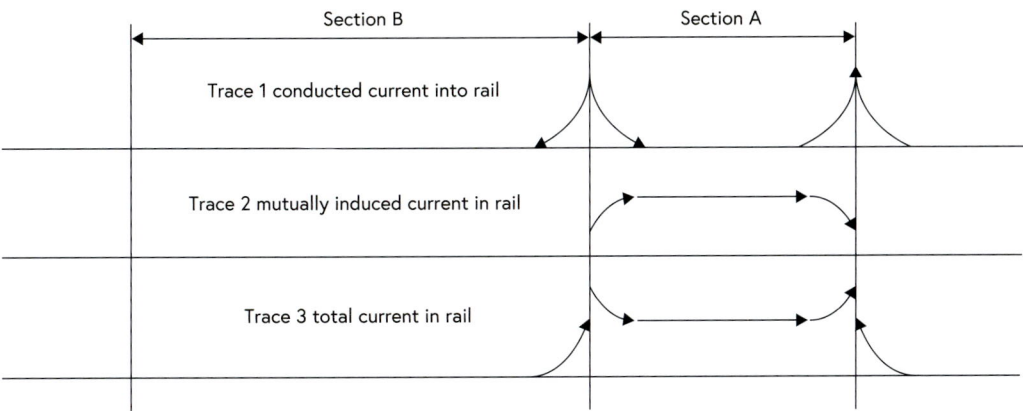

Between the train and the midpoint connection: The traction current that is conducted into the rail conductively and typically leaves the rail within 1 to 2 km (Trace 1 in Figure 3.6 and 3.7). The current that remains in the rail is held there by mutual coupling between the overhead line 25 kV conductor and the rails (Trace 2). The total current in the rail is approximately 70 % of the traction load current (Trace 3),

Section 3 – Mass of Earth as part of a traction return circuit

Figure 3.7 Booster transformer with forced return system

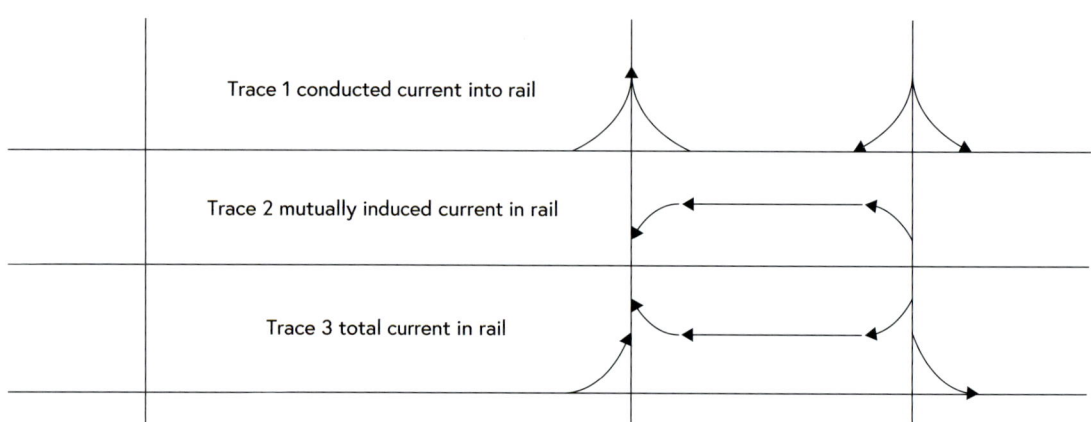

and comprises both a conductive element (Trace 1) and a mutually induced element (Trace 2). The current in the ground is approximately 30 % of the load current and is responsible for electromagnetic induction into lineside cables.

Between the mid-point connector and the feeder station the current in the return conductor is equal and opposite in phase to the current in the contact wire.

3.6 Traction earth return system – electromagnetic field

The AC electrification current flows in the overhead contact line system (OCLS) and returns through the rails and the distributed earth return system. This current flow is responsible for producing magnetic flux lines that cut lineside cables, inducing a voltage into copper data and communications conductors.

The induction into the railway communication system and can be safety-related and performance affecting. This disturbance must be addressed to ensure the railway is both safe and reliable.

The following sections show the different path of the 'out' and 'return' current for rail return and forced traction return circuits. With rail return, the current is towards the feeder station, whereas with forced return systems, the current in the rail can flow towards the feeder station as well as away

Section 3 – Mass of Earth as part of a traction return circuit

from it. This characteristic behaviour significantly increases the complexity of the analysis of electromagnetic induction.

In more recent years, the induction into lineside cables has been mitigated by the introduction of optical fibre communication systems. However, it is still necessary for some railway and non-railway applications to have final copper circuits at the receiving end. These non-railway applications can include telephone companies and the highway's data communication systems. Civil structures that are not adequately grounded can also be affected by induction, including handrails and fences.

Faraday's Law of electromagnetic induction

The magnetic fields of 'p' separate conductors add together by the principle of superposition (equation 3.6). Therefore, where the current is in phase opposition this reduces the total level of the magnetic flux ($\emptyset_{total}$).

Principle of superposition

$$\emptyset_{total} = \sum_{(n=1)}^{p} \emptyset n \tag{3.6}$$

Faraday's Law of electromagnetic induction (equation 3.7) states that an induced voltage is generated where there is a flux linkage (resultant) between the 'out' and 'return' currents and lineside conductors.

Faraday's Law of electromagnetic induction

$$E = -N \frac{d\emptyset_{total}}{dt} \tag{3.7}$$

N the number of turns

$\emptyset$ magnetic flux

3.6.1   Induced voltage and the earth return system

The level of induced voltage into lineside cables can be controlled by creating magnetic fields from the current in both the contact wire and the return circuit that are in phase opposition. This balance is needed to be achieved in both space and time. In an AC railway, to minimize the resultant 50/60 Hz magnetic field, the return conductor should ideally be placed concentrically with the contact wire, however, to do this is impractical. Therefore, the return conductors are mounted on the masts at a height similar to the contact wire to keep the induced voltage to a minimum.

Where the return current is in deep earth, it is then responsible for the flux cutting and subsequent induction of 60 Hz, 50 Hz, 25 Hz or 16.7 Hz voltages into lineside conductors and cables. For safe operation of the railway, and to protect maintainers and the public from electric shock, the induced voltage level should be controlled' Limits are detailed in EN 50122-1 (Section 9.2) and IEC 62128-1 and ITU-T Vol VI: *Danger, damage and disturbance*[1], where national regulations are used these may vary from the limit stated above.

---

[1]ITU-T *Directives concerning the protection of telecommunication lines against harmful effects from electric power and electrified railway lines – Volume VI: Danger, damage and disturbance.*

Section 3 – Mass of Earth as part of a traction return circuit

3.6.2 Traction rail return system

The rail return system distributes the return current between all the rails through rail-to-rail and track-to-track cross-bonding. The proportion of current flows in the deep earth is typically 30–40 %, which occurs between the traction load and feeder station (see Figure 3.8). The depth of the earth return current depends on the Earth's resistivity: the higher the resistivity, the deeper the path.

With 30–40 % of the current in the earth return path, the rail return system has the most significant current-flux imbalance, with no return conductor, and therefore the highest magnetic flux cutting and induced voltage. A 6 kA 50/60 Hz rail return railway will notionally need to limit lineside conductors and cables to less than 1 km to comply with the protective provision for humans detailed in EN 50122-1 (Section 9.2) and IEC 62128-1, in Table D.3.

Figure 3.8 Magnetic field – AC traction rail return system

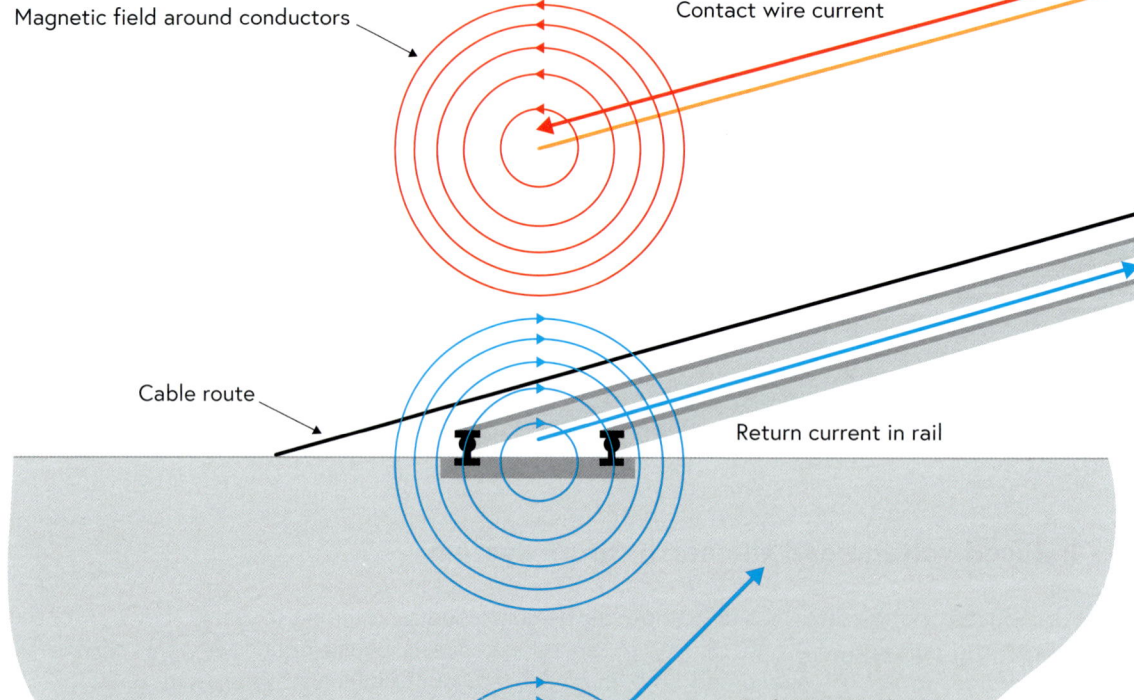

3.6.3 Traction forced return system

3.6.3.1 Booster transformer system

The booster transformer forced return system (see Figure 3.9) has a return current conductor mounted on the overhead line structure. This location is good for the current-flux balance with the contact wire. In this configuration, the traction return current has three possible return paths:

1. the rails;
2. the return conductors; and
3. deep earth.

The current in deep earth between the traction load and the mid-point connection is mainly responsible for determining the level of imbalance in the resultant magnetic field. This resultant field creates flux cutting of lineside conductors and the subsequently induced voltage into lineside communications cables.

With booster transformers located at typically 3.2 km, this system controls the magnetic flux cutting and induced voltage. Typically, in a 6 kA railway this forced return system will notionally need to limit lineside conductors and cables to less than 10 km to ensure the induced voltage is below the safety limits of EN 50122-1 (Section 9.2) (IEC 62128-1 (Section 9.2)).

3.6.3.2 Autotransformer system

The autotransformer forced return system (see Figure 3.9) has a −25 kV auto feeder conductor mounted on the overhead line structure and this location is good for the magnetic flux balance. In this configuration between autotransformer sites, the traction return current has three possible return paths:

1. the autotransformer feeder conductor;
2. the rails; and
3. deep earth.

The current in deep earth between the traction load and the autotransformer site is mainly responsible for determining the level of imbalance in the resultant magnetic field. This resultant field creates flux cutting of lineside conductors and the induced voltage into lineside communications cables.

With autotransformers located at 5 km, this system can control the magnetic flux cutting and induced voltage. Typically, in a 12 kA railway this forced return system will notionally need to limit lineside conductors and cables to less than 5 km to ensure the induced voltage is below the safety limits of EN 50122-1 (Section 9.2) (IEC 62128-1 (Section 9.2)).

Figure 3.9 Magnetic field – forced return systems (return conductor and autotransformer)

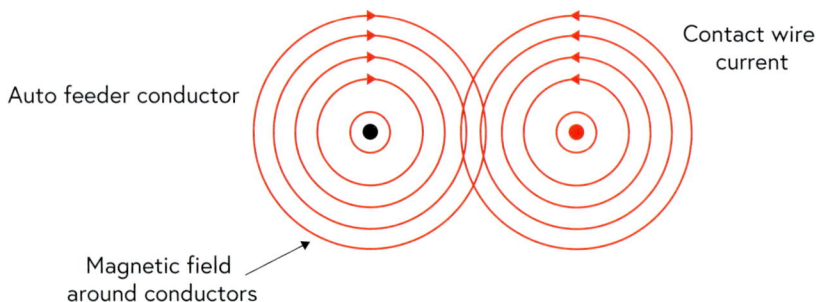

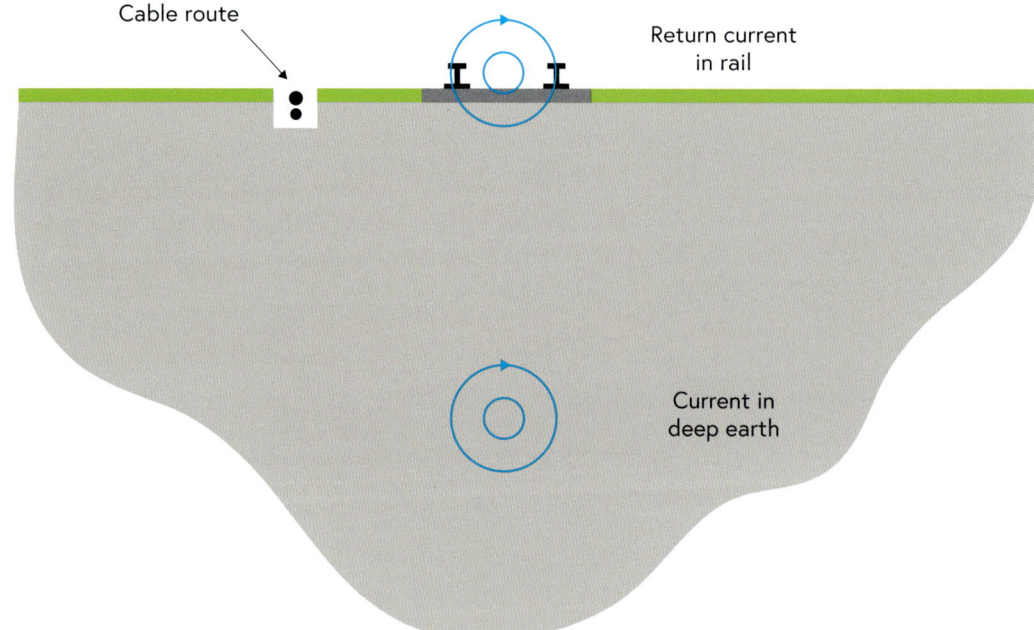

3.7   Modelling AC railway electrification systems

Modelling examples are given in the following sections for the short-circuit characteristic of a 25 kV 'single-phase rail return', a 'single-phase booster transformer' and an 'autotransformer' distribution systems. The three main system characteristics that have been analyzed include the magnitude of the short circuit, the rail potentials at the location of the earth fault and the induced voltage due to the changing magnetic flux linkage.

The following examples are for a 6 kA (AC single-phase) and 12 kA (AC autotransformer) systems and the results should not be treated as representative of any particular railway.

3.7.1   Modelling a 25 kV 50 Hz single-phase rail return 6 kA railway

Figure 3.10 shows the results for the modelling of a 6 kA short-circuit characteristic for a 25 kV single-phase rail return system. The model depicts:

Section 3 – Mass of Earth as part of a traction return circuit

(a) a 25 kV electrified railway from 0 km (traction substation) to 9.6 km (traction switching station);
(b) the short circuit is applied at each cross-bond, and mid-points between cross-bonds;
(c) the railway modelled is a two-track railway with single-rail return and 320 m cross-bonding;
(d) a mutual screening conductor has been included; and
(e) copper telecommunications cable is positioned between 0 km and 2 km.

The traces in Figure 3.10 graphically depict:

(a) the fall in the magnitude of the short-circuit current with distance from the substation;
(b) the rail potentials at the point of the short circuit – the voltage is recorded is at each cross-bond and mid-location between cross-bonds; and
(c) the induced voltage for a cable exposed along the 2 km route.

This induced voltage characteristic of the rail return system is the worst-case scenario as the magnitude of the induced voltage rises in a near-linear fashion with distance from the traction substation and is approximately 300 V per km with a mutual screening conductor.

Figure 3.10 6 kA short-circuit characteristic with rail return

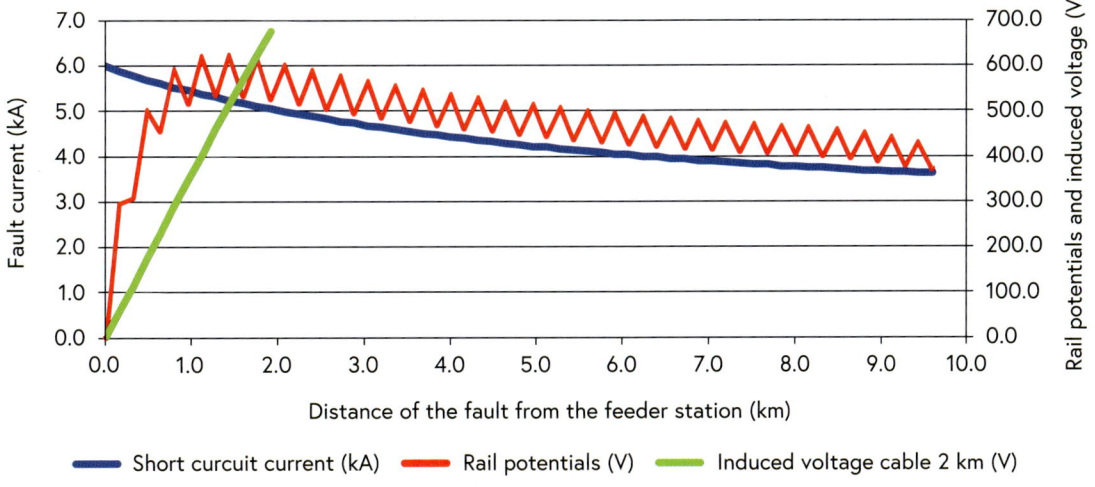

3.7.2 Modelling a 25 kV 50 Hz single-phase booster and return conductor

Figure 3.11 shows the results for the modelling of a 6 kA short-circuit characteristic for a 25 kV single-phase booster transformer and return conductor system. The model depicts:

(a) a 25 kV electrified railway from 0 km (traction substation) to 9.6 km (traction switching station);
(b) booster transformers are located at 1.6 km, 4.8 km and 8 km;
(c) the short circuit is applied at each cross-bond, and mid-points between cross-bonds;
(d) the railway modelled is a two-track railway with single-rail return and 320 m cross-bonding;
(e) a mutual screening conductor has been included; and
(f) copper telecommunications cable is positioned between 0 km and 9.6 km.

Section 3 – Mass of Earth as part of a traction return circuit

Figure 3.11 6 kA short-circuit characteristic booster transformer with return conductors

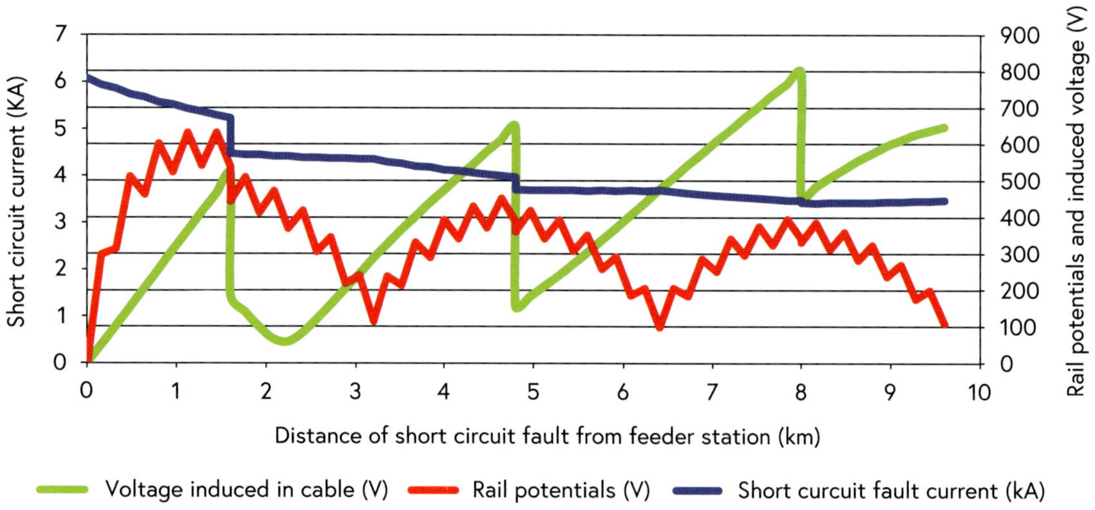

The traces in Figure 3.11 graphically depict:

(a) the fall in the magnitude of the short-circuit current with distance from the substation;
(b) the rail potentials at the point of the short circuit – the voltage recorded is at each cross-bond and mid-location between cross-bonds;
(c) the induced voltage for a cable exposed along the 9.6 km route (note the 645 V limit of EN 50122-1 is marginally exceeded); and
(d) the short circuit applied at each cross-bond, and mid-points between cross-bonds.

The induced voltage characteristic is a typical sawtooth-shape. This is because a short circuit located at the mid-point connector causes a balance in the currents of the return current conductor and the contact wire.

3.7.3   Modelling a 25 kV-0-25 kV autotransformer 12 kA short circuit

Figure 3.12 shows the results for the modelling short-circuit characteristic of an autotransformer. The model depicts:

(a) an autotransformer-fed electrification system between 0 km (autotransformer substation) and 5 km and 10 km (autotransformer sites);
(b) a short circuit is applied along the railway at each cross-bond and mid-location between cross-bonds;
(c) the railway modelled is a two-track railway with double-rail return (500 m cross-bonding);
(d) overhead line masts are bonded to AECs;
(e) a buried earth conductor also acts as a screening conductor; and
(f) copper telecommunications cable is 10 km long.

The traces in Figure 3.12 graphically depict:

(a) the fall in the magnitude of the short-circuit current with distance from the substation;
(b) the rail potentials at the point of the short circuit – the voltage recorded is at each cross-bond and mid-location between cross-bonds;

(c) the rail potentials at cross-bond and mid-location between bonds; and

(d) the induced voltage for a cable exposed along the 10 km route.

Figure 3.12   12 kA short-circuit characteristic 25 kV-0-25 kV autotransformer

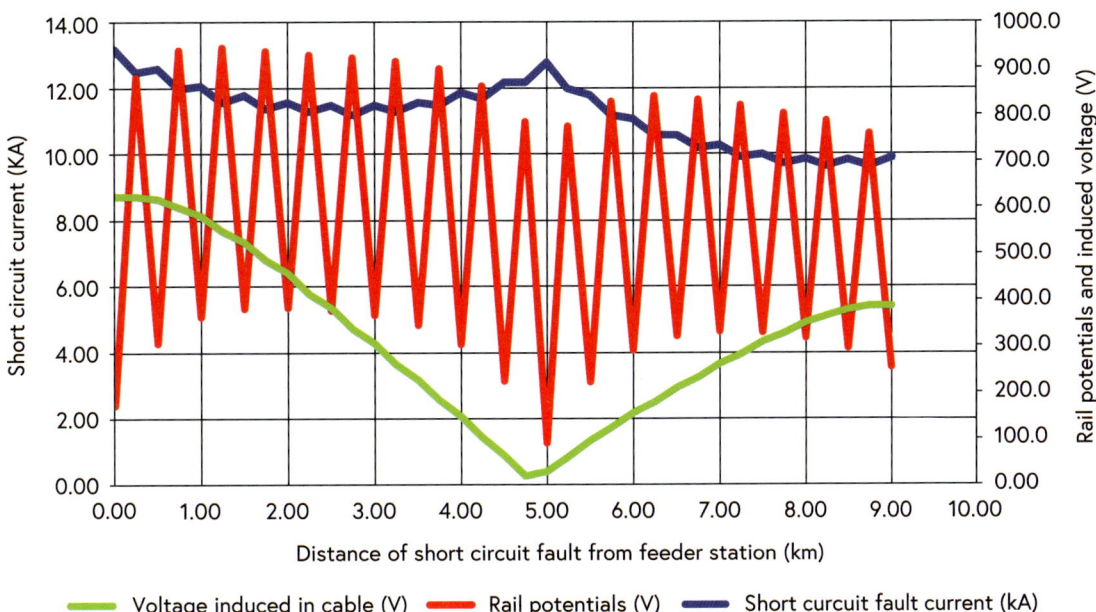

Where the short circuit is located at the mid-section between the autotransformer sites the currents are approximately equal in magnitude and opposite in phase, hence nearly cancelling each other out when calculated for a cable exposed along the 10 km. This creates an induced voltage with a typical V-shape

3.8   Electromagnetic interference to non-traction systems[2]

The alternating current that flows in the contact and catenary wire creates a longitudinal EMF in conductors that are parallel to it. Typical parallel conductors in an electrified railway are communications cables, traction rails, AECs installed on structures, cables (for example, lead sheath, steel armour or copper strands), and metal pipes. The induced EMF increases from the location where the conductor is earthed. The magnitude is a function of the length of the parallelism and other factors, such as the separation distance. If the parallel conductor's remote end is also connected to the earth, then a current can circulate through it, this is in the opposite direction to the inducing current.

The AC electrified railway is an earthed system, and therefore the return current flows in the earthed rails, earthed conductors and deep earth. Additionally, this return path may include third-party parallel earthed conductors and third-party utilities. This stray current is likely to cause interference to these systems, including danger (touch voltage and malfunction), damage (overheating or fire)

[2]NR BR 13422: *50Hz Single Phase AC Electrification, Immunisation of Signalling & Telecommunications Systems Against Electrical Interference*

and disruption (service). This current may affect the rating of the conductor and its operational temperature rise.

3.8.1 Electromagnetic disturbance

Where railway overhead lines are parallel with lineside insulated conductors (railway) or buried pipelines (third party), the AC currents of the electrification system induces a voltage into these assets. The lineside insulated conductors and assets experience an EMF that can result in touch voltages from short-term (earth faults) and long-term (train load) events. These induced voltages may represent a danger to personnel touching the asset[3]. This subject is addressed in detail in Section 5.

Such inductive effects are due to time-varying changes in the 50/60 Hz line current that is flowing in overhead line conductors. The current is in the contact wire and in the return circuit which is shared between earth and return conductors. These conductor currents are responsible for creating a time-varying magnetic field. The resultant 50/60 Hz magnetic field is shown in Figure 3.10 and 3.11. Ideally, for no induction, the out and return circuit should be coincident and, in this case, the resultant flux would approach zero.

3.8.2 Calculation of the electromagnetic inductive coupling

The electromagnetic interaction between the overhead lines and lineside conductors is due to Faraday's Law of electromagnetic induction. This induced voltage into insulated conductors is dependent on several parameters, including[4]:

(a) type of traction power supply system used:
  (i) rail return;
  (ii) booster transformer system; or
  (iii) autotransformer system;
(b) additional return conductors;
(c) geometry and distance between the electrification conductors and the affected conductor;
(d) length of parallelism;
(e) frequency of the power system (60 Hz, 50 Hz, 25Hz or 16.7 Hz);
(f) the level of the inducing traction current;
(g) soil resistivity;
(h) length of parallelism of the copper cable;
(i) mutual coupling between power circuit and copper cables;
(j) separation of the feed and return conductors (see Figures 2.7 and 2.16);
(k) earth proximity effects; and
(l) intentional and unintentional screening effects by other metallic structures or effects of conductance to earth of the running rails etc. (civilization factor).

---

[3]*ITU-T Directives concerning the protection of telecommunication lines against harmful effects from electric power and electrified railway lines – Volume VI: Danger, damage and disturbance*
[4] *ITU-T Directives concerning the protection of telecommunication lines against harmful effects from electric power and electrified railway lines – Volume II: Calculating induced voltages and currents in practical cases*

Section 3 – Mass of Earth as part of a traction return circuit

3.8.2.1   Calculation for parallel conductors with a common earth return

The most significant form of inductively coupled telecommunications systems is where copper circuits are laid adjacent to AC electrification systems. Where there is no forced return, the mutual coupling is high, with a large inductive loop being set up between the traction overhead (outgoing feed) and the earthed rail return system. The analysis used is normally based on ITU-T Vol II which is defined in terms of the mutual coupling per unit length with the Carson-Clem equation.

This method relies upon the use of Carson's expressions for the computation of conductors' self and mutual impedances in the presence of semi-infinite earth. That is, the calculated impedances include the earth, acting as the return path of the traction return circuit. Carson's results were expressed in terms of convergent infinite series, however, (for the sake of convenience) simplified expressions are normally used.

The per-unit-length self-impedance Z of an above-soil conductor with earth return can be expressed as per the equations 3.8a, 3.8b and 3.8c for the self-impedance of conductor with earth return (refer to ITU-T *Directives - Volume II: Calculating induced voltages and currents in practical cases*).

(This is based on ITU-T Directive Vol II Equation 4.1-16[5]): Function for calculating the self-impedance per metre

$$(Z_{\text{si}}) = \left[R_i + 0.99 \times 10^{-3} f + j\omega \times 2 \times 10^{-4} \left(\ln \frac{r_i}{r_i{}'} + \ln \frac{D_e}{r_i} \right) \right] \Omega/\text{km} \tag{3.8a}$$

Function for calculating the mutual impedance per metre between conductors

$$(Z_{\text{Si}}) = \left[R_i + 0.99 \times 10^{-3} f + j\omega \times 2 \times 10^{-4} \ln \frac{D_e}{r_i{}'} \right] \Omega/\text{km} \tag{3.8b}$$

Equivalent depth of hypothetical earth return

$$D_e = \frac{1.852}{\alpha} = 659\sqrt{(\rho \setminus f)} \tag{3.8c}$$

The induced voltage into a data lineside conductor is based on the summation of the mutual impedance of each of the conductor currents as shown in equations 3.8a–c. The number of conductors in an electrified railway is typically 12 conductors for a double-rail two-track autotransformer railway and 20 conductors for a double-rail four-track autotransformer railway. This includes contact, catenary, rails, aerial earth conductors, buried earth conductors and autotransformer feeder conductors. To calculate the resultant induced voltage, it is necessary to undertake a summation of all the voltages included by each separate conductor, as shown in equation 3.9.

---

[5]ITU-T *Directives concerning the protection of telecommunication lines against harmful effects from electric power and electrified railway lines – Volume II: Calculating induced voltages and currents in practical cases.*

Section 3 – Mass of Earth as part of a traction return circuit

Induced voltage

$$(V_{\text{induced}}) = \sum_{k=1}^{n} I_n \cdot XM_n \cdot \text{length}(\text{km}) \cdot J \qquad (3.9)$$

where:

> V is the induced voltage (V)
> I is current (A)
> XM is the mutual reactance (Ω)
> J is a complex operator
> n is an integer.

The calculation is undertaken as part of a multi-conductor model of the electrified railway, where the 'branch currents' is multiplied by the mutual impedance matrix, as given by equation 3.10.

Matrix induced voltage calculation

$$\begin{pmatrix} v(1) \\ v(2) \\ v(n) \end{pmatrix} = \begin{pmatrix} XM_1 \times \text{length} & 0 & 0 \\ 0 & XM_2 \times \text{length} & 0 \\ 0 & 0 & XM_n \times \text{length} \end{pmatrix} \begin{pmatrix} i(1) \\ i(2) \\ i(3) \end{pmatrix} \qquad (3.10)$$

where:

> length is the length of the conductor that is parallel to the electrification conductors
> $v(1)$ is the induced voltage due to the current in conductor (1)
> $i(1)$ is the current in conductor (1).

3.8.3   Study of induced voltage calculation

A simple study of the induction defined in ITU-T Vol II is provided in 50122-3 Annex A and EN IEC 62128-1 Annex A: this annex gives information on how to determine the dimensions of the zone of mutual interaction. The purpose is not to calculate accurately the voltage coupled into the DC system due to an adjacent AC system.

IEC 62128-3 and EN 50122-3 Example 1 shows the calculation of the distance between the tracks of an AC railway and a DC railway for the zone of mutual interaction in the case of a system with return conductors. This example indicates that the electrified railway will create significant interference (induced voltage) into third-party systems.

3.8.4   Psophometric weighting

The purpose of psophometric weighting is to emphasize the parts of the audible spectrum that ears perceive most readily, and attenuate the parts that contribute less to the perception of loudness, to get a measured figure that correlates well with subjective effect. A major use of noise weighting is in the measurement of residual noise in audio equipment, usually present as hiss or hum in quiet moments of programme material.

Section 3 – Mass of Earth as part of a traction return circuit

3.8.4.1 Psophometrically weighted primary current

Psophometric weighting refers to any weighting curve used in the measurement of noise. In the field of audio engineering, it has a more specific meaning, referring to noise weightings used especially in measuring noise induced in telecommunications circuits.

Consultative Committee for International Telephony and Telegraphy (CCITT) and C-message weighting filters are bandpass filters used to measure audio-frequency noise in telephone circuits. The ITU-T filter is used for international telephone circuits. The C-message filter is typically used for North American telephone circuits.

The frequency response of the psophometric and C-message weighting filters are specified in the ITU-T O.41 standard and Bell System Technical Reference 41009, respectively. Figure 3.13 shows the relative attenuation defined for the psophometric and C-message weighting filters.

Psophometric weighted current is defined as the rms addition of all the harmonic currents in the traction units primary current wave-shape, each harmonic first being attenuated in accordance with the

Figure 3.13   Comparison between psophometric and C-message average weighting

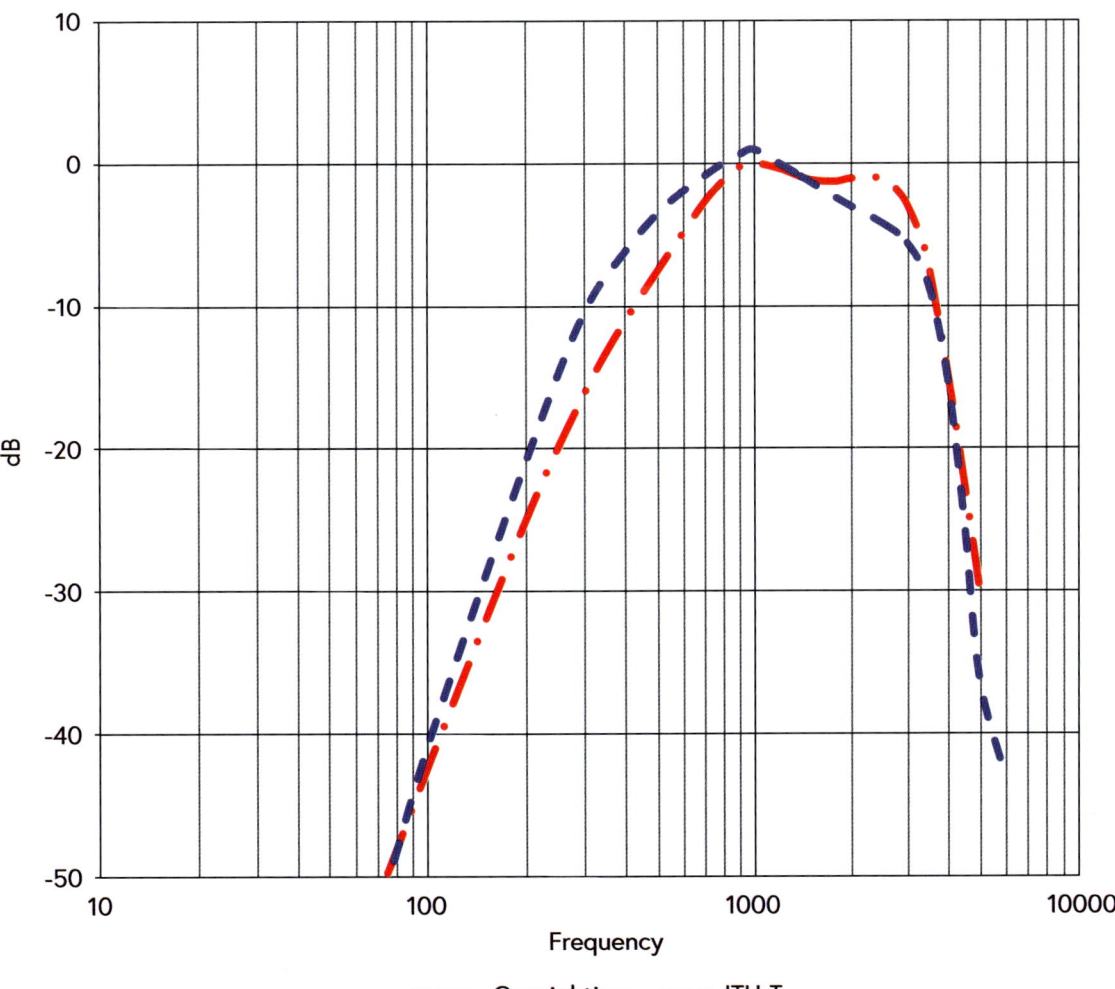

appropriate ITU-T weighting curve, which is defined in Figure 3.13. The psophometric traction unit primary current, therefore, is an indication of the level of interference that will be produced at that point in time for the traction unit. The interference mechanism is by electromagnetic induction from the high traction power into the low-power telecommunications networks. Only conventional telecommunications networks are affected, digitized and optical links are inherently immune from such interference. The relationship between measurements made with the North American noise meter and the ITU-T psophometer, as shown in Figure 3.13, is dependent on the frequency spectrum of the noise being measured (see Equation 3.11).

Psophometrically weighted primary current

$$\text{Ipso} = \sum_{1}^{n} I_n^2 \times P_n^2 \qquad (3.11)$$

where:

  n is the harmonic number
  p is the psophometric weighting factor of the nth harmonic

3.8.5   Screening and the screening conductor

The principle of electromagnetic screening of low voltage data and communications cables is based on Lenz's Law, which is named after Emil Lenz, and says:

  An induced electromotive force (EMF) always gives rise to a current whose magnetic field opposes the change in original magnetic flux. The direction of the induced EMF is always such as to result in opposition to the change producing it.

The difference between Faraday's Law and Lenz's Law is that Faraday's Law of Electromagnetic Induction tells us the magnitude of the EMF produced, whereas Lenz's Law tells us the direction in which that current will flow. Therefore, when a screening conductor is located in the troughing or on a cable tray within close proximity to data and voice cables, the screening conductor will conduct a portion of the traction return current. This is due to both mutual and conductive coupling, and this current produces a magnetic field in opposition to the original field.

There are different arrangements for screening conductors:

(a) a mutual screening conductor which is independent of the traction return circuit is located in the troughing and is earthed every 1 km with a typical earth electrode of 4 Ω;
(b) a return current screening conductor which is located in the troughing and bonded to the rails at every track-to-track cross-bond; or
(c) a buried earth conductor which is located beneath the troughing and is bonded to the rails at every track-to-track cross-bond.

Screening factor is the ratio of the induced voltage in a conductor in the presence of the screening conductor to the induced voltage which would be induced in the conductor in the absence of the screening conductor.

The level of reduction in electromagnetic induction is related to the spatial positioning and the phase displacement and magnitude of the traction return current flowing through the screening conductor. This behaviour is relative to conductors in the contact line system and is depicted in Figure 3.14. The reduction in the resultant magnetic field provides a reduction in the voltage (sometimes called immunization) that is induced into the data and voice communications cables. Typically, a 'mutual screening conductor' will give a screening factor of between 0.6 and 0.7. A 'return current screening conductor' gives a screening factor of less than 0.5.

A screening factor of 0.25 would imply that the screening reduced the induced EMF by 75 %. It should be noted that the lower the screening factor, the better the screening and this is always less than unity.

Figure 3.14   Principle of a screening conductor

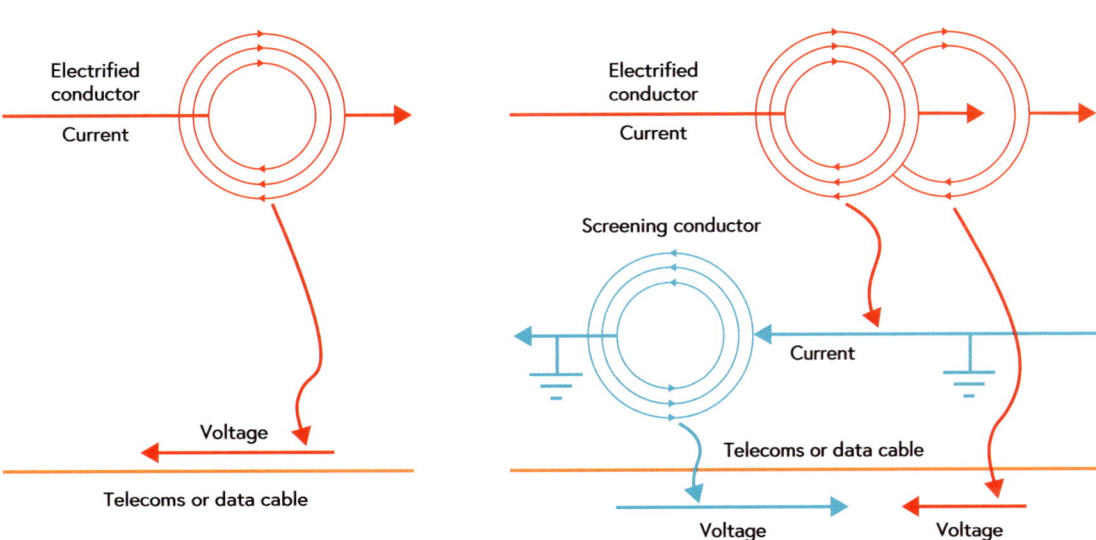

The induction from electrified conductor into telecoms cable

The current induced from contact wire and screening conductor cancel in the telecoms cable

3.9   Electrostatic interference

The electric fields emanate from any HV AC conductor (11 kV, 15 kV, 25 kV, 275 kV or 400 kV) and the lines of the field travel from the conductor which is energized and terminates on any conducting surfaces (see Figure 3.15). The electric field creates a standing voltage on any unearthed conductive surface that is mounted between the ground and the live conductor, including fences, metal gutters and unearthed overhead line conductors.

The electric field produced can be visualized by lines whose direction at each point is the same as the field's, which was a concept introduced by Michael Faraday. The strength of the electric field is proportional to the density of the lines, and the electric field lines represent paths that a positive charge would follow as it is forced to move within the field, as shown in Figure 3.15. Field lines have several important properties:

(a) they always originate from a positive charge and terminate at a negative charge;
(b) they enter all conductors at right angles; and
(c) they never cross or close in on themselves.

Section 3 – Mass of Earth as part of a traction return circuit

Figure 3.15   Electric field

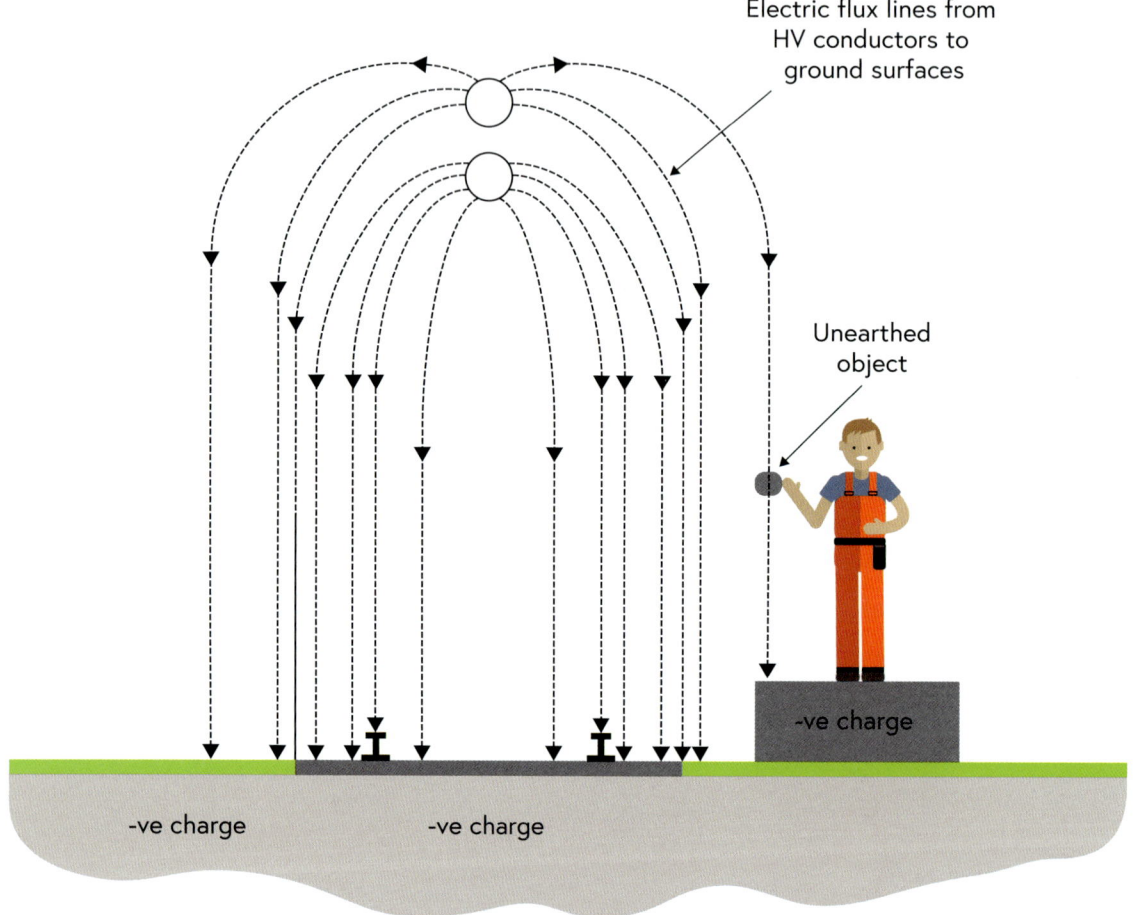

When conductors are laid out in parallel, the capacitance which exits between the two conductors is proportional to the separation and the length of the parallelism. For human protection, there is a requirement for any exposed conducting surfaces to be bonded when they are in the presence of the electric field of an AC railway. An unearthed conductive structure may give an electric shock to persons touching exposed unearthed surfaces shown by the human figure in Figures 3.15 and 3.16.

The voltage that stands on a conductor is proportional to the capacitance (distance) between the 25 kV conductor, the unearthed conductor and the ground, as shown in Figure 3.16.

3.9.1   Operation and maintenance of the overhead lines

Induction problems can occur where there are unearthed overhead line conductors suspended in air, particularly when railway overhead conductors are being installed, maintained, or removed in close proximity to other live AC conductors or live HV power lines crossing the railway.

The construction and maintenance procedures for the AC railway should include mitigating voltages induced (electric and magnetic fields) on unearthed conductors due to the close proximity of overhead

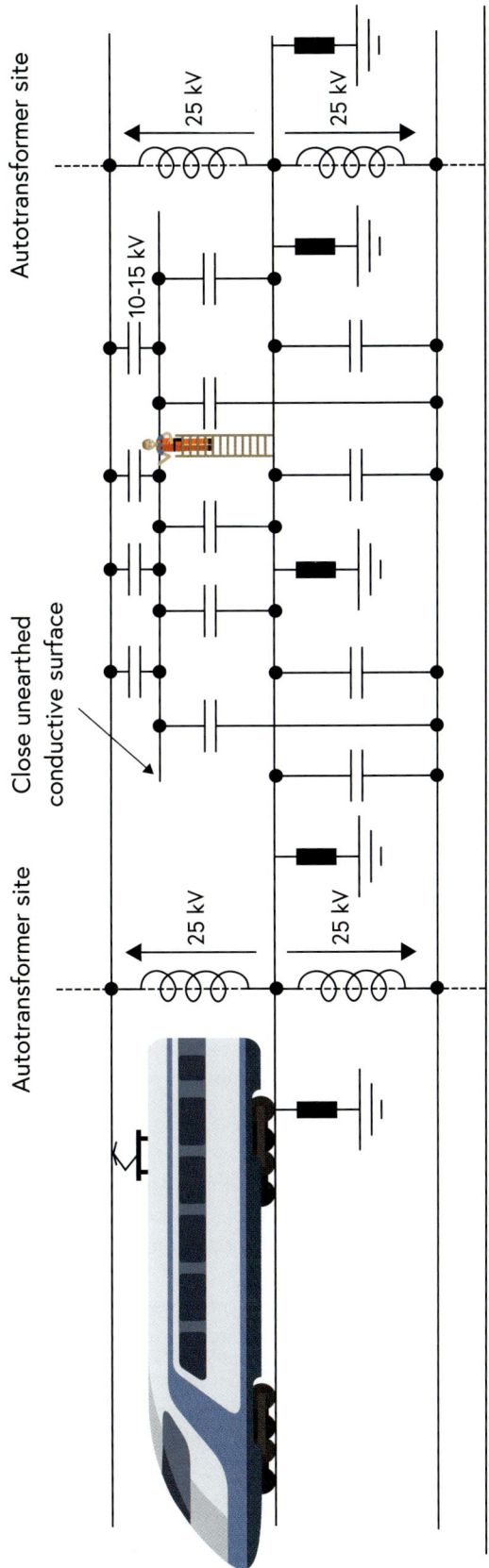

Figure 3.16 Capacitive coupling between conductors

live conductors. Where AC overhead lines are de-energized, they are required to be earthed to prevent any standing voltages, due to the electric or magnetic fields.

Where AC overhead lines are bonded to earth, these parallel conductors can become a legitimate AC return path, due to the mutual coupling during both train load and an overhead line short circuit. All necessary precautions must therefore be taken when applying and removing temporary and permanent earth bonds.

This matter is addressed in more detail in Section 11.

3.10   Pipeline electromagnetic interference

Pipelines interference by electromagnetic induction is addressed in the standard EN 50443[6].

This European standard deals with the maximum tolerable limits of the interference effects, taking into account the AC electrification system's behaviour, both in normal operation and during faults. The presence of AC electrification traction systems may cause voltages to build up in pipeline systems, running parallel, due to one or more of the following mechanisms:

(a) inductive coupling;
(b) conductive coupling; or
(c) capacitive coupling.

The build-up of AC interference voltages may cause danger to persons, damage to pipelines, or disturb the electrical/electronic equipment connected to the pipeline.

3.10.1   Interference distance (based on EN 50443)

Interference distance is defined in EN 50443 as "the maximum distance between the pipeline system and AC traction supply for which an interference shall be considered". The AC traction supply is likely to cause interference and the objective of interference distance is to identify any potential interfering systems to be assessed. It is necessary to consider the following distance between the railway and the metallic pipeline:

(a) in rural areas, for a soil resistivity below 3,000 Ω.m, an interference distance of 1,000 m between the AC railway line and the metallic pipeline system should be considered. In cases of soil resistivity values greater than 3,000 Ω.m, the interference distance value, in metres, should be equal to the soil resistivity value, in Ω.m, divided by three.
(b) in urban areas, the previous interference distance may be decreased, taking into account the environmental reducing factor of the metallic structures existing in these areas. The interference distance should not be less than 300 m.

NOTE: Typical values for the environmental reducing factor are 0.1 to 0.7 (refer to ITU-T K.68:2008 Appendix II).

---

[6]EN 50443 *Effects of electromagnetic interference on pipelines caused by high voltage AC electric traction systems and/or high voltage AC power supply systems*

Section 3 – Mass of Earth as part of a traction return circuit

3.11 Conductive interference due to AC return currents in the ground

The AC electrified railway is earthed along its entire length, this is achieved through the rail leakages and the overhead line mast structure foundations. A percentage of the current conducts to earth through these paths. This leakage current can become large and exceed tens of amps over a length of 2 to 3 km.

This current is able to use various unintended paths, including insulated cable screens, armours, protective earth and conductive civil structures, including metal overline or underline bridges. The current can be responsible for damage to conductive utilities as well as the cable screens and armours due to overheating or fire. It can also cause damage due to the possibility of mechanical vibration and AC corrosion.

HV electrical faults can create EPR at many locations including electrical substations, switching sites, and where the railway is close to HV transmission lines. When an earth fault occurs, the short-circuit current flows through the structure or equipment and into the earth electrode. The fault current and the earth's resistance produces a potential rise with respect to remote earth. The resulting EPR can cause hazardous voltage, many hundreds of metres away from the actual fault location.

The fault type (short-circuit), fault location, return paths and typical ground potentials are given in Table 3.2.

3.11.1 Earth potential rise at National Grid sites

At substations switching sites and near tower lines, the EPR is a particular concern because the high potential may be a hazard to both maintainers, passengers and equipment. A potential gradient in the ground may be high enough that a person or animal is injured due to the voltage gradient (step potential) that develops between their feet. A person can also receive an electric shock due to the potential difference (touch potential) between the ground they are standing on and an exposed-conductive-part.

Hazards exist where utilities are connected to the earth of an HV substation. The exposed parts of utilities (for example, telephones, fences, piping) may become energized during an HV earth fault. This can then cause a transferred potential to maintainers who are remote from the substation.

3.11.2 Modelling of earth potential rise

Modelling the electrical conductivity of the foundations of a transmission tower and also the conductivity of a substation earth mat has been undertaken. This parametric study assesses the EPR and is based on varying the following parameters:

(a) homogeneous resistivity (Ω.km); and
(b) distance from the earth fault (m).

EPR has been analyzed using the modelling as specified in ENA Engineering Recommendation EREC S34 Issue 2 *A guide for assessing the rise of earth potential at electrical installations*[7]. This outlines the methods used in the power industry to assess EPR profiles at substation sites. The models are based on fault current passing to the ground via the earth electrode at the site.

---

[7]Engineering Recommendation EREC S34 Issue 2, November 2018 *A guide for assessing the rise of earth potential at electrical installations.*

Table 3.2   Typical earth faults on 25 kV electrified railways

Fault type	Fault location	Typical fault level	Return path	Typical ground potential	Likelihood of an earth fault
Tower line earth fault					
Grid HV earth fault	Tower line mast	10/20 kA	Earth foundations and tower chain earth	<10 kV	Low
Substation earth fault					
Transformer HV (bushing) to transformer case (earth fault)	Grid transformer	10/20 kA	Grid site earth mat and tower chain earth	<1 kV	Low
Transformer HV winding fault to transformer case (earth fault)	Grid transformer	10/20 kA	Grid site earth mat and tower chain earth	<1 kV	Low
25 kV winding fault to transformer case (earth fault)	Grid transformer and dis-connector compound	6/12 kA	Transformer neutral	<645 V	Low
25 kV fault to neutral/earth (earth fault)	Grid transformerand dis-connector compound	6/12 kA	Transformer neutral	<645 V	Low
Railway earth fault					
25 kV line fault to neutral/earth	25 kV feeder station	6/12 kA	Traction negative return at feeder station (the % into the ground is low)	<645 V	Low
25 kV line fault to neutral/earth	OCLS Line to earth fault	6/12 kA	Traction negative return and earth return (30/40 % of 6 kA enters the ground)	<645 V	High (De-wirement)
25 kV line fault to rail (traction return/earth) due to dewirement	25 kV traction switching station	3/6/9 kA	Traction return conductors (the % into the ground is low)	<645 V	Low
Non-conductive structure (non-conductive bridge without mitigation)	Line of route	<6 kA	Earth return (100% of kA enters the structure)	>1 kV	High (Bird strike)

Section 3 – Mass of Earth as part of a traction return circuit

The following are the parametric variation which are specified in the modelling method that is described in more detail in the following sections:

(a) each fault is modelled for varying soil resistivities from 10 – 2,000 Ω.m, which covers most earth types of soil as shown in BS 7430:2011+A1:2015 Table 1;
(b) soil resistivity is a key parameter as it directly affects the level of earth potential rise (EPR) and its profile; and
(c) the modelling assumes that soil resistivity is homogenous with distance.

Typical ground potential due to 400 kV, 132 kV and 25 kV high voltage (HV) earth faults are provided in Table 3.2 Typical earth faults on 25 kV electrified railways.

3.11.3 Modelling of EPR: HV tower line structure

HV transmission towers are responsible for creating an EPR due to an insulator flashover or a conductor earth fault. The value of the EPR is dependent on the structure footing and the ground resistivity. The earth fault current enters the earth via the tower structure foundations and also through the tower line chain impedance. This impedance of the chain (or ladder) network of an overhead line earth wire with its connections to earth via metal lattice towers along its route, or an insulated cable's sheath that has connections to earth via installations along its length.

The magnitude of the EPR depends on several factors, including the design of the tower footing, the resistivity of the ground and the magnitude of the earth fault current. A parametric variation for EPR on a tower line structure foundation has been undertaken using a proprietary mathematical package. The calculation was based on the formulae given in EREC S34 'Formula P6 Voltage Profile around Earth Electrode: B3.7: Column P 6.3'.

The earth electrode used in the modelling is based on a reinforced concrete tower footing. The shape of the earth electrode at the site determines the current in the ground and the current density away from the electrode. This then determines the characteristics of the drop of EPR with distance from the site.

Modelling and assumptions

In the calculation for EPR the following assumptions have been made:

(a) phase-to-earth fault current is 20 kA;
(b) a 10 % proportion of the fault current at a pylon site goes to earth via the tower footings; and
(c) the pylon site is assumed to be a single mast in a line of transmission towers, the foundations and earth wire forming a parallel earth network.

The tower footing is modelled as a hemispherical earth electrode, which is indicated in Section B3.7 Column P6.1 of EREC S34[8] issue 2.

The EPR profile has been modelled in a mathematical modelling tool. Figure 3.17 shows the variation in ground voltage in volts/kA with respect to distance from the tower footing. Soil resistivity is a key parameter as it directly affects the level of earth potential rise (EPR) and its profile. The modelling assumes that soil resistivity is homogeneous with distance.

---

[8]Engineering Recommendation EREC S34 Issue 2, November 2018 *A guide for assessing the rise of earth potential at electrical installations.*

Section 3 – Mass of Earth as part of a traction return circuit

Each fault is modelled for varying soil resistivities from 10–2,000 Ω.m, which covers most earth types of soil as shown in BS 7430:2011+A1:2015 Table 1.

For reference, purpose lines are drawn for 430 V[9], 645 V (IEC 62128-1&EN 50122-1 effective touch voltage limit) and 1,000 V (cable insulation).

Figure 3.17   Modelling of transmission tower EPR

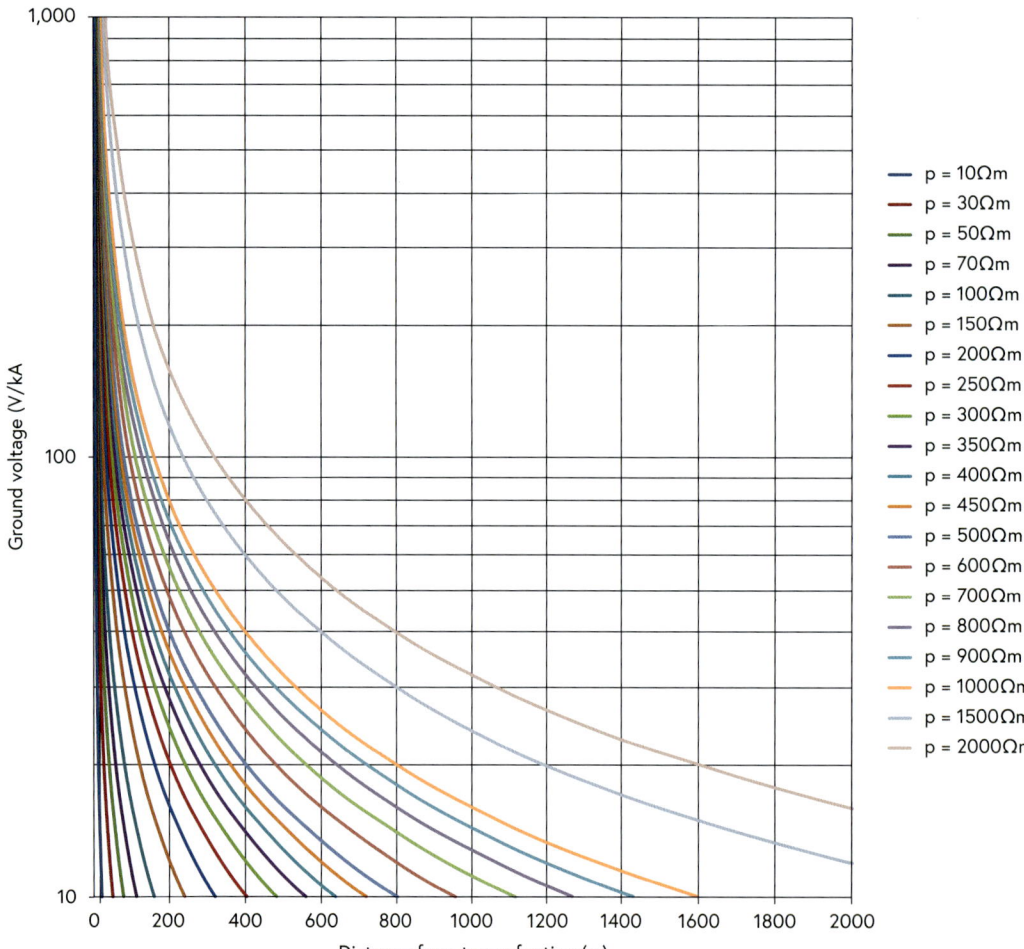

3.11.4   Modelling of EPR: electrical grid substation

HV/LV equipment located at an electrical substation is mounted on an electrical earth grid (mat). When there is an earth fault, the grid (earth mat) of a substation is responsible for creating an EPR due to the conduction of the earth fault current. The value of this EPR is dependent on the grid earth and the ground resistivity.

---

[9]ITU-T *Directives concerning the protection of telecommunication lines against harmful effects from electric power and electrified railway lines – Volume VI: Danger, damage and disturbance* Table 1/5

Section 3 – Mass of Earth as part of a traction return circuit

When there is an earth fault in the substation compound the fault current enters the earth via the substation earth mat (grid) and the transmission tower line chain impedance. The magnitude of the EPR depends on several factors, including the design of the earth mat, the ground's resistivity and the magnitude of the earth fault current.

A parametric variation for EPR has been undertaken using a proprietary mathematical package. The calculation was based on the formulae given in EREC S34 Issue 2[10] (reference Formula P6 Voltage Profile around Earth Electrode: B3.7: Column P 6.3)

Modelling and assumptions

In this calculation for the substation, EPR has been modelled with the following assumptions:

(a) an earth fault current of 20 kA;

(b) a 'tower line chain' impedance of 2 Ω; and

(c) the earth electrode is taken to be the buried earth mat which is approximated to an equivalent circular plate electrode of resistance 0.4 Ω (base case).

The substation is assumed to be connected to a tower line which includes an alternate earth path to the earth through each tower footings. The earth resistance value of this alternate path is assumed at around 2 Ω resistive (Table G.1 EREC S34 Issue 2.).

The modelled EPR profile has been carried out using a mathematical modelling tool. The modelled ERP profile for the grid site earth is taken from the edge of the earth mat. The modelled ERP profile is highly sensitive to the grid site earth mat resistance value. This is demonstrated by modelling the EPR profile with various resistance values. The influence of the earth mat resistance value is very significant near the edge of the earth mat (see Figure 3.18). Also the value of earth mat resistance is a function of soil resistivity.

NOTE: The value of earth mat resistance is a function of soil resistivity.

The graph shows the variation in ground voltage in volts/kA with respect to distance from the edge of the earth mat. Resistivity is varied from 10 to 2,000 Ω.m. For reference, purpose lines are drawn for 430 V[11], 645 V (IEC 62128-1 and EN 50122-1 effective touch voltage limit) and 1,000 V (cable insulation).

---

[10]Engineering Recommendation EREC S34 Issue 2, November 2018 *A guide for assessing the rise of earth potential at electrical installations.*

[11]ITU-T *Directives concerning the protection of telecommunication lines against harmful effects from electric power and electrified railway lines – Volume VI: Danger, damage and disturbance.* Table 1/5

Section 3 – Mass of Earth as part of a traction return circuit

Figure 3.18 Modelled substation earth mat EPR

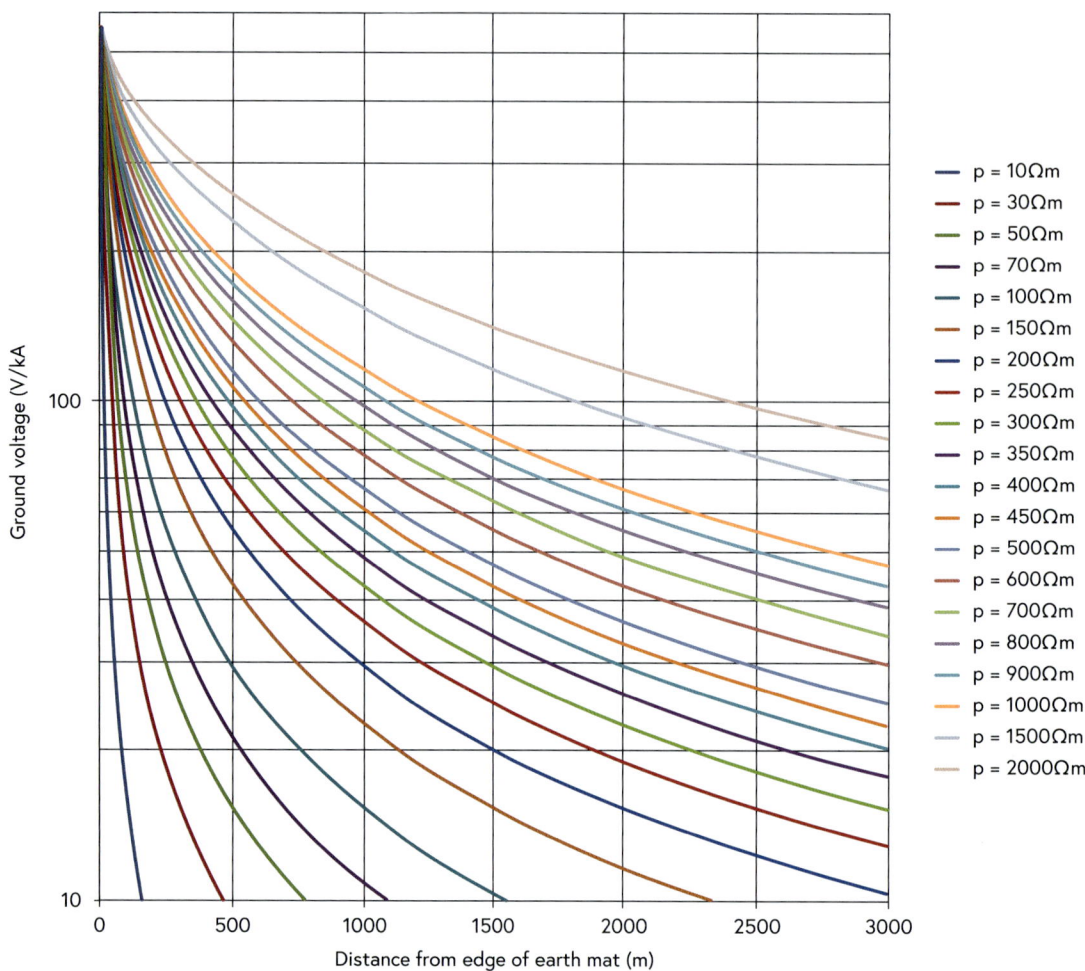

© The Institution of Engineering and Technology

 **Section 4**

Traction return requirements and circuit configuration

The traction return circuit comprises the running rails, traction return bonds and the foundations of the overhead line masts. It is, electrically, the negative return of the 25 kV distribution system and is required to provide the return circuit for both the traction current and to act as a protective conductor during a 25 kV short circuit. The traction return circuit is also required to maintain the rail potentials within the limits prescribed in national regulations and international standards. This Section identifies the elements of the traction return circuit, its configuration and the functional requirements.

4.1 Earthing of AC electrified railways

The complexity of integrating an electrified railway with various electrical systems and exposed-conductive-parts means that it is impossible to prescribe one traction return design solution that addresses the needs of every AC electrified railway.

Since the early twentieth century, different traction return systems have evolved and there are now several system designs that have been developed, including those with a rail return system and a 'forced' return system, utilizing either booster transformers or autotransformers. In all cases, however, the railway earthing design is based on the fact that AC railways should be 'earthy' and that the return circuit conductors are required to provide many features, including a traction return path, 25 AC fault return paths and a protective conductor. This arrangement can be described as a TN-C arrangement based on IEC 60364-1 (and its equivalent CENELEC harmonized document HD 60364-1).

'T' — direct connection of a point with earth (French: *terre*).

'N' — the neutral connection is supplied by the electricity supply network (French: *neutre*), either separately to the protective earth (PE) conductor or a protective earthed neutral (PEN), is a single conductor that has the combined function of providing the neutral and protective earth conductor.

TN–C – this supply has a combined PEN where the conductor fulfils the functions of both a protective earthing conductor and a neutral conductor.

4.2 Earthing and bonding of the traction return circuit

The traction return circuit's earthing is primarily made up of two main elements as shown in Figures 4.1 to 4.4:

1. the leakages to earth of the running rails; and
2. the resistance of the concrete or pile mast foundations.

The arrangement is bonded together and produces a distributed earthed system that provides a path for the traction return current to the traction switching stations, and controls the traction return system's voltage. The return circuit is required to have a high level of integrity, such that if one part (or bond) becomes disconnected the remainder of the system shall still maintain safe voltages on the interconnected infrastructure.

4.2.1 Bonding of the overhead line masts

Masts and the running rails are generally bonded together in one of two ways. Traditionally overhead line masts and all extraneous earth connections are bonded directly to the rails (see Figure 4.1). The downside of this arrangement is twofold: the sheer number of bonds required (as the rail is required

Section 4 – Traction return requirements and circuit configuration

to be drilled before the attachment of a bond), and the tamping of the railway ballast can cause damage to the bonds that are lying on the ballast.

More recently it has become common practice to install aerial earth conductors (AECs) on both sides of the railway to bond overhead line masts, signal gantries and steel overbridges (see Figures 2.7, 2.16 and 3.1). The AECs and rails are then bonded together at a track-to-track cross-bond location (see Figures 4.2 and 4.3). Where the mast is made of non-conducting material then a bond is required between the AEC and the earth plate. Where the mast is made of steel, the provision of a bond between the AEC and the earth plate (also called spider plate) is necessary where the mast or fittings (earth plate) or fixings (AEC) are not designed to pass traction load current (see Figure 4.4).

4.2.2 Conductance and leakage of the running rails

The conductance and leakage of the running rails have been addressed in detail in Section 3.

4.2.3 Bonding of the traction running rails

The feeding length of AC electrification schemes can vary from anywhere between 20 km and 80 km and is usually dependent on the level of the load current and the availability of the high voltage (HV) electrical infeed.

For over 100 years, signal engineers have utilized the running rails for train detection. With the single-rail track circuit (Figure 4.1), one rail (the signalling rail) is used solely for train detection, while the other rail (the traction rail) is used for train detection and as part of the traction return system.

More recently with increased train loads, it has become common to use both rails for traction return:

(a) Figure 4.2 depicts the traction bonding for double-rail traction return and there are no track circuits.
(b) Figure 4.3 depicts the traction bonding for the double-rail traction return and double-rail track circuit. In this configuration both rails are used for train detection; both rails are bonded together through an

Figure 4.1   Single-rail return and single-rail track circuit

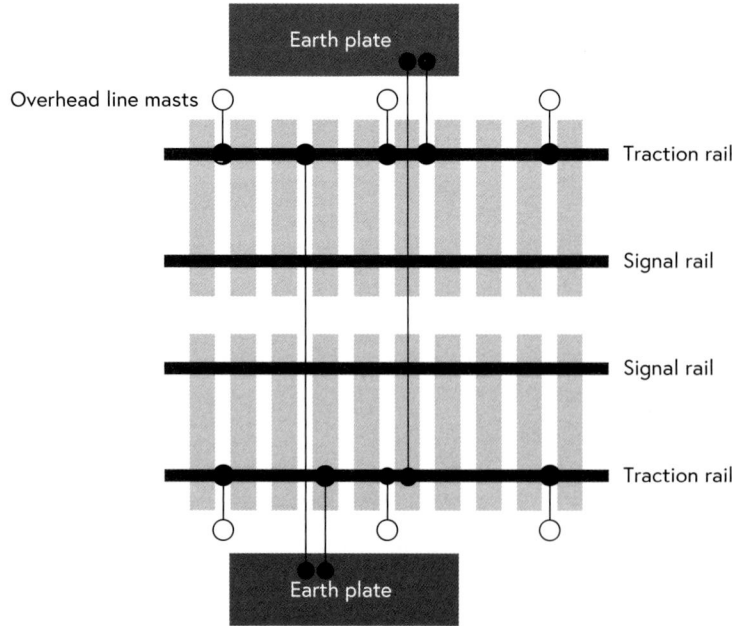

Section 4 – Traction return requirements and circuit configuration

Figure 4.2   Double-rail traction return (no track circuits)

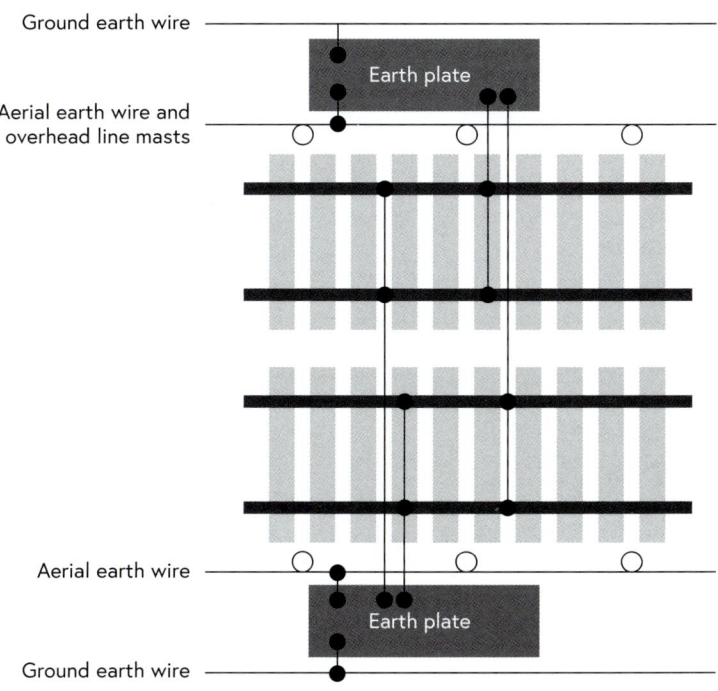

Figure 4.3   Double-rail return and double-rail track circuit

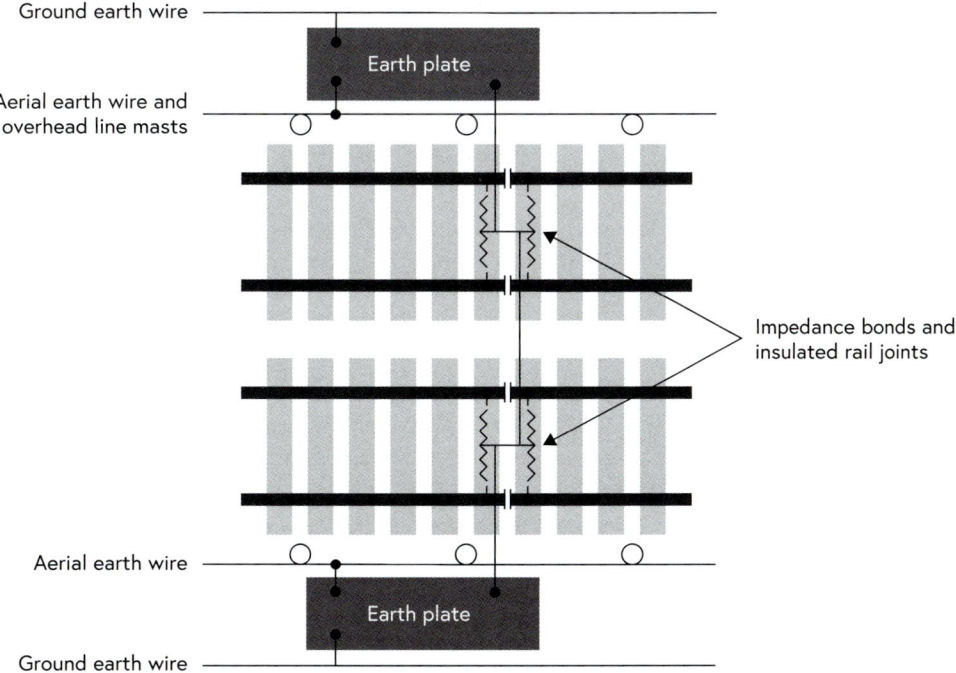

impedance bond. Impedance bonds used for AC track circuits consist of two low resistance windings wound in opposite directions on a laminated iron core. Each winding is connected across the rails on either side of the track, and centre taps from each winding are connected together. This interconnection of the rails and tracks is called cross-bonding.

Figure 4.4   Overhead line mast bonding

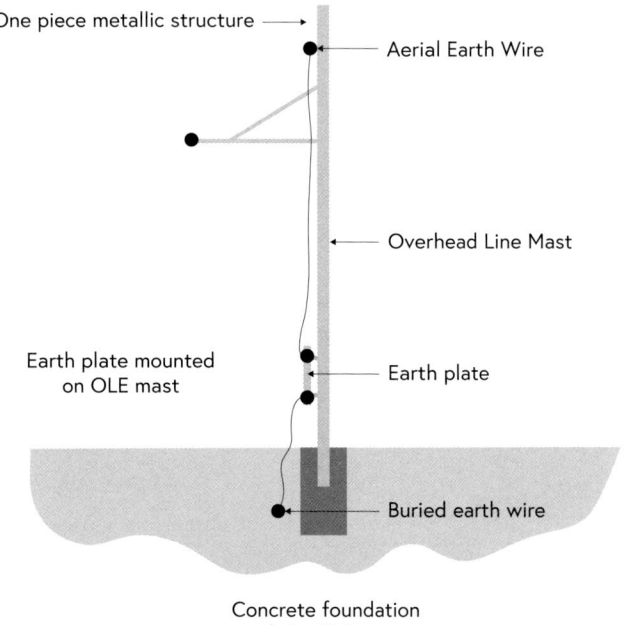

When compared to a single-rail return, the double-rail return has a reduced impedance of the traction return circuit. The benefit of this is twofold: first, the rail potentials are reduced and, second, the double rails provide an increased screening effect on lineside conductors. This increased screening effect on lineside conductors will reduce the induction into those cables. Where there are multiple tracks, the overall impedance of the traction return circuit can be further reduced by bonding all the rails together at regular intervals.

4.3   Design requirements for the traction return system

The rails of the traction return system run the length of the railway and create a type of earth grid which is made up of the rails, sleepers, track-bed and anything else that is bonded to the rails. The rails are connected longitudinally throughout their length with continuity bonds. The rails are then regularly cross-bonded to each other to form a lattice network.

In an AC railway, insulated rail joints (IRJs) should not be installed as this creates a discontinuity in the traction return path, and the return current is then required to find an alternative path around the IRJ, which inevitably causes a higher rail potential. IRJs, however, are required at specific locations, for example, AC/DC interfaces. In these locations it is necessary to undertake a project-specific 'risk assessment' (see Section 10).

4.3.1   Functional requirement for touch potential

Effective touch potential of the rails or exposed-conductive-parts is the main protective characteristic that is required to be controlled. The rails' potential should be limited to 60 V root mean square (rms) under long-term (train operation) and 645 V for < 200 ms under short-term (fault) conditions[12].

---

[12]EN 50122-1 *Railway applications. Fixed installations. Electrical safety, earthing and the return circuit. Part 1. Protective provisions against electric shock*

Section 4 – Traction return requirements and circuit configuration

The effective touch potential is the voltage that is experienced by a person, and this is further detailed in Section 5.

4.3.2 Functional requirements of the traction return system

The railway environment is large, encompassing several other non-traction assets (for example, signalling and telecommunications, low voltage (LV) supplies) and third-party assets (for example, bridges, utilities and services).

The traction return system (rails and overhead line masts) is required to also provide protective provisions and protective functions that relate to non-traction power supplies, civil structures and third-party assets. These requirements are specified in national wiring regulations and various national and international standards, and are referenced in Sections 5 and 6.

The protective functions and characteristics of the traction return circuit are as follows:

(a) Traction negative return bonding

The traction continuity and cross-bonding create a return path for traction load and fault current. The traction bonding controls the current return path and ensures accessible touch and step potentials, in line with the national railway earthing standards and codes of practice.

(b) Protective earthing (25 kV HV)

Protective earthing is required where conductive surfaces and electrical systems could become 'live' following a dewirement. The equipotential bonding provides a return path for fault current enabling the operation of the 25 kV protection systems.

(c) Equipotential bonding

Equipotential bonding is required to ensure that non-continuous metallic surfaces and conductive surfaces (for example, structures, handrails, equipment cabinets) are all held at the same potential to minimize the risk of electric shock.

(d) Common earth network

The rails and the overhead line mast foundations are a lattice network (see Figures 4.1–4.4) that create the common earth to be used by the 25 kV traction return, conductive structures and non-traction electrical systems.

(e) Functional earth (HV and LV earth)

Functional earthing is required to provide a return current path to enable regular operation of a device. This is a low impedance path to connect the equipment to the reference earth.

(f) Protective earthing (lightning)

National regulations require that protection from a direct or indirect lightning strike is provided for railway civil structures, including stations, viaducts and bridges.

Where electrified railways are in exposed locations the 25 kV overhead lines are at an increased risk of being struck by lightning. In this case the overhead lines act as the air termination, the masts act as down conductors, the traction return system the earth termination. Lightning that strikes the overhead lines can thereby discharge into the earth via the structural foundations and the running rails. This will provides protection from a direct lightning strike to trains, track-based electrical equipment and humans.

(g) Suppression of electrical noise

Railway traction earth is notoriously noisy due to the high levels of harmonic currents and also due to the conduction path (rail) being exposed to the 25 kV current and lightning transients.

Section 4 – Traction return requirements and circuit configuration

Earthed systems must therefore be designed to ensure compatibility with the noise, in particular 25 kV switching transients and lightning transients.

4.4 Traction return circuit conductors

Conductors form an integral part of the traction return system, enabling both the continuity of traction current and providing protective earthing.

4.4.1 Traction return bonds

Traction return bonds across discontinuities are used along the length of the railway, providing a continuous path for the return traction current to the 25 kV feeder stations. Rail-to-rail and track-to-track cross-bonding is also required at regular intervals.

A typical example of conductors that are used include aluminium conductors and steel-reinforced (ACSR) conductors 19/3.25 – 157.6 mm^2.

4.4.2 Protective earth wires and earth conductors

Earth wires are used to bond together all the overhead line masts and overline structures to the traction return system at each track-to-track cross-bond, and are typically bare or PVC sheathed with aluminium 19/3.25 – 157.6 mm^2 or with copper conductor 90 mm^2.

Return conductors: the AC single-phase booster transformer system has return conductors mounted on the masts and is insulated from them with a low voltage insulator. The conductor is then connected to the rails typically every 3.2 km. The return conductor may be used on its own, but it is more familiar with a booster transformer which acts as a forced return system.

Return conductors are typically a bare all-aluminium conductor (AAC) or all-aluminium alloy conductor (AAAC) (19/4.22 – 265.7 mm^2 or 19/3.25 – 157.6 mm^2).

Screening conductor: this is located next to lineside data conductors and is located in the lineside trough, or on conductor trays in tunnels or viaducts. It uses the principle of Lenz's Law and provides screening from 50/60 Hz induction of the overhead conductors and the return rails.

50/60 Hz screening conductors are typically AAC or AAAC (19/4.22 – 265.7 mm^2 or 19/3.25 – 157.6 mm^2 XLPE with sheath rated to 250 °C).

Buried earth conductors: this is bonded together with the overhead line mast and the running rails at the track cross-bonding location. They are typically 35 mm^2 or 50 mm^2 bare copper conductors. As the conductor is buried it has the added benefit of providing a fast transient lightning earth. It can be utilized on one or both sides of the railway and is most commonly found on high speed lines. The buried earth conductor provides an earth and a protective conductor for low voltage track systems.

Supplementary ground and earth conductors: on viaducts and in tunnels it is normal to install either a supplementary earth conductor or an earth bar along the length of the tunnel or viaduct. This conductor provides an earth and a protective earth for low voltage systems. On viaducts this earth conductor can

Section 4 – Traction return requirements and circuit configuration

become exposed to lightning surges as it is interconnected to the mast down conductor and the lightning earth pit.

4.4.3 Operational requirements for conductors

The 25 kV earthing system, its components and bonding conductors should be specified such that they are capable of distributing and discharging the 25 kV fault current without exceeding the thermal and mechanical design limits of the conductors. The calculation should be based on the operating time of the backup protection.

The railway substation earth typically shall withstand up to 12 kA for 3 s. This rating is specified for the main earth bar, earth electrodes, conductors and connections within the substation and is based on the backup protection of the public electricity supply.

Traction return current bonding shall be designed to withstand 6 kA on a 25 kV single-phase or 12 kA on an autotransformer supply for 1 s. This rating is based on the backup electrical protection of the railway feeder station.

The following conductors and bonding equipment are required to be assessed for adequate load current and fault current rating:

(a) traction return rails and bonding conductors;
(b) AECs and auxiliary/supplementary earth conductors;
(c) electrical connections to the conductors, for example, earth wire clamps, bonding lug connections; and
(d) switching station return bonds and bonding conductors, terminal plates and return current busbars.

The earthing system should also be designed to maintain a high level of integrity for the expected installation lifetime with due allowance for corrosion, damage and mechanical constraints.

4.4.4 Method of calculating the conductor short-circuit current rating

An adiabatic calculation occurs without including any transfer of heat between a thermodynamic system and its surroundings. This method of calculating the short-circuit rating is based on the assumption that the heat is normally retained inside the conductor for the duration of the short circuit (i.e. adiabatic heating). However, some heat transfer occurs into the adjacent materials during the short circuit and advantage can be taken of this.[13]

4.4.4.1 Calculation of adiabatic short-circuit current rating

The general adiabatic temperature rise equations 4.1a and 4.1b applies for calculating current-carrying capacity, which is applicable to an initial temperature, is without the transfer of heat between a thermodynamic system and its surroundings:

$$I^2_{\text{AD}}.t = K^2 S^2 \ln\frac{(\Theta_{\text{f}} - \beta)}{(\Theta_{\text{i}} + \beta)} \tag{4.1a}$$

---

[13]BS 7454:1991 & IEC 60949:1988+A1:2008 *Method for calculation of thermally permissible short-circuit currents, taking into account both adiabatic and non-adiabatic heating effects*

$$K = \sqrt{\frac{\sigma_c(\beta + 20).10^{-12}}{\rho_{20}}} \qquad (4.1b)$$

where:

I_{AD} is the short-circuit current (rms over duration) calculated on an adiabatic basis (A)

t is the duration of a short circuit(s)

K is the constant depending on the material of the current-carrying component ($As^{½}/mm^2$) (Table I of BS 7454)

S is the geometrical cross-sectional area of the current-carrying component (mm^2) (for conductors specified in IEC 228 it is sufficient to take the nominal cross-sectional area)

Θ_f is the final temperature (°C)

Θ_i is the initial temperature (°C)

β is reciprocal of the temperature coefficient of resistance of the current-carrying component at 0 °C (K)(Table I of BS 7454)

ln is $\log_e$

σ_c is the volumetric specific heat of the current-carrying component at 20 °C ($J/K.m^3$): (Table I of BS 7454)

ρ_{20} is the electrical resistivity of the current-carrying component at 20 °C ($\Omega.m$) (Table I of BS 7454).

4.4.4.2 Calculation of non-adiabatic short-circuit current rating

The adiabatic equation assumes no heat is dissipated from the conductor during a fault. While putting the calculation on the safe side, in some situations, particularly for longer fault duration, there is the potential to utilize a smaller cross-section. In these instances, it is possible to use the non-adiabatic which is a more accurate calculation.

The non-adiabatic calculation occurs with the inclusion of transferring heat between a thermodynamic system and its surroundings which are only valid for short-circuit durations. When compared to the adiabatic method, the non-adiabatic calculation allows for significant increases of the permissible short-circuit currents in the case of screens, sheaths and small conductors of less than 10 mm^2 (mainly when used as screen wires).

The following calculation is for thermally permissible short-circuit currents, taking into account non-adiabatic heating effects, and is based on BS 7454 and IEC 60949.

The method adopted is to use the adiabatic equation (4.2) and apply a factor to cater for the non-adiabatic effects of the maximum fault current:

$$I_{s/c} = \varepsilon\, I_{AD} \qquad (4.2)$$

where:

$I_{s/c}$ is the permissible short-circuit current (A or kA)

I_{AD} is the adiabatic calculated permissible short-circuit current (A or kA)

ε is the factor to allow for heat dissipation from the conductor (see BS 7454 Section 5).

4.4.5 Method of calculating the load current rating – non-adiabatic

This section is based on the requirements as detailed in IEC 60287-1-1:2006 *Electric cables – Calculation of the current rating – Part 1-1: Current rating equations (100 % load factor) and calculation of losses – General.*

Section 4 – Traction return requirements and circuit configuration

AC losses occur within conductors, which are able to simultaneously produce heat into the surrounding materials which can be non-homogeneous and have boundary and external temperature limitations.

IEC 60287-1-1:2006 addresses the rating, including the following losses associated with operational load currents:

(a) AC resistance of conductors including conductor losses due to resistance (I^2R losses), skin effect, proximity effect;
(b) dielectric losses of conductor insulation; and
(c) circulating current losses of screens and armours.

As the losses within a conductor will create heat, the methodology necessary to size conductors means treating the issue as a thermal problem. Depending on the installation and the environment, the heat will ultimately be dissipated to its surrounding at a given rate.

The insulation and environment of the conductor determine the maximum temperature and this needs to be factored into the current-carrying capacity. The maximum temperature for thermoplastic insulation is 70 °C and for thermosetting insulation is 90 °C.

The current rating depends on how this heat can be dissipated through the conductor surface and into the surrounding environment. As the conductor heats, there is a thermal equilibrium that occurs when the heat is dissipated to the environment at the same rate at which it is generated (due to losses in the conductor). At a specific current level, the conductor will reach this thermal equilibrium. This is the designed maximum allowable temperature for the insulation, and also the maximum current-carrying capacity of the conductor. Where multiple conductors are located, it is also necessary to take these and the containment into account.

The current-carrying capacity of a conductor is based on achieving the thermal equilibrium, but in a railway environment, the load current varies due to the movement of the trains. Therefore, the thermal equilibrium is assumed to be the averaged current that can flow for a period of time, without exceeding its insulation or sheath temperature rating. The time taken to reach equilibrium is typically assumed to be one hour of the ultimate train full-service timetable.

As traction return bonding conductors are required to pass current continuously, it is necessary to consider using XLPE. This can run at a higher average current due to its maximum working temperature of 90 °C.

4.4.6 Calculation conductors directly exposed to solar radiation

Where the track-bed is an open route, the traction return conductors are mainly located on the track-bed or suspended on overhead structures exposing them to solar radiation. Where the track is underground, the traction return conductors are not exposed to solar radiation.

From IEC 60287-1-1:2006 *Electric cables - Calculation of the current rating - Part 1-1: Current rating equations (100 % load factor) and calculation of losses - General*. Taking into account the effect of solar radiation on a conductor, the permissible current rating is given by the equation in 1.4.4.2.

4.5 Traction power system modelling – conductor current rating

Railway operators and infrastructure owners are required to design the railway to specific national and international technical and safety performance standards. These standards and codes of practice provide the basis for a railway company's codes of practice, which detail the design methodology, application and system installation. The modelling of AC railway networks should cover all types of AC feeding arrangements, including the rail return system, the booster transformer system and the autotransformer system. The separate modelling of multi-conductors in an AC power network – instead of a simplified lumped analysis – enables more accurate calculations of induced voltage, electromagnetic compatibility (EMC) analysis, positive and negative energy consumptions, and calculation of conductor losses.

The operation of the 25 kV single-phase rail return systems, 25 kV single-phase booster transformer systems and autotransformer systems have been covered in Section 2.

4.6 Traction return bonding

As part of the traction system design, it is necessary to undertake traction power system modelling to determine how frequently the rails are required to be bonded to each other to control rail potentials. The interval of this cross-bonding is determined by the traction load and the short circuit (fault current) for each particular electrification system. Typically, the cross-bond spacing on a 12 kA autotransformer railway is between 400 m and 500 m, while the cross-bond spacing on a 25 kV single-phase 6 kA railway is typically 400 m, but can be longer. In some specific instances, close to grid supply points, it may be necessary to reduce the cross-bond spacing further in order to control the high rail potential.

4.6.1 Conductor rating

Traction power modelling, including the proposed electrical loads and proposed timetable, is used to determine the operational behaviour of the 25 kV system and the rating of conductors. Where the project requires the calculation of the rms current in each of the 25 kV electrical sections, 25 kV conductors, overhead lines and the traction return bonds, it is necessary to prepare a multi-conductor traction power model.

The electrical load characteristic is usually at its highest value where it is closer to the feeder station, as the load current falls away over distance. Similarly, the short-circuit characteristic will be at its highest value where it is closer to the feeder station and also reduce characteristically in the same way.

The load currents in these circuits will vary over a 24-hour period due to the operational timetable. The rating of cables and conductors is therefore generally based on a one-hour period of a busy intercity service (peak) with a specified headway. During this one hour, conductors will usually have reached their highest temperature due to the internal thermal capacity and environmental conditions.

The power and distribution engineer has the responsibility of specifying adequately sized overhead line conductors (contact and catenary) and traction return bonds at the feeder station. It is also necessary to provide redundancy for the traction return bonds to mitigate any damage to those bonds from rail tamping machines or a derailed train.

The traction power modelling should additionally address any degraded mode of operation that may occur. Degraded modes include the re-energization of the 25 kV system following the loss of power to the overhead line or the complete outage of an electrical supply point. Following a 25 kV outage, there may be a requirement for the electrical control room (ECR) to schedule a restart of the traction power system. The restart is necessary to ensure that multiple trains do not cause the section breaker to trip due to overcurrent, or the conductors exceed their operational working temperature.

4.7 Traction return circuit configuration

Historically, to prevent disturbance of the track-based LV signalling and communication systems by earth potential rise (EPR) from the 25 kV system, the railway companies have opted for segregated earth bonding and connections for the two systems.

Signalling and telecommunications engineers believed that a segregated system would protect their equipment from voltages induced in the ground by 25 kV railway earth faults. This arrangement is depicted in Figure 4.5. The LV systems, however, were still subjected to ground potentials and transients induced by the 25 kV earth fault and lightning strikes.

To overcome this ground voltage difference, it has now become standard on new railway electrification schemes to integrate the earth. This design is not proposed in national or international standards but is made a requirement through codes of practice that have been developed by railway companies, and these are quite specific to the particular nature of their electrification system (for example, SNCF EF4D1n°1 *Ligne Electrifiees En Courant Alteratif Monophase*; UK High-Speed 000-GDS-LCEEN-00041-05 *2x25kV Earthing and Bonding Principles*; and Network Rail NR/L2/ELP/21085 *Earthing and Bonding on AC Electrified Railways*).

4.7.1 Railway segregated earthing

The earth of the LV signalling and telecommunication systems is segregated by creating local earth as shown in Figure 4.5. It was formerly believed that 25 kV earth faults would not disturb LV systems on a railway with a 'segregated' earth. However, they are still susceptible to transients, particularly where the ground resistivity is high or the ground is 'made up'. To comply with the protective limits detailed in Section 5, there is an additional requirement to keep the mandatory physical separation of 2.5 m between these two separately earthed systems.

It should be noted that in segregated railways it is still a requirement to have a common earth system between HV electrical systems (25 kV HV grid supplies) and civil structures, which include conductive structures within stations and bridges that could be endangered by the electric traction system. In these cases, the conductive structure is bonded directly to the traction return system, and this occurs along the railway.

With the segregated earthing system, the rating of the LV earth circuit protective conductor (cpc) is solely determined by the LV fault current. This makes the LV design more straightforward and it is not required to take the conducted currents, including the (HV) traction return system, into consideration.

Where HV earth fault currents are conducted into the ground, the EPR generates contours of equal voltage, spreading out from the point of the earth fault. Signalling systems are known to be susceptible to HV earth faults from nearby tower lines or substations. The EPR at an HV tower line can be as high as 5,000 V. This is due to the percentage of the fault current (approx. 10 %) which is conducted into the ground through the tower footings. In this case, the railway signalling or control system may see the EPR, which may then require the LV system to be reset (see Section 3.11.3).

4.7.2 Railway integrated earthing

With the introduction of new electrification schemes and the increase in fault levels, it has become more popular to implement an integrated earthing design ensuring compliance with the protective limits for

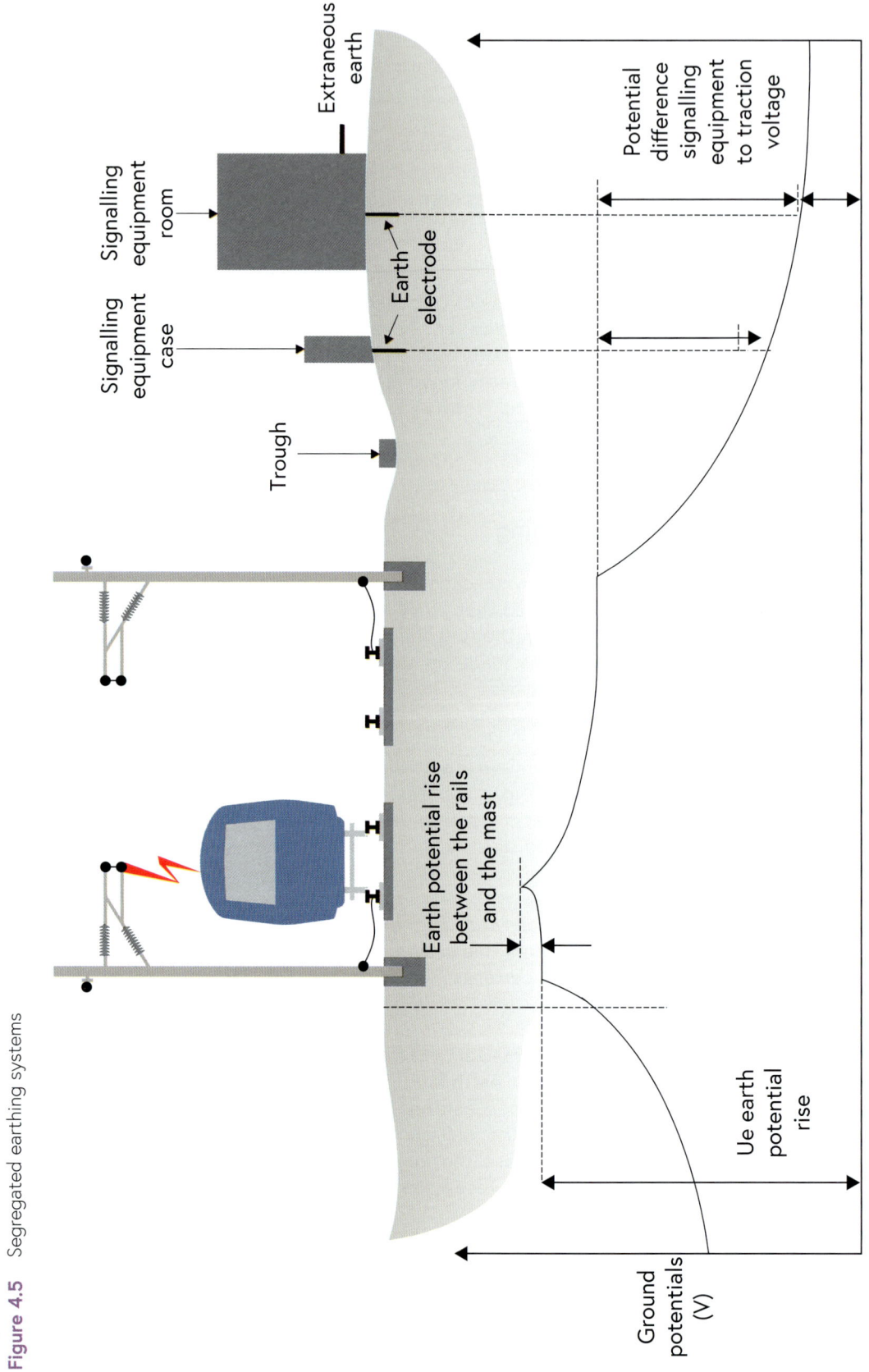

Figure 4.5   Segregated earthing systems

Section 4 – Traction return requirements and circuit configuration

safety as detailed in Section 5. Also, the application of a common-bonding of LV systems provides greater immunity from the EPR generated by 25 kV, HV supplies and lightning transients.

The integrated earthing design is called a common earthing system or integrated earth. This applies to all railway earthed systems (25 kV, HV, MV and LV), signalling, telecommunications systems, civil structures and non-railway auxiliary supplies.

The advantage of having multiple paths with an integrated railway earth, is that the earth voltages across the railway are low. However, the 25 kV load current, fault current and lightning currents can choose to flow in the LV earth conductor (for example, cpc, screens and armour). The rating of the LV cpc should therefore take into consideration a component of the 25 kV traction load. The size of the LV earth conductor can be greater than that used in standard practice and defined in the following national rules (IEE Wiring Regs, Australian/New Zealand AS/NZS 3000:2018 Electrical installations, known as the 'Wiring Rules'). This is further developed in Section 7.

Figure 4.6   Integrated earthing systems

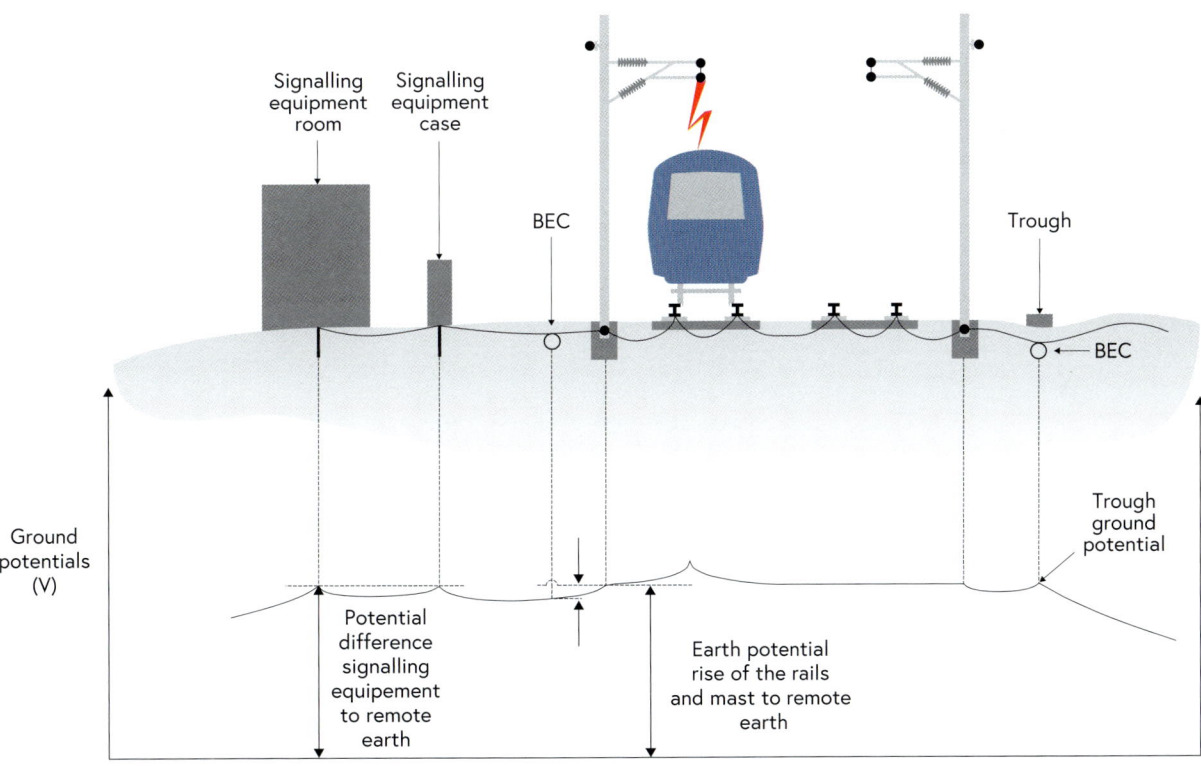

4.8   Criteria of the failure modes for LV systems

When the LV systems (signalling or data control system) are in close proximity to 25 kV and HV power systems (for example, 132 kV and 400 kV), they are likely to experience earth transients. The railway standard EN 50121-1 *Railway applications. Electromagnetic compatibility. General* 'Section 4 Performance' Criteria specifies the following failure modes which may occur:

Section 4 – Traction return requirements and circuit configuration

(a) **Performance criterion A:** The apparatus shall continue to operate as intended during and after the test. No degradation of performance or loss of function is allowed below a performance level specified by the manufacturer when the apparatus is used as intended. The performance level may be replaced by a permissible loss of performance.

(b) **Performance criterion B:** The apparatus shall continue to operate as intended after the test. No degradation of performance or loss of function is allowed below a performance level specified by the manufacturer when the apparatus is used as intended.

(c) **Performance criterion C:** Temporary loss of function is allowed, provided the function is self-recoverable or can be restored by the operation of the controls.

 # Section 5

Protection against electric shock

5.1 Fundamental objective

The need to protect people (for example, workers, passengers and the general public) many of whom have, at best, a limited understanding of its danger, has been as a fundamental objective throughout history as the use of electricity has become widespread. This objective has been codified in national and international standards and codes of practice, and adopted into national statutory legislation. It is defined in the international electrotechnical standard IEC 61140 *Protection against electric shock – Common aspects for installation and equipment as a fundamental rule of protection against electric shock*. The standard states that:

"Hazardous-live-parts shall not be accessible and accessible conductive parts shall not be hazardous live either

1. under normal conditions (operation in intended use and absence of a fault) or
2. under single fault conditions."

Although the definition is straightforward, it employs accepted terminology which can make it difficult to read. An explanation of these is provided below and will be used throughout this Guide.

Conductive part: This is a part that can carry electric current. Its applicability is not limited to electrical installations. It includes those parts that are designated to carry current, such as overhead contact lines, and conductors within busbars, transformers and switchgear. It also includes fences, metal components within bridges, buildings and other structures, fences, handrails, platforms. Within the scope of this section, a conductive part needs to be categorized as live, exposed or extraneous, and whether or not it is accessible in order to understand the electric shock hazard and determine appropriate mitigations.

Accessible part: This means that a conductive part, whether a live part or not, can be touched by persons or animals. The description 'accessible conductive part' is often used but in most cases requires qualification to provide a complete understanding of the electric shock hazard. Examples include 'accessible exposed-conductive-parts', and 'accessible extraneous-conductive-parts'.

Hazardous-live-part: This is a conductive part intended to be energized under normal conditions (the live part) and can, under certain conditions, give a harmful electric shock (i.e. hazardous). A live part includes a neutral conductor or mid-point conductor, but should these be connected to earth to provide the function of a protective earthing conductor then by convention they are not considered to be a live part. For an AC electrified railway, the rails usually perform the functions of a combined protective earthed neutral (PEN) conductor, but the nature of the electric traction system and of a train load that varies by magnitude, time and location mean that the rails need to be considered as a live part.

The term 'hazardous live' is used to provide an overarching description of any situation that could lead to a harmful electric shock. The various conditions are dependent on how an electrical system is designed, constructed, operated and maintained. Such a situation may occur under normal conditions where voltages and currents are sufficiently large to create a hazard through electrostatic and electromagnetic induction between live parts and accessible conductive parts.

Exposed-conductive-parts: These are parts of the electrical equipment or installation that can conduct electricity and can be touched by persons or animals. Unlike live parts they are not intended to be energized under normal conditions, but they can become live under fault conditions. Examples are the external surfaces of metal-clad switchgear or transformers, or metallic overhead contact line system (OCLS) supporting structures in an electric traction system. Because these parts can be touched, IEC 61140 makes clear that these parts must not be a danger to persons under both normal and fault conditions.

Extraneous-conductive-part: This is a conductive part that does not form part of the electrical installation but can introduce an electrical potential into it, normally from a local earth that is remote to the installation. Examples of extraneous-conductive-parts are metallic water or gas pipes, or the metallic sheaths of telecommunication cables that are run into a traction switching station. On electrified railways, metallic fences or LV protective conductors that are bonded to the traction return circuit can also be extraneous-conductive-parts as they may transfer rail potential from the railway to third-party installations.

Normal conditions include the various alternative configurations for which an electrical system has been designed and built. For a railway electric traction system, alternative configurations can include:

- HV switching to ensure the continuation of supplies during plant outages;
- HV switching to transfer load from alternative traction supply sources; and
- isolation of part of the system for maintenance while keeping other parts energized to support the train service.

The type of single fault cannot be specifically defined as it depends on the type and configuration of the electrical system. Instead, IEC 61140 imposes the following test to establish whether a fault needs to be considered for fault protection:

- **Cause a:** an accessible, non-hazardous-live-part to become a hazardous-live-part (for example, due to failure to limit the steady state touch current and charge);
- **Cause b:** an accessible conductive part which is not live under normal conditions to become a hazardous-live-part (for example, due to insulation failure); or
- **Cause c:** a hazardous-live-part to become accessible (for example, by mechanical failure of an enclosure).

There are two further terms that are not included within this definition but are related and are frequently used:

- **direct contact:** this is the electric contact of persons or animals with live parts as stated in IEC 60050 *International Electrotechnical Vocabulary (IEV) – Part 195: Earthing and protection against electric shock*. In respect of the fundamental rule of protection in IEC 61140, this relates to the statement that "hazardous-live-parts shall not be accessible". It is worth noting that in LV systems, direct contact represents actual physical contact with live parts. For HV systems where the electric field around live conductors is strong, a space around each live part is defined as a danger zone. As stated in IEC 61140, entry into the danger zone is equivalent to touching live parts (i.e. direct contact).
- **indirect contact:** this is the electric contact of persons or animals with exposed-conductive-parts which have become live under fault conditions as stated in IEC 60050. In respect of the fundamental rule of protection in IEC 61140, this relates to the statement that "accessible conductive parts shall not be hazardous live".

5.2 Electric shock and hazardous-live-parts

Electric shock occurs when current flows through the human body. How much current flows is a function of the voltage that is applied across the body and the body impedance at that voltage between the points where the current enters and leaves the body. Direct contact with hazardous-live-parts will normally result in death or cause severe injuries such as burns that in many cases can also be fatal. Accordingly, such hazardous-live-parts should not be accessible.

Section 5 – Protection against electric shock

However, HV and high current systems can cause accessible conductive parts to be become hazardous live due to:

- **Hazard (a):** electrostatic induction in normal conditions;
- **Hazard (b):** electromagnetic induction in normal and fault conditions;
- **Hazard (c):** earth potential rise (EPR) in normal and fault conditions; and.
- **Hazard (d):** insulation failure or mechanical failure of support arrangements for bare live conductors.

Hazards (a) and (b) are described in Section 3. In general, they are controlled by equipotential bonding and earthing, and, additionally, for hazard (b), by limiting the length of the part exposed.

Hazard (c) occurs when current flows in the earth, which results in the voltage of the earth at that point to be increased such that there is a voltage difference between that point and the nominal voltage (0 V) of the mass of Earth. In general, for AC railways this situation is typically more significant under fault conditions, but it may need assessment where there is significant traction return current flowing back to the traction switching station and substation via earth. The calculation of the EPR is covered by various standards and textbooks, such as EN 50522 and IEEE Standard 80, but in essence reflect the magnitude of the earth fault current, the design of the earth electrode and the ground resistivity. The earth fault current shall be the maximum value of all the allowable configurations of the electric traction system. Within a substation or switching station, an earth electrode comprising a buried mesh with earth rods is usually provided. This rises to the value of the EPR, but its design is such that voltage differences across the meshed earth electrode do not cause hazardous voltages. Beyond the edge of the meshed earth electrode, the EPR decays to zero as a function of the inverse of the distance from the centre of the electrode and is dependent on the shape and type of earth electrode. However earth electrodes generally produce the typical shape of the decay in EPR with distance that is shown in Figure 5.1. Section 6 provides more information on the design of earth electrodes.

Lightning surge current into the ground also creates an EPR and this is addressed in Section 9.

Hazard (d) is controlled by equipotential bonding and automatic disconnection of the supply (see Section 3).

5.3   Touch, step, mesh and transfer voltage

How the current flows through the body is dependent on the activity when the hazard occurs. It follows that the current could enter at any point on the body and leave from another. By convention, the voltages a human may experience are placed into two categories that generally reflect the body's position in most activities:

1. touch voltage; and
2. step voltage.

The touch voltage, or more specifically the effective touch voltage, is the voltage across the body when conductive parts are touched simultaneously. The definition of effective touch voltage is used to differentiate this from the prospective touch voltage, as both terms are used in the electric shock model described in Section 5. It can be further divided into hand-to-hand and hand-to-feet touch voltages. Hand-to-hand touch voltage occurs when two conductive parts with different voltages can be touched simultaneously (simultaneously accessible parts). Current will flow from one hand through the body to the other hand. The distance between two accessible exposed-conductive-parts that can be touched simultaneously. is, by convention, specified as 2.5 m or less between the two accessible exposed-conductive-parts. Hand-to-hand touch voltages need to be assessed where simultaneous touching distances exist.

Section 5 – Protection against electric shock

Hand-to-feet touch voltage occurs when a person is touching a conductive part with their feet on the ground, which, by convention, are 1 m away from the conductive part (IEEE Standard 80: *IEEE Guide for Safety in AC Substation Grounding*). The current flows from the hands through the body and out through the feet into the ground.

Step voltage is the voltage difference in the ground between two points as seen by the feet while walking or stepping. This step distance is defined in IEC 60050 as 1 m for people, but it will be much greater for some four-legged animals because their front and back legs can be more than 1 m apart and they do not have foot coverings that can provide the additional insulation that shoes provide for people. This larger distance mean that these animals experience a greater voltage difference between their front and rear legs arising from the earth potential rise. Where shock risk to animals is important, step voltages may be as significant as touch voltages.

There are two further categories that are often described:

1. transfer voltage; and
2. mesh voltage.

Both may present a touch or step voltage hazard but are identified as particular situations because of their prevalence. Transfer voltage describes the situation where a conductive asset, at which the earth potential locally at that asset has been raised due to a fault, could transfer that raised earth potential to a remote location where the earth potential is nominally zero. Examples of conductive assets that could transfer voltage are cable screens/armour, telecommunication cables, pipes and fences. The voltage transferred will subject persons or animals at the remote location to touch voltages that may be hazardous because the local earth would be at a different potential to that of the asset transferring the voltage from the location of the high EPR.

Traction switching stations and substations usually have an earth electrode formed by a combined earth grid with a mesh arrangement and earth rods at the perimeter of the earth grid. An earth grid does not create a uniform voltage with respect to remote earth across its area, with peak voltages occurring at the edge of the earth grid and lower voltages at other points within the grid and also within the edge of a mesh within the grid and its centre. These voltage differences may be sufficient to cause a touch or step voltage hazard. The mesh voltage is defined in IEEE Standard 80 as "the maximum touch voltage

Figure 5.1 Touch, step and transfer voltages

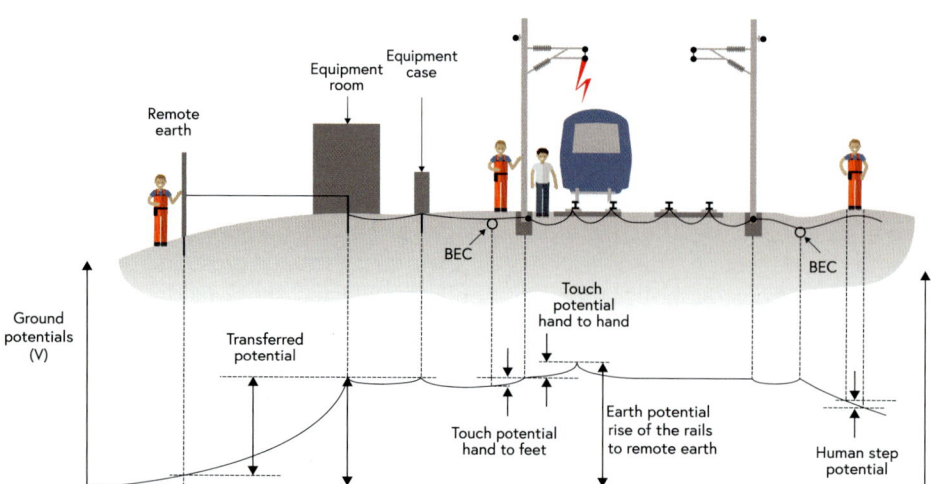

between the centre and edge of a mesh" and as maximum voltages occur at the edge of the earth grid, the mesh used for the mesh voltage calculation is normally a corner mesh. Calculation of the mesh voltage is a requirement of IEEE Standard 80 but is not explicitly a requirement in EN or IEC standards, though the methodology has been adopted by various national utilities to address this issue.

Touch, step and transfer voltages are illustrated in Figure 5.1 (a cross section across an electrified railway during an electric traction fault).

5.4 Nature of electric shock: physiological effects

The human body is sensitive to electric current passing through it and this can result in either fatal or non-fatal effects. Harmful effects include burns, neurological damage (including seizures) and cardiac abnormalities (including arrhythmia, and in particular ventricular fibrillation), which may lead to death. There are also many long-term harmful effects. Non-fatal and potentially less serious effects include muscle spasms, startle reactions, or other effects that cause distress and pain. IEC 60479-1 *Effects of current on human beings and livestock – Part 1: General aspects* provides basic guidance on the effects of current on persons and livestock and illustrate the sensitivity of the human body to electric current. Table 5.1 is derived from IEC 60479-1 and from IEEE Standard 80 and shows body current threshold values relating to the following non-harmful effects. The effects are formally defined in both IEC 60479-1 and IEEE Standard 80 but have been simplified to assist in understanding their essential meaning. It needs to be stressed that they are not absolute values and should be used for guidance only as the actual values must be calculated in accordance with these standards:

(a) **Perception threshold current:** minimum value of electric current through the body of a person or animal which causes any sensation for that person or animal.
(b) **Threshold of reaction current:** minimum value of body current which causes involuntary muscular reaction.
(c) **Let-go threshold current:** maximum value of electric current through the body of a person at which that person can release himself or herself from a live object.
(d) **Immobilization threshold current:** also known as the tentanization or freezing current. For a stated frequency and waveform, this is the minimum value of electric current for which an insuperable, involuntary, sustained muscular contraction is produced. This muscular reaction means that the person cannot move voluntarily as long as the current flows.

Table 5.1 Non-harmful electric shock currents

Effect	50 Hz body current (mA) (IEC 60479-1)	60 Hz body current (mA) (IEEE Standard 80:2013)
Perception threshold	0.5 mA	1 mA
Threshold of reaction	>0.5 mA to 5 mA (independent of time)	1 mA to 6 mA
Let-go threshold	10 mA for adult males 5 mA for entire population	6 mA to 9 mA
Immobilization threshold	5 mA to < 40 mA	9 mA to < 60 mA

The International Commission on Non-Ionizing Radiation Protection (ICNIRP) published *Guidelines for Limiting Exposure to Time-Varying Electric and Magnetic Fields (1 Hz–100 kHz)*, which states perception threshold limits that align with those of the IEC 60479 and IEEE Standard 80. Because of the scope of the ICNIRP's work, different limits have been stated for occupational exposure (i.e. exposure involving electrically skilled or instructed persons) or general public exposure (i.e. ordinary persons).

The values in Table 5.1 should never be considered as absolute limits and must not be considered as free of pain and discomfort. IEC 60479, IEEE Standard 80 and the ICNIRP guidelines are clear that the human body and the conditions in which electric shock occurs introduce significant variance. These standards therefore state threshold values that are generally accepted as unlikely for the physiological reactions to occur. Above the perception threshold, currents can cause significant discomfort, although not fatal, and through instinctive muscular reaction to break clear of the live part, can lead to consequential accidents, such as falls, head and limbs striking objects and letting tools fall on other persons. Within the range of immobilization, extreme discomfort will occur with the body unable to move voluntarily and there could be breathing difficulties. These are not permanent and usually disappear when the current is interrupted. Above the immobilization threshold, it is highly probable that the electric shock will result directly in death through ventricular fibrillation.

Under normal conditions it is expected that body current will be below the threshold of perception. Under fault conditions it is not practicable to ensure that the body current can be controlled to below the threshold of perception current level. For HV systems, the maximum permissible body current is determined by the threshold of ventricular fibrillation as this is the fatal condition that is directly caused by electric shock. This does not mean other, potentially life changing, consequences such as burning may not occur, but protective provisions can be implemented to control such consequences in addition to limiting the body current. In practice, body current is difficult to measure on site and therefore normally the touch voltage experienced by the body is used as the limiting value. The limiting touch voltage takes account of the body impedance, duration of the fault current and additional resistances such as footwear and standing surfaces. The determination of the limiting body current and the touch voltage is described in two standards that are applicable to HV systems: IEC 60479-1 and IEEE Standard 80. Both utilize the same electric shock model but adopt different methodologies to determine the limiting values of body current and touch voltage.

5.5 The electric shock model

Both IEC 60479-1 and IEEE Standard 80 use a similar electric shock model, which is shown in Figure 5.2. The prospective touch voltage U_{tp} is the voltage between the structure and the point at which the person is in contact with earth in the absence of contact. The effective touch voltage U_{te} is the voltage across the body after taking into account the other resistances that may be present. These act as a resistance voltage divider circuit which can result in an effective touch voltage that is less than the prospective touch

Figure 5.2 Electric shock model

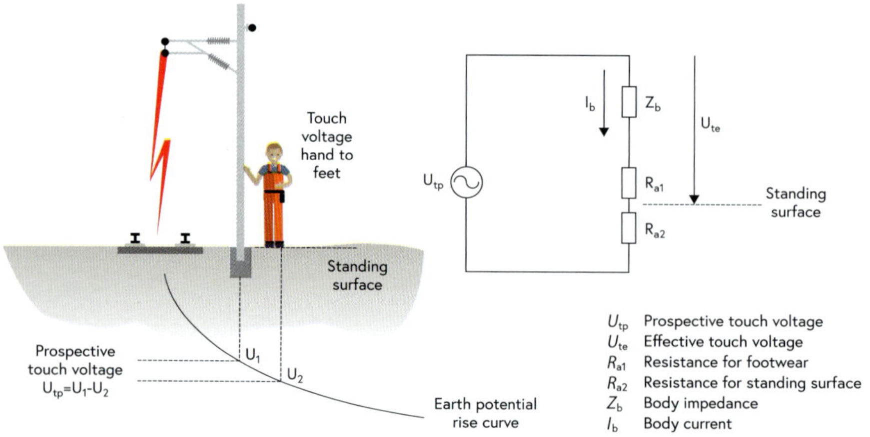

U_{tp}	Prospective touch voltage
U_{te}	Effective touch voltage
R_{a1}	Resistance for footwear
R_{a2}	Resistance for standing surface
Z_b	Body impedance
I_b	Body current

voltage. In Figure 5.2, the hand-to-feet touch voltage scenario is illustrated in which the prospective touch voltage is the voltage difference along the EPR curve between the structure and the person's feet. For step potential the prospective touch voltage would be determined by the position of the feet.

The methodologies for determining the body impedance Z_b and permissible body current I_b are defined in IEC 60479-1 and IEEE Standard 80, but are different because IEEE Standard 80 provides information that may be used directly by designers of HV AC substations, whereas IEC 60479-1 provides data that then has to be developed by national and international technical committees that prepare standards for LV ($\leq$ 1,000 V AC) and HV > 1,000 V AC) systems, equipment and products to define the permissible effective touch voltages. These standards reflect the conditions under which systems, equipment and products are designed, constructed, maintained and operated and whether the persons that will be using and operating such products or systems are electrically skilled or instructed, or are ordinary persons as defined in IEC 60050.

The electric shock model includes additional resistances that reflect those for footwear R_{a1} and the standing surface R_{a2}. IEC 60479-1 provides no guidance on these additional resistances, whereas IEEE Standard 80 provides data and formulae that can be used by designers.

5.6   Maximum permissible body current

5.6.1   IEC 60479-1: methodology

The value of fatal body current is dependent on the duration for which the current flows and IEC 60479-1 provides current/time zones for these various physiological conditions (see Figure 20 and Table 11 of IEC 60479-1:2018).

The current path is important when using this information. The left hand to both feet represents current through the heart and is the path for which the least current will be likely to cause ventricular fibrillation (zone AC-4). If other current paths need to be assessed, then IEC 60479-1 provides heart current factors to be applied to the zone AC-4 left hand to both feet currents for the current path.

5.6.2   IEEE Standard 80: methodology

The permissible body current to which exposure would not lead to ventricular fibrillation and death in 99.5 % of all persons, is determined by equation 5.1 that reflects the magnitude and duration of the current:

$$I_b = \frac{\sqrt{S_B}}{\sqrt{t_S}}$$

(5.1)

where:

I_b is the permissible body current
t_s is the duration of the current
S_B is a value that reflects that the electric shock energy that can be survived by 99.5 % of persons for a given body weight. Its value is a function of body weight such that for a given current duration the permissible body current reduces as body weight reduces, as shown in Table 5.2. For a 50 kg person, $S_B = 0.0135$ and for a 70 kg person $S_B = 0.0246$. Body mass rather than body weight is the more accurate term, but body weight is used in IEEE Standard 80 and accordingly has been used in this description.

Table 5.2 Maximum permissible body currents by body weight and duration

Fault duration (seconds)	Body weight (kg)	Maximum permissible body current (mA)
1	50	116
1	70	157
0.1	50	367
0.1	70	316

Table 5.2 illustrates that the ventricular fibrillation current increases with increasing body weight for a given fault duration. The lower body weight of 50 kg is considered appropriate for areas accessible to the general public because this value represents the lower end of adult weight. The permissible safety voltage limits are lower because the maximum permissible body current is lower, and this represents a conservative approach that experience has shown to be acceptable. The 70 kg body weight is typically used in substations and other installations not accessible to the public and where other protective provisions can be used.

5.6.3 The importance of rapid disconnection of faults

Both the IEC and IEEE standards illustrate how the ability of the body to withstand electric shock is improved as the time taken to disconnect faults is made smaller. Considering the significance of fault duration, rapid disconnection of faults can be advantageous for two reasons:

1. the probability of exposure to electric shock can be reduced by a fast fault clearing time, in contrast to situations in which fault currents could persist for several minutes; and
2. tests and experience show that the chance of severe injury or death can be reduced if the duration of a current flow through the body is very brief.

5.7 Human body impedance

5.7.1 Elements of body impedance

The physiological effects are caused by the magnitude and duration of current that flows through the human body. From a practical perspective that is not particularly helpful as the voltage across the human body will be known rather than the current flowing through it. This voltage is the touch voltage and to determine the body current, the body impedance needs to be determined.

The human body impedance depends on various factors, including current path through the body, duration of current flow, frequency, skin moisture, contact area and pressure. It consists of two main components:

1. internal impedance; and
2. skin impedance.

Skin can be considered as a network of resistances and capacitances, the impedance of which falls as the current increases. It is dependent on contact area and moisture as well as the skin type. For lower touch voltages, there is wide variance; but for higher touch voltages, the impedance decreases considerably and becomes negligible when the skin breaks down due to the current flow.

Section 5 – Protection against electric shock

The internal impedance is mainly resistive, its value dependent on the current path through the body. There is a small capacitive component, but for the conditions relevant to an AC electrified railway it can be ignored.

The total body impedance varies with the touch voltage mainly due to the skin impedance. As the touch voltage increases, the total body impedance is less influenced by the skin impedance and its value approaches that of the internal impedance.

5.7.2   IEC 60479-1 body impedance values

IEC 60479-1 describes how the internal impedance is distributed between the limbs, trunk and head. For hand-to-hand or one hand to one foot current paths the partial internal impedances are mainly located in the arms and legs, and the trunk partial internal impedance is small in comparison. IEC 60479-1 also provides a simplified representation that for many applications is sufficient for technical committees to develop their limits of permissible effective touch voltage. This simplified representation assumes that the arm and leg partial impedances are all equal and means that the hand-to-hand internal impedance is the same as that for the one hand to one foot path (see Figure 3 of IEC 60479-1). The simplified circuit also enables factors to be applied should an assessment of other current paths be required. For example, the internal impedance from one hand to both feet is 75 % of the internal impedance between hand-to-hand.

IEC 60479-1 details the hand-to-hand total body impedance for large, medium and small contact areas, and for dry, water-wet and saltwater-wet conditions. IEC 60479-1 illustrates that the total body impedance is not linear but tends towards a limiting value at higher touch voltages that correspond to the internal body impedance. It also shows that skin contact conditions have no significant variation at a touch voltage of 400 V and above. The various conditions are:

(a) a large contact area is defined as 0.01 m^2 and can be roughly considered to be representative of the size of an adult human hand;

(b) dry skin condition: condition at the contact area with regard to humidity of a person at rest under normal indoor environmental conditions;

(c) water-wet skin condition: condition at the contact area being exposed for 1 min to water of public water supplies (average resistivity $\rho = 35$ Ω.m, pH = 7 to 9); and

(d) saltwater-wet skin condition: condition at the contact area being exposed for 1 min to a 3 % solution of NaCl in water (average resistivity $\rho = 0.3$ Ω.m, pH = 7 to 9) – IEC 60479-1 assumes that the saltwater-wet condition simulates the condition of the skin of a sweating person or a person after immersion in seawater.

IEC 60479-1 states that for the conditions shown above, the total body impedance for children is expected to be somewhat higher but of the same order of magnitude. In general, the values are sufficiently conservative for them to be applied to 'normal' persons, including children irrespective of age and weight.

IEC 60479-1 represents the best available knowledge of body impedance, but this has to be considered in the context of a population whose physical condition varies with age and health. Body impedance will vary across a population and to assist in understanding the range, IEC 60479-1 provides body impedance for 5 %, 50 % and 95 % of a population for each of the three skin conditions. Table 5.3 of IEC 60479-1:2018 shows the range of body impedances for large contact area in dry skin conditions, but the notes illustrate the level of uncertainty around the data, and which can be considered when touch voltage limits are being developed.

Section 5 – Protection against electric shock

Table 5.3   Total body impedances by population percentile (Table 1 of IEC 60479-1:2018)

Touch voltage V	Values for the total body impedances Z_T (Ω) that are not exceeded for		
	5% of the population	50% of the population	95% of the population
25	1 750	3 250	6 100
50	1 375	2 500	4 600
75	1 125	2 000	3 600
100	990	1 725	3 125
125	900	1 550	2 675
150	850	1 400	2 350
175	825	1 325	2 175
200	800	1 275	2 050
225	775	1 225	1 900
400	700	950	1 275
500	625	850	1 150
700	575	775	1 050
1 000	575	775	1 050
Asymptotic value = internal impedance	575	775	1 050

Notes: 1. Some measurements indicate that the total body impedance for the current path hand to foot is somewhat lower than for a current path hand to hand (10% to 30%).
2. For living persons the values Z_T correspond to a duration of current flow of about 0, 1 s. For longer durations Z_T values can decrease (about 10% 20%) and after complete rupture of the skin Z_T approaches the internal body impedance Z_i.
3. For the standard value of the voltage 230 V (network-system 3N ~230/400 V) it can be assumed that the values of the total body impedance are the same a touch voltage of 225 V.
4. Values of Z_T are rounded to 25 Ω.

As electric shock is a function of the current that flows through the body, the body impedance limits the body current for a given touch voltage. Selection of a 5 % body impedance to determine the maximum permissible touch voltage is the most conservative approach as the majority of the population is expected to have a body impedance in excess of that value and therefore will experience lower body currents for the same touch voltage. The converse applies if a 95 % value is used.

5.7.3   IEEE Standard 80 body impedance values

IEEE Standard 80 specifies a fixed value of 1,000 Ω, irrespective of the body voltage U_b and this covers both children and adults. It makes it clear that this value relates to the current path between one hand and one or both feet. It does note that other paths, such as from foot to foot, can be less hazardous although still painful and may result in in other types of injury, such as those associated with falling or from involuntary reaction. The standard does allow higher body impedance values but does not specify what these should be.

The value of 1,000 Ω corresponds to the 95 % probability asymptotic values for large contact areas in dry, wet and saltwater-wet conditions described in IEC 60479-1 (see Table 1 of IEC 60479-1:2018).

Section 5 – Protection against electric shock

5.8 Permissible effective touch voltages

5.8.1 IEC 60479-1

For these values, IEC 60479-1 remains silent as its purpose is to provide core information and data. The values for permissible effective touch voltage are left to appropriate national technical committees to determine. Because these committees are specific to an electrical domain, such as power or telecommunications, different permissible effective touch voltage values have been developed that are appropriate to the technology and working practices. IEC 61936-1 *Power Installations Exceeding 1 kV AC – Common Rules* and EN 50522 *Earthing of power installations exceeding 1 kV AC* set limits for HV installations and systems. The ITU-T *Directives concerning the protection of telecommunication lines against harmful effects from electric power and electrified railway lines – Volume VI: Danger, damage and disturbance* sets limits for telecommunications systems.

For electric railways the limiting value of the effective touch voltage U_{te} is defined in EN 50122-1 *Railway applications. Fixed installations. Electrical safety, earthing and the return circuit. Part 1. Protective provisions against electric shock* such that the magnitude and duration of current flowing through the body does not cause ventricular fibrillation of the heart.

Figure 5.2 shows that the effective touch voltage U_{te} takes account of any additional resistance that a person may have through shoes or gloves. Figure 5.2 also illustrates how insulation of the standing surface may allow the touch voltage between the conductive part and earth (referred as the prospective touch voltage U_{tp}) to be greater than the effective touch potential by increasing the resistance of the standing surface R_{a2}. For electrified railways, increasing the standing surface resistance is not practicable along its line of route but can be used at traction switching stations or other locations that are fenced and where access is limited to authorized and competent persons.

The equations to calculate the effective (equation 5.2) and prospective (equation 5.3) touch voltage limits are:

$$U_{te} = I_b(t_f)\left(\frac{1}{HF} \cdot Z_b(U_b) \cdot BF + R_{a1}\right) \tag{5.2}$$

$$U_{tp} = U_{te} + I_b(t_f) \cdot R_{a2} \tag{5.3}$$

where:

> $I_b(t_f)$ is the limiting body current as a function of the fault duration for the left hand-to-feet current path
>
> HF is the heart current factor detailed in Table 12 of IEC 60479-1:2018. The heart current factor is used to calculate the limiting body current for current paths through the body other than left hand-to-feet path that represent the same danger of ventricular fibrillation
>
> $Z_b(U_b)$ is the body total impedance at the applied body voltage U_b

NOTE: EN 50122-1 uses Z_b and U_b in lieu of the Z_T and U_T used in IEC 60479-1.

> BF is body factor – this adjusts the body impedances values from the hand-to-hand body impedance values provided in Table 6 of IEC 60479-1:2018 to reflect different current paths through the body (for example, the body impedance from one hand to both feet is approximately 75 % of the hand-to-hand body impedance). The factors are provided as a note to Figure 3 of IEC 60479-1:2018. Should a body impedance be required for more complex current paths, partial impedances are provided in Figure 2 of IEC 60479-1:2018 that allow specific calculations.

Section 5 – Protection against electric shock

Table 5.4 Maximum permissible effective touch voltages as a function of time duration (from EN 50122-1:2011 +A4:2017 Table 4 and Section 9.2.2.3)

Current time duration (seconds)	$U_{te,max}$ long-term (volts)	$U_{te,max}$ short-term (volts)
> 300	60	Not applicable
300	65	Not applicable
1	75	Not applicable
0.9	80	Not applicable
0.8	85	Not applicable
0.7	90	Not applicable
< 0.7	Not applicable	155
0.6	Not applicable	180
0.5	Not applicable	220
0.4	Not applicable	295
0.3	Not applicable	480
0.2	Not applicable	645
0.1	Not applicable	785
0.05	Not applicable	835
0.02	Not applicable	865

EN 50122-1 defines how the various parameters covered in IEC 60479-1 are to be used. These are summarized below and the resulting maximum permissible effective touch voltages $U_{te,max}$ are shown in Table 5.4:

(a) Body impedances are for a large surface area in dry skin conditions and are not exceeded by 50 % of the population (see Table 5.5).

   (i) The choice of dry conditions for the railway environment is essentially historical. The first issue of EN 50122-1 in 1997 used body impedances from document IEC 479-1:1994 (a predecessor to IEC 60479-1:2018). The quoted body impedances did not consider skin contact conditions. The effect of skin contact conditions was published in IEC 60479-1:2005 which produced new sets of body impedance values that differed from those stated in IEC 479-1:1994. Although not clear, it would appear that experience of the touch voltage limits derived from IEC 479-1:1994 indicated that these provide acceptable safety limits against electric shock. It would therefore appear reasonable to use the body impedances in IEC 60679-1:2005 that most closely matched those in IEC 479-1:1994 (i.e. the 50 % probability values for dry conditions). It is worth noting the development of touch voltages in EN 50122-1 followed very closely the approach used in the HV electrical systems as defined in HV standards such as EN 50522 and IEC 61936-1: *Power installations exceeding 1 kV AC and 1.5 kV DC - Part 1: AC.*

   (ii) The rationale behind the choice of a body impedance value not exceeded by 50 % of the population is not explained. It means that 50 % of the population will experience higher body currents for a given touch voltage and does raise questions as to why this is acceptable. However, it is a value used in the ITU-T CCITT directives and European standards covering HV systems such as EN 50522, although not in the UK HV utility sector where a body impedance not exceeded by 5 % of the population is used (see Table 5.7).

(b) Current path is left hand to both feet. This results in a body factor of 0.75 in the 50 % population body impedances stated in Table 6 of IEC 60479-1:2018. Selected values of the factorized body impedances are shown in Table D1 of EN 50122-1:2011+A4:2017.

(c) The limiting values for body current I_b are taken from curve c1 of IEC 60479-1 Figure 20 and marks the upper limit of zone AC-3. Although this represents 0 % probability of ventricular fibrillation, there are possibly severe physiological effects as described in Table 5.3. The limiting values are time-dependent and selected values are provided in Table D3 of EN 50122-1:2011+A4:2017 for AC electrified railways.

(d) Heart current factor set at 1. This is because the current path chosen represents one of the current paths that is most hazardous.

(e) An additional 1,000 Ω resistance R_{a1} is added to represent 'old wet shoes' for short-term conditions which have durations less than 700 ms. A key consideration in making a distinction between long- and short-term conditions is to control the rail potential. The rails form part of the traction return circuit and can be considered to be a live part. They are not insulated and cannot be protected by barriers to prevent direct contact, which means that they present an effective touch voltage to persons in both normal and fault conditions. Long-term conditions therefore reflect normal conditions in the context of electric shock protection. Exclusion of R_{a1} resistances for long-term conditions means that the limiting value of effective touch potential cannot be greater than the maximum body voltage U_b. It assumes that persons have bare feet and hands but this cautious assessment is considered appropriate because members of the public, who are considered as ordinary persons in the context of knowledge and awareness of electrical hazards, can come in contact with the rails at, for example, level crossings or from contact with trains when boarding and alighting. For short-term conditions EN 50122-1 makes a key assumption that the inclusion of R_{a1} at 1,000 Ω means that the maximum permissible body voltage U_b is never exceeded.

(f) No additional resistance R_{a2} is included for the standing surface. Where it is considered reasonable to do so and where the resistance can be guaranteed over the life of the asset, then it can be added to determine the prospective touch potential. This can be reasonably assured within restricted areas or compounds, such as traction switching stations, where environment and access is controlled but is more difficult in open areas of the railway corridor.

There are workshops, depots, or similar where work is undertaken on trains or on-track plant when the rails remain connected to the traction return circuit even though the overhead contact line system (OCLS) is isolated and earthed. Even though the work activities are normally carried out by competent persons and in controlled areas, the nature of the work requires lower values of permissible touch voltages as specified in EN 50122-1 and summarized in Table 5.5.

The maximum voltage of 25 V is specified for long-term conditions because the limiting value is not set by curve c1 of IEC 60479-1 Figure 20 but to within the area AC-2 by limiting the body current to less than 5 mA. It is selected to mitigate against injuries caused by uncontrolled reaction to touch voltages and reflects the working practices within depots where persons work in more restricted spaces. The restricted space can mean that the current path is between head-to-hand or back-to-foot rather than the default hand to both feet path. While EN 50122-1 provides a limiting value for these special conditions based on hand to both feet current path, it may be necessary to assess different safe values to reflect these different body current paths, contact areas and wet conditions. IEC 60479-1 provides methodologies to support such assessments.

It should be noted that even lower values may be desirable, particularly where the public has regular access. An example is where platform screen doors are provided and there will be a voltage between the platform screen door and the train, both of which can be touched as persons alight and board the train. In such a situation, reducing the touch voltage between the train and platform screen door to values (< 2 V) that are below the threshold of perception under normal conditions may be considered desirable to reduce the risk of a touch voltage creating a startle reaction that may cause a fall or sudden movement. Under fault conditions the short-term touch voltage limits defined in EN 50122-1 would be used, as it would be difficult to provide measures to hold the touch voltage to such low values.

Section 5 – Protection against electric shock

Table 5.5   Maximum permissible effective touch voltages as a function of time duration for workshops, depots [from EN 50122-1:2011+A4:2017 Section 9.2.2.3]

Current time duration (seconds)	$U_{te,max}$ long-term (volts)	$U_{te,max}$ short-term (volts)
> 300	25	Not applicable
300	25	Not applicable
1	25	Not applicable
0.9	25	Not applicable
0.8	25	Not applicable
0.7	25	Not applicable
< 0.7	Not applicable	155
0.6	Not applicable	180
0.5	Not applicable	220
0.4	Not applicable	295
0.3	Not applicable	480
0.2	Not applicable	645
0.1	Not applicable	785
0.05	Not applicable	835
0.02	Not applicable	865

5.8.2   IEEE Standard 80

IEEE Standard 80, unlike IEC 60479-1, does provide the means to calculate the permissible touch voltages (hand-to-feet and hand-to-hand) and step voltages. In part, this is because IEEE Standard 80 is applicable to outdoor AC substations, either conventional or gas insulated, including distribution, transmission and generating plant substations. It therefore incorporates the derivation of body resistances and currents into one document, together with how to apply the additional resistances that may be applicable, such as footwear and the additional insulation provided by standing surfaces. The equivalent circuit is similar to that in Figure D1 of EN 50122-1, but the difference is that IEEE Standard 80 provides significantly more information on how to calculate the standing surface resistance R_{a2}. The equations for touch and step potential take account of the different body weights described in Section 5.6.2. The equations 5.4 and 5.5 for a 50 kg person are (reproduced from IEEE):

$$E_{touch50} = (1,000 + 1.5 C_s \rho_s) \left(\frac{0.116}{\sqrt{t_s}} \right)$$
(5.4)

$$E_{step50} = (1,000 + 6 C_s \rho_s) \left(\frac{0.116}{\sqrt{t_s}} \right)$$
(5.5)

The terms in these equations are:

1,000 represents the body impedance described in Section 5.7.3

$\sqrt{s_B} = 0.116$ for a 50 kg person described in Section 5.6.2

$\sqrt{s_B} = 0.157$ for a 70 kg person described in Section 5.6.2

1.5 (or 6) $C_s \rho_s$ represents the impedance of two feet together and the standing surface where ρ_s represents the resistivity of the standing surface layer. C_s is a factor that is applied to take account of the different resistivity of the standing surface additional layer and the underlying ground resistivity. If there is no additional standing surface layer applied, then $C_s = 1$ and ρ_s takes the resistivity of the underlying ground

t_s represents the duration of the fault current.

The insulation of any footwear is excluded from these equations but may be added as an additional resistance.

An important point to note is that the calculations are specified as suitable for fault durations between 0.03 seconds and 3 seconds. This means that IEEE Standard 80 cannot be used to set the permissible touch voltages for the long-term conditions that are addressed in EN 50122-1. Accordingly, for some railways in the US A, such as the California High Speed Rail project, the limiting touch voltages described in EN 50122-1 have been adopted.

5.9 Comparison across standards applicable to high-voltage systems

Table 5.6 illustrates how different national and international technical committees have provided different values of maximum permissible effective touch voltage for HV systems, and how each standard has defined maximum permissible effective touch voltages that reflect the working methods, equipment, and environment applicable to the standard. It is therefore essential that the basis on which these maximum permissible effective touch voltages have been calculated is fully understood so that they remain applicable to real life situations and, importantly, where there is a need to apply these to apparently similar situations that are not covered by a particular standard.

When designing railway systems in which the HV electric traction system and LV systems co-exist, questions are often asked as to why the LV 50 V touch voltage limit used in the LV standard IEC 60364 (and its equivalent CENELEC HD 60364 harmonized document), and the EN 50122-1 60 V touch voltage limit, are so close and why they are they not the same. The description above of how the information contained in IEC 60479 is used by various technical committees to specify the different touch voltage limits listed in Table 5.6 helps explain how the 50 V limit used in LV systems has been developed and incorporated into many national standards, such as BS 7671 in the UK. For the LV IEC 60364 limits, the technical committee selected a current duration/current magnitude that is within the AC3 zone but below the c1 curve shown in Figure 20 of IEC 60479-1 . A hand-to-foot body impedance for dry conditions that is not exceeded by 95 % of the population was selected from Table 5.3 and modified by a body factor to reflect the body impedance for a two-hands-to-two-feet current path. A combined footwear and standing surface impedance of 1,000 Ω was added. The committee also considered that as LV equipment and systems are used mainly by ordinary persons as defined in IEC 60050 (IEV ref 195-04-03), and the wide range of fault clearance times found in LV systems, the permissible touch voltage limit should not be dependent upon the fault clearance time. Within these parameters, the LV permissible touch voltage with no duration limit is specified as 50 V. (Refer to the IET's *Commentary on IET Wiring Regulations (17th Edition)*.)

Table 5.6 Comparison of maximum permissible effective touch voltage $U_{te,max}$ for different standards

Fault duration t (seconds)	EN 50122-1:2011 +A4:2017 (long-term)	EN 50122-1:2011 +A4:2017 (short-term)	EN 50522:2010 Figure 4 IEC 61936-1 Figure 12 no additional resistances	EN 50522:2010 (UK Annex N) no additional resistances	ENA TS 41-24 Issue 2 Note 1	ITU-T Directive Volume 6:2008 - Typical Situations (Short Duration) Note 2	ITU-T Directive Volume 6:2008 - Typical Situations (Long Duration) Note 2	IEEE Std 80:2013 (50 kg body weight) Note 3	IEEE Std 80:2013 (50 kg body weight) Note 4
>300	60		80		57	Not applicable	60	Not applicable	Not applicable
300	65		80		57	Not applicable	60	Not applicable	Not applicable
1	75		117	68	80	430	Not applicable	116	Not applicable
0.5		220	220	135	166	650	Not applicable	164	328
0.2		645	537	320	407	1,500	Not applicable	259	519
0.1		785	654	405	521	2,000	Not applicable	367	734
Basis of assessment									
Probability of ventricular fibrillation	0 %	0 %	5 %	5 %	5 %	5 %	0 %	0.5 %. See Section 0	0.5 %. See Section 0
IEC 60479-1:2018 Figure 20 Time/current curve	C_1	C_1	C_2	C_2	C_2	C_2	B	Not applicable. See Section 0	Not applicable. See Section 0
Current path	One hand to both feet	One hand to both feet	Weighted average of left hand to both feet, right hand to both feet, both hands to both feet, hand-to-hand	Weighted average of left hand to both feet, right hand to both feet, both hands to both feet, hand-to-hand	Weighted average of left hand to both feet, right hand to both feet, both hands to both feet, hand-to-hand	Hand-to-hand and one hand to both feet	Hand-to-hand and one hand to both feet	One hand to one foot and one hand to both feet	One hand to one foot and one hand to both feet
Contact area (as defined in IEC 60479-1:2018)	Large	Large	Large	Large	Large	Large	Large	Not applicable	Not applicable

Table 5.6   cont.

Fault duration t (seconds)	EN 50122-1:2011 +A4:2017 (long-term)	EN 50122-1:2011 +A4:2017 (short-term)	EN 50522:2010 Figure 4 IEC 61936-1 Figure 12 no additional resistances	EN 50522:2010 (UK Annex N) no additional resistances	ENA TS 41-24 Issue 2 Note 1	ITU-T Directive Volume 6:2008 - Typical Situations (Short Duration) Note 2	ITU-T Directive Volume 6:2008 - Typical Situations (Long Duration) Note 2	IEEE Std 80:2013 (50 kg body weight) Note 3	IEEE Std 80:2013 (50 kg body weight) Note 4
Degree of moisture (as defined in IEC 60479-1:2018)	Dry	Dry	Dry	Dry	Dry	Dry	Dry	Not applicable	Not applicable
Probability of exceeding assumed body impedance	50 %	50 %	50 %	5 %	5 %	50 %	50 %	Not defined	Not defined
Source impedance (ohms)	0	0	0	0	0	180	180	0	0
Contact area resistance (ohm)	0	0	0	0	150	0	0	0	0
Shoe resistances (ohm)	0	1,000	0	0	0	3,000	3,000	0	1,000 (long-term only >0.7 seconds)

Notes: 1. ENA TS 41-24 is the UK Electricity Network Association Technical Standard 41-24. It is based upon the UK Annex N of EN 50522:2010.
2. ITU-T is the Telecommunication Standardization Sector of the International Telecommunications Union. The values of body impedance are those used in the document IEC TS 60479-1:1994 (alternative reference CEI IEC 479-1 third edition 1994) and not those in the latest version of IEC 60479-1.
3. The values are based on no contact or additional resistances to enable comparison with EN 50522:2010, IEC 61936-1:2010 and EN 50522:2010 Annex NA.
4. The value includes a 1,000 Ω shoe resistance for the long-term conditions to allow comparison with EN 50122-1.

Section 5 – Protection against electric shock

5.10 Protective measures and protective provisions

5.10.1 Basic and fault protection

The fundamental rule described in IEC 61140 requires provision of protective measures to reduce the risk of electric shock by preventing direct contact with hazardous-live-parts and accessible conductive parts from becoming hazardous live under normal conditions and under fault conditions.

Normal conditions are not formally defined by the IEC but can be understood as a condition that encompasses all of the following:

(a) when there is no fault (i.e. it is fault-free);
(b) functions as intended over the specified range of steady state and transient currents and voltages for which it has been designed, manufactured and installed; and
(c) operation in intended use (IEC Guide 51:2014 Clause 3.6). For an electric traction system, this includes the various configurations for which the system has been designed and typically includes:

    (i) switchgear, transformers etc., taken out of service for maintenance repair or renewal;
    (ii) non-availability of electrical supplies resulting in alternate traction feeding configurations; and
    (iii) isolation of OCLS electrical sections for maintenance, repair and in response to emergencies.

Protection against electric shock under fault-free conditions is described as basic protection. Protection against electric shock under single fault conditions is described as fault protection. It is considered unlikely that two or more faults will occur simultaneously that result in an electric shock hazard worse than the most onerous single fault. Accordingly, fault protection provisions are developed for the single fault condition that has the highest fault current and longest fault duration time from all of the permissible electrical configuration arrangements.

The terms 'protection against direct contact' and 'protection against indirect contact' are sometimes used in lieu of normal and fault protection. These terms do not adequately cover the scope of the fundamental rule because the electric shock risk and, particularly in HV systems, may not only involve direct contact with live parts. Touch voltages above the threshold of perception can occur on accessible conductive parts under normal conditions through inductive and capacitive coupling from live parts, a condition that is very typical of AC electric traction systems. These may not be fatal but do represent an electric shock hazard. Furthermore, the term 'indirect contact' is a rather oblique way of describing the contact of persons or animals with exposed-conductive-parts which have become live under fault conditions. The term does not make it clear that hazardous voltages under fault conditions may appear as a step or touch voltage due to high EPR or through the transfer of voltages from remote locations by extraneous-conductive-parts.

Basic and fault protection is achieved by one or more protective measures, each of which is built up from one or more protective provisions. Some protective provisions can be used as part of the protective measures for both basic and fault protection. An example of a common protective provision on HV systems is protective earthing and protective equipotential bonding.

It is important to appreciate that IEC 61140 applies to all electrical systems and equipment from extra-low voltage (< 50 V AC) through low-voltage ($\leq 1{,}000$ V AC) to high-voltage ($> 1{,}000$ V AC). It describes a wide range of protective provisions but not all of these are applicable or suitable for every electrical system or equipment. Specific standards are therefore required to define the appropriate protective measures for particular systems and equipment. For example, IEC 61936-1 and EN 50522 describe the protective measures and provisions specific for power installation exceeding 1 kV AC (HV), and EN 50122-1 describes the protective measures and provisions for railway electric traction systems. The suite of

Section 5 – Protection against electric shock

IEC 60364 standards, and its equivalent CENELEC suite of HD 60364 harmonized document, describes the protective measures for LV AC and DC electrical installations.

Basic and fault protection measures described will consider only those protective provisions most applicable to AC railway electric traction systems.

5.10.2 Basic protection

The protective measures for basic protection employ all or some of the following protective provisions:

Basic insulation: This is provided to prevent contact with hazardous-live-parts. The insulation type may be liquid (oil), solid or gas (including air). For accessible live parts within HV installations that are air insulated, basic insulation is an imaginary boundary of the danger zone around live parts and in the context of basic protection entering the danger zone is equivalent to touching live parts. To provide a margin of safety, and to reduce the risk of entering the danger zone, railway infrastructure owners often define a larger vicinity (or proximity) zone within which entry is prohibited. Section 11 describes the danger and vicinity zones in more detail.

It is worth noting that basic insulation using solid rather than air insulation for HV systems will prevent contact with hazardous-live-parts, but through capacitive or inductive coupling they may have touch voltages on the accessible surface of the solid insulation or on the metal containment of the gas or liquid insulation. In such cases, additional protective provisions, typically earthing and equipotential bonding, are used to control the touch voltages to below the threshold of perception.

The term 'basic insulation' is to make it clear that the insulation is suitable for normal conditions. It makes a distinction from supplementary insulation that provides a fault protection protective measure when combined with basic insulation. IEC 61140 requires that supplementary insulation shall be dimensioned to withstand the same stresses as specified for basic insulation under the same environmental conditions. The addition of supplementary insulation provides double insulation. Reinforced insulation provides the equivalent of double insulation, but its construction tends to involve several layers that do not allow it to be tested in separate parts as basic and supplementary insulation. In general, supplementary or reinforced insulation is not economic for use in HV installations and therefore in practice it is used more commonly in LV installations.

(Electrically) protective barriers or enclosures: This provision provides protection against contact of live parts or entering the danger zone by persons, including ordinary persons (i.e. those who neither possess knowledge or understanding of electrical systems and their hazards), and animals. Although similar, the terms 'barriers' and 'enclosure' have important differences. A barrier provides protection against contact with a live part or entering the danger zone from any usual direction of access, whereas an enclosure prevents access from any direction. It should be noted that the definition of a barrier implicitly assumes that it is a protective measure against both unintentional and deliberate action. An example of a barrier is a robust high fence, often with anti-climbing measures to make it very difficult to have direct contact with live parts given that the usual direction of access is from ground level. An enclosure prevents contact with live parts or entering the danger zone from any direction and is also often designed to protect against environmental effects such as water and dust and sometimes to provide a level of vandal resistance. Examples of enclosures include metal-clad switchgear, distribution boards and electrical appliances used within a domestic environment. IEC 61936-1 describes the requirements for barriers and enclosures for HV installations.

(Electrically) protective obstacles: Although a protective provision against contact similar to barriers, an obstacle is a significantly different provision. An (electrically) protective obstacle prevents unintentional contact or entry to the danger zone, but does not prevent contact from happening deliberately. Unlike an (electrically) protective barrier the definition of an (electrically) protective obstacle makes no mention of usual direction of access and, in comparison to a barrier, diminishes its effectiveness by allowing contact or entry to the danger zone by deliberate action. A fence is an example of an obstacle. When employed as an obstacle the design prevents unintentional contact but will not have the anti-climbing and other measures to prevent intentional access from any usual direction of access that is required from a barrier. It could be climbed deliberately and thereby give access to live parts or the danger zone. IEC 61140 makes clear that (electrically) protective obstacles should be provided only where electrically skilled or instructed persons will be present and should not be provided for where ordinary persons will be present.

Placing out of arm's reach: Where basic insulation is provided by air and it is not practicable to place the live parts behind barriers or within enclosures, the live parts are placed out of arm's reach to prevent contact with live parts or enter the danger zone. Examples of such a measure include power distribution overhead lines that carry bare conductors. On an electrified railway, the overhead contact line is placed out of arm's reach at station platforms, road and footpath at-grade crossings and other locations where workers, passengers and the public may be present. Although IEC 61140 describes this measure, it is left to sector-specific technical committees to establish the distance below which a live part is no longer considered to be out of arm's reach. Section 6.2.4.1 describes how this is applied by EN 50122-1.

Protective earthing and protective equipotential bonding: At HV installations, there may be hazardous voltages on accessible surfaces of basic insulation by inductive and capacitive coupling, as described previously, or from an EPR due to currents that return to their source via the installation earth electrode. At high voltages, basic insulation alone may not avoid touch voltage hazards due to insulator leakage current flowing through an accessible conductive part and provide the intended insulation stress grading and performance under normal conditions. In these situations, the accessible conductive surfaces and non-live end of HV insulators will be bonded together and to the installation earth electrode to form an earthed equipotential zone.

5.10.3   Fault protection

The protective measures for fault protection employ all or some of the following protective provisions in addition to those required for basic protection:

Automatic disconnection of supply: In the event of a fault, the fault current is detected, and a protective device operated to disconnect that part of the installation. Section 5.6.3 explains how the rapid disconnection of faults can reduce the chance of injury and death.

Protective earthing and protective equipotential bonding: All accessible conductive surfaces are equipotentially bonded together and connected to earth to avoid dangerous touch and step voltages. This will include accessible conductive surfaces that do not form part of the installation, such as fences and gates, and other assets that are at risk of becoming hazardous live as a result of the fault. Protective earthing at HV installations is provided by an earth electrode which could be a number of earth rods or an earth grid, or a combination of both. The design of the earth electrode aims to limit the EPR such that step, touch and mesh voltages within the switching station or substation and outside its perimeter are within safe limits.

Potential grading: The EPR can vary between different points within an installation or immediately outside its boundary and these could result in hazardous touch and step voltages. In such circumstances, additional earth electrodes would be provided and connected to the installation protective equipotential bonding arrangement.

Insulation of standing surfaces: The insulating layer provides an increased insulation from earth such that a higher prospective touch voltage U_{tp} will not create a hazardous effective touch voltage U_{te} when a person touches an earthed conductive part that has become hazardous live. The use of granite chippings or similar can provide additional insulation, but in practice its application is normally restricted to secure areas such as within the fence of traction switching stations where only competent electrically skilled or instructed persons have access.

5.11 Application of protective measures and protective provisions to an AC electric railway

5.11.1 Electrification system hazards

Section 2 describes the parts that form the electric traction system. The protective measures and provisions applied reflect how the electric traction system is configured and the hazards that are present. These can be summarized as follows:

(a) The rails form part of the return circuit for traction current. They cannot be insulated and therefore they present a touch voltage hazard. The rails are normally cross-bonded together at regular intervals along the length of the railway and to other conductors that form part of the return circuit and to earth. This is an equipotential bonding protective provision that seeks to maintain the rail potential to within the limiting voltages defined in EN 50122-1. Should this not be possible, EN 50122-1 describes a methodology by which higher rail potential may be acceptable by the application of the 'a' factor, the concept of which is based on the following:

  (i) the permissible touch voltage (and hence the rail potential) is stated with reference to a nominal zero-volt earth reference that is remote from the rails;

  (ii) touch voltages arise from normal and fault current flowing from the rails into earth to create an EPR around the point where the current flows to earth. As described in Section 3, the EPR decays in with increasing perpendicular distance from the track for homogeneous soil resistivity; and

  (iii) the distances for touch voltage and step voltage hazards are such that the voltage imposed (the prospective touch potential) is the distance between two points on the EPR curve. It shows that the voltage between where a person stands and the rail (or an accessible conductive part connected to the rail) will always be less than the full rail-to-remote-earth potential. The example described in Section 3 and EN 50122-1 shows that the voltage between the rail and a point 2 m away is generally only 50 % of rail potential to remote earth.

EN 50122-1 uses this concept to specify two values for the 'a' factor. The first value allows the touch voltage (and rail potential) to be twice the maximum permitted effective touch voltage (i.e. a = 2). If this approach were to be adopted, the following would need to be considered:

  (i) ground resistivity: if this is very high or is not homogeneous, the decay of the EPR may be considerably steeper such that 50 % reduction may occur closer than 2 m.

  (ii) transfer voltages: this approach must ensure that there are no assets that permit transfer voltages, and particularly to those locations with assets at a nominal zero-volt earth where they could be touched simultaneously. The adoption of an integrated earthing system as described in Section 4 can mitigate against the resulting hazards within the railway boundary, but there may be unacceptable transfer voltages to outside the railway. A segregated earthing system would not be practicable because of the different earths and the difficulty of ensuring the equipment separation of the two segregated systems in all circumstances.

Section 5 – Protection against electric shock

EN 50122-1 defines a second 'a' factor value of 3.3 but the implementation of this requires significant protective measures, such as standing surface insulation, use of voltage limiting devices (VLDs), and access restrictions, all of which can be difficult to install and which impose a significant burden on the asset management regime to maintain these over the life of the railway.

The application of the 'a' factor requires care and thought to make sure that the implications of higher rail potentials across the line of route, including transfer potentials to third parties, are understood. In practice, many railway infrastructure owners do not allow the use of the 'a' factor except for specific locations where the environment is known and can be controlled, and where the 'a' factor can then be used as part of hazard risk reduction approach.

(b) The overhead contact line system (OCLS) has live conductors at a nominal voltage of 25 kV. Insulation of the live conductors is by air except at support locations where solid (typically ceramic or polymeric type) insulators are provided at the OCLS masts and at the support and registration assemblies. The position of the live conductors is determined by the railway structure gauge and the effective and reliable contact between train pantograph and contact wire. The OCLS must pass beneath bridges, through tunnels, in cuttings and through stations. Hazards include:

   (i) breakdown of the air insulation and flashover to railway infrastructure structures and fixed equipment, and to rail vehicles (corresponds to cause **(b)** in Section 5.1).

  (ii) persons either coming into contact with live parts or approaching the danger zone (corresponds to cause **(c)** in Section 5.1).

 (iii) windborne debris caught in the OCLS, or water or ice from overline structures, or bird strikes that bridge the live and earthed parts and particularly those that are of high impedance such that the fault is not detected by the electric traction protection system (corresponds to cause **(b)** in Section 5.1).

 (iv) vehicles and plant coming into contact with live parts, particularly at at-grade crossings accessible to the public (corresponds to cause **(c)** in Section 5.1).

  (v) insulation breakdown due to lightning and other transient overvoltages (corresponds to cause **(b)** in Section 5.1).

 (vi) the OCLS breaks, and live conductors fall onto structures infrastructure fixed equipment, and rail vehicles which then become live and create and electric shock hazard (corresponds to cause **(b)** in Section 5.1). The zone in which a broken overhead line contact wire is likely to fall is the overhead contact line zone (OCLZ) which is defined by EN 50122-1. It should be noted that the extent of the OCLZ shown is based on an overhead contact line that is positioned above the track centre line and takes account of contact wire stagger but does not intend to reflect the complexity of an OCLS with its overlaps, jumpers, section insulators and neutral sections, etc. Dimension *X* is left for national railway authorities to define as the knowledge of the structure gauge, pantograph type and practical experience for their railways is required to establish practical values for the dimensions. In determining dimension *X*, national railway authorities would need to consider the following:

- the likely fall area of the overhead contact line in the event of a breakage of the contact wire or other overhead contact line support arrangement. The OCLZ is an important concept because any asset or equipment within that zone will require bonding to the traction return circuit and to be rated for traction fault current and fault clearance duration.
- the overhead contact line is formed of discrete lengths (known as tension lengths) that require anchoring at each end on a structure and which may require part of the overhead contact line to be out of running (i.e. it runs at a height and position where it is not intended to be directly used for current collection). For some overhead line systems, the out of running parts are live. Dimension *X* needs to take account of the fall area of a broken out of running live conductor.

- for autotransformer systems with autotransformer feeder conductor or in single-phase systems with aerial along track or reinforcing feeders these may be placed inside or on top or on the outside of OCLS structures. EN50122-1 in defining the OCLZ considers only those overhead contact lines that are primarily at risk from damage or breakage by the train pantograph (i.e. the contact wire and the messenger (catenary) wire) and therefore these other aerial conductors are not considered in defining X. Railway infrastructure owners would need to understand the risk of these conductors breaking and how they would fall to the ground. It may be decided that the risk is sufficiently low to be acceptable. If not, either the X dimension could be increased to take account of the likely fall area or alternative protective provisions provided.
- it can be argued that the location of overhead line contact structures and their spacing along the track means that they could be considered as a bare protective part that is connected to the traction return as defined in EN 50122-1. It therefore could be considered that the limit of the OCLZ should not be extended beyond the line of the overhead contact line structures.

(c) Broken train pantographs that may raise above the OCLZ while remaining live and corresponding to cause (b) in Section 5.1. These may strike structures such as overbridges and make then live. The zone in which such an event is likely is the current collector zone (CCZ). EN 50122-1 leaves it to national railway authorities to determine the dimensions of Y and Z as these are strongly influenced by the railway structure gauge, the mechanical properties of the OCLS and the pantograph types permitted to run.

(d) Traction switching stations contain transformers and switches to control the distribution of energy to the OCLS. Traction current returns to these switching stations via the earth and the rail which result in EPR and touch potential hazards within the switching station.

5.11.2   Protective measures and provisions

The application of protective measures and provisions is best considered in two parts, although they are linked as many of the protective provisions are applicable to both. The basic and fault protective provisions for the electric traction system should be considered as one part. Assets that do not form part of the electric traction system but would become hazardous live under an electric traction system normal and fault conditions should be the other part. Examples of these assets include bridges, viaducts, non-traction HV and LV equipment, train command and control equipment, and telecommunications equipment. Section 6 describes the protective provisions appropriate for the electric traction system, and Section 7 describes those for assets not forming part of the traction return system. However, at this point it is worthwhile summarizing that basic and fault protection for AC electrified railways is provided by two protective measures: equipotential bonding and earthing, and automatic disconnection of the supply. Tables 5.7 and 5.8 illustrate these protective measures and the various protective provisions that make up the protective measures used for AC electric traction systems.

Table 5.7 Equipotential bonding protective measure

Protective measures	Protective provisions	
	Basic protection	Fault protection
Protection by equipotential bonding	Basic insulation using solid, gas (including air) and liquid (for example, oil) insulation used for cables and in electrical equipment such as transformers, switchgear, busbars, and insulators used at live overhead contact lines support structures in accordance with product standards.	Equipotential bonding of rails and other parts of the traction return circuit (for example, earth wires, return conductors) and to earth to maintain the rail potential and effective touch voltage to limits defined in EN 50122-1 for short-term conditions.
	Basic insulation by air by providing clearance distances to earthed conductive parts in accordance with EN 50124-1 and defining the boundary of the danger zone (refer to Section 11) plus the following provisions described in EN 50122-1: • provision of obstacles to prevent contact with or approaching near live parts (for example, overbridge parapets); • provision of barriers such as fences around traction switching stations; and • placing out of reach for example at station platforms, at-grade road and footpath crossings and other locations where workers, passengers and the public may be present.	Equipotential bonding to traction return of the accessible conductive parts of assets and equipment not part of the electric traction system that may become live due to: • failure of basic insulation; • OCLS conductor breakage within the overhead contact line zone; and • pantograph failure within the current collector zone.
		Equipotential bonding normally is a direct conductive connection. Where direct connection is not permitted, a voltage limiting device – fault (VLD-F) as defined in EN 50122-1 is provided and connected between the part to be protected and the traction return circuit.
	Equipotential bonding of rails and other parts of the traction return	Equipotential bonding, earthing and connection to the traction return circuit of accessible exposed-conductive-parts at switching stations and substations, and of the OCLS (for example, OCLS support structures).
		Rating of protective conductors within the electric traction system to carry traction

Table 5.7 *cont.*

Protective measures	Protective provisions	
	Basic protection	Fault protection
	circuit (e.g., earth wires, return conductors) and to earth to maintain the rail potential and effective touch voltage to within limits defined in EN 50122-1 for long-term conditions.	fault current for the specified fault clearance time.
	Earthing and equipotential bonding of accessible conductive parts of high voltage installations that are not live parts but under normal conditions unacceptable touch voltages may occur though inductive or capacitive coupling (for example, gas insulated switchgear) or at substations or at substations where traction return currents flowing through the earth electrode produce hazardous unacceptable touch and step voltages.	Rating of protective conductors within electrical installations that do not form part of the electric traction system that are made live through a fault on the electric traction system.
		Potential grading within and around traction switching stations and substations to limit touch and step voltages to within limits defined in EN 50122-1.
		Insulation of standing surfaces
		Provision of surge divertors connected between normally live parts and the traction return circuit.

Table 5.8 Automatic disconnection of supply protective measures

Protective measures	Protective provisions		
	Basic protection		Fault protection
Protection by automatic disconnection of supply	Basic insulation using solid, gas (including air) and liquid (for example, oil) insulation used for cables and in electrical equipment such as transformers, switchgear, busbars, and insulators used at live overhead contact lines support structures in accordance with product standards.	Basic insulation by air by providing clearance distances to earthed conductive parts in accordance with EN 50124-1 and defining the boundary of the danger zone (refer to Section 11) plus the following provisions described in EN 50122-1: • provision of obstacles to prevent contact with or approaching near live parts (for example, overbridge parapets); • provision of barriers such as fences around traction switching stations; and • placing out of reach for example at station platforms, at-grade road and footpath crossings and other locations where workers, passengers and the public may be present.	Automatic disconnection of the supply in the event of short circuits and overloads within the electric traction system: • protective devices to detect the fault; and • circuit-breakers to open and disconnect the supply.

 **Section 6**

Electrification system assets excluding the traction return

6.1 Non-traction return assets

Section 4 describes the traction return system and its various parts. This section discusses other assets within the electrification system that are not considered part of the traction return system but form part of the electric traction system. These assets are:

(a) the overhead catenary system;
(b) 25 kV cables;
(c) switching stations; and
(d) feeder switching stations and substations.

6.2 The overhead contact line system

6.2.1 Description

An overall description of the overhead contact line system (OCLS) is detailed in Section 2. Electrically, it can be considered as consisting of two parts:

1. parts that are intended to be live at the nominal traction system voltage; and
2. parts that are connected to the traction return and earth.

The live parts are attached to the OCLS support structures by insulators. The support arrangements ensure that the live conductors fit within the railway structure gauge and the contact wire provides effective and reliable contact with the static and dynamic movement of the train pantograph. Between the support structures the live conductors must pass beneath bridges, through tunnels, in cuttings and through stations where the electrical insulation is provided by air. Similarly, at the OCLS structures, insulators hold the live conductors away from any part that is connected to the traction return and earth.

EN 50119 *Railway applications. Fixed installations. Electric traction overhead contact lines* describes the requirements for the OCLS. It covers the electrical and mechanical requirements of the overhead system and the structural requirements for support structures and foundations. Although a European standard, it is used widely throughout the world, albeit supplemented with national standards and codes of practice, particularly in respect of structural requirements where Eurocodes may not be applicable.

6.2.2 Insulators

Insulators need to satisfy a number of general requirements. They:

(a) must have the mechanical strength to support and hold the conductors in their required position under the environmental conditions in which the OCLS will operate.
(b) should be sufficiently durable to withstand vandalism.
(c) shall have a rated impulse voltage appropriate for the electric traction system voltage. The impulse voltage is higher than the nominal or rated voltage of the electrification. Impulse voltages are of high magnitude but short-term and can be caused by switching transients, harmonics and lightning

strikes. Impulse voltage is an important consideration in achieving insulation coordination throughout the electrification system. EN 50124-1 *Railway applications. Insulation coordination – Basic requirements. Clearances and creepage distances for all electrical and electronic equipment* provides a range of rated impulse voltages for various nominal electric traction system voltages.

(d) have a creepage path across its surface between the live end of the insulator and the earthed end appropriate for the atmospheric environment in which it is to be installed. EN 50124-1 provides the creepage distance requirements for various degrees of pollution. EN 50119 cross-references other European standards that cover various types of insulators. Most of these are applicable to any HV system, although a railway specific standard EN 62621 *Railway applications. Fixed installations. Electric traction. Specific requirements for composite insulators used for overhead contact line systems* has been written for composite insulators.

6.2.3 Air clearance insulation

Air clearance insulation is easily provided if space is available and, if sized correctly to take account of the environmental and electrical insulation requirements, will need practically no maintenance other than regular visual inspection to check that the air clearance is not compromised by debris, vegetation, water leakage or stalactites.

EN 50124-1 provides air clearance insulation dimensions for various electric traction system nominal voltages that take account of the environmental conditions and overvoltage withstand, and these are the dimensions that provide basic insulation. It should be noted that EN 50119 specifies different air clearances to those stated in EN 50124-1. This is because EN 50119 considers only technical and economic factors based on operational experience that have provided acceptable levels of performance and reliability and do not claim that that they are acceptable for human safety (basic insulation) purposes. Railway infrastructure owners should consider the impact on human safety as well as reliability and performance by providing the clearances stated in EN 50119 in lieu of those stated in EN 50124-1. The air clearance insulation dimensions specified in EN 50119 are listed in Table 2 of EN 50119:2020, together with the dimensions for basic insulation as specified in EN 50124-1.

Air clearance insulation does not provide protection against direct contact and, therefore, additional protective provisions, such as obstacles, barriers and placing out of reach, are needed.

Because the OCLS has to thread its way within the railway structure gauge, there may be locations where the air clearances in Table 2 of EN 50119:2020 cannot be achieved. The preferred solution would be to move or modify the asset to improve the air clearance, but this may be impracticable for economic or technical reasons. Additional secondary insulation could be provided in addition to the air clearance; but while such arrangements may provide adequate insulation for the expected highest permanent voltage, and the highest non-permanent voltages and long-term overvoltage as defined in EN 50163 *Railway applications. Supply voltages of traction systems*, they may not be sufficient to accommodate the impulse voltage withstand specified for the electrification system as defined in EN 50124-1. Non-linear surge arrestors provide an established and reliable means of mitigating the human safety and asset damage consequences of air clearances that are non-complaint to EN 50119. They are typically connected between the live catenary system and earth to protect equipment from the effects of lightning and switching overvoltages.

Surge arrestors typically comprise a core of metal oxide with a non-linear characteristic that is encased in a porcelain or silicone rubber type housing, the choice being dependent on the application.

During normal operation, the surge arrester should have no adverse effect on the electric traction system. When there are overvoltages, excess energy can quickly be removed from the electric traction system by a high discharge current through the surge arrestor. The protection relays will detect this as fault and operate circuit-breakers to isolate the section in which the surge arrestor has operated.

Section 6 – Electrification system assets excluding the traction return

Section 9 provides further detail about surge arrestors and how they fit within the overall insulation coordination of the electric traction system.

Surge arrestors offer an opportunity to apply reduced air clearances as a basis of design when electric traction systems are being planned. Such an approach requires careful attention to detail in their placement and their overall impact on insulation coordination and electric traction system reliability, particularly in parts of the world where lightning is frequent.

6.2.4 Protection against direct contact

Air clearances alone cannot prevent persons entering the danger zone (see Section 5.10.2), which is the same as direct contact with hazardous-live-parts. Additional protective provisions to prevent direct contact with hazardous-live-parts are required to fulfil the requirements for basic protection. The two methods are:

1. placing conductors out of reach; and
2. the provision of obstacles.

6.2.4.1 Placing out of reach

EN 50122-1 provides the minimum distances required to satisfy the protective provision of 'placing out of reach' of live parts. In this context live parts include the live parts of train pantographs.

This distance is made up of the arm's reach dimension plus a safety margin. The arm's reach dimension is independent of the nominal voltage of the electric traction system and is specified in IEC 60364-4-41 or the CENELEC harmonized document HD 60364-1. To this, an additional safety distance dimension is added, but how this has been derived is not explained. These distances vary depending on the position of the live parts in relation to the standing surface.

The minimum dimensions to live parts is shown in Figure 4 of EN 50122-1:2011+A4:2017. This shows a minimum distance of 3.5 m between a standing surface and live parts of either rail vehicles or OCLS at locations where the live parts are above the plane of the standing surface and where passengers or the public may be present. The build-up of this distance is shown in Table 6.1.

Table 6.1   Derivation of placing out of reach overhead dimensions in public areas

Arm's reach dimension from IEC 60364-4-41 (CENELEC harmonized document HD 60364-1) (m)	2.50
Additional safety distance (m)	1.00
Placing out of reach distance (m)	3.50

EN 50122-1 also specifies placing out of reach for restricted areas. The dimensions are significantly less than those for the public areas but their provenance is not explained within the standard. The overhead direction clearance restricted area clearance is stated as 2.75 m, compared to 3.5 m for public areas. Accordingly, the use of the restricted area dimensions should be limited to controlled areas to which only electrically skilled or instructed persons have access and in which there are strict work practices and procedures. National regulations may vary these minimum distances from those stated in EN 50122-1.

It should be noted that the placing out of reach distances specified in EN50122-1 represent minimum distances, and where the clearance is to the OCLS, the position of the live conductors has to be determined under those environmental and electrical load conditions in which they are closest to the standing surface. As a protective provision, the design and installation of electric traction systems should aim to maximize these distances rather than design to them.

The form of the standing surface assumes that it is not compliant to the obstacle requirements in EN 50122-1 for a standing surface above live parts and that no fluid is being used on the standing surface that could drop onto the live parts. If the standing surface is compliant to the standing surface obstacle requirements, the dimension can be reduced.

EN 50122-1 also describes the clearances for protection by placing out of reach at locations where road vehicles are likely to be present, such as road/rail crossings at-grade or where vehicles are parked for loading and unloading. . EN 50122-1 provides minimum distances, although these may be revised to suit national requirements because the maximum height and type of road vehicles permitted and the highway standards are defined on a national basis.

6.2.4.2   Protection by obstacles

This protective measure is to prevent direct contact with live parts. For strict compliance with IEC 61140 *Protection against electric shock – Common aspects for installation and equipment*, these should be barriers rather than obstacles, unless their use is restricted to controlled areas accessible only to electrically competent persons. There are many locations where protection by clearance is not sufficient to provide basic protection. Examples include bridges over the railway and working platforms for equipment within the current collector zone, such signal posts and gantries. In such circumstances, EN 50122-1 requires obstacles to be provided.

The design and installation of obstacles needs to take account of the particular locations and should consider:

(a) the location of the standing surface relative to the adjacent live parts;
(b) whether parts of the obstacle are inside the overhead contact line zone (OCLZ) or current collector zone (CCZ);
(c) the distance between the obstacle and the live parts;
(d) whether the standing surface is a public area or a restricted area; and
(e) the presence of openings in the obstacle and the characteristics of these openings.

Although obstacles are designed to suit the particular location, they all need to consider the following hazards:

(a) approaching the boundary of the air clearance distance surrounding a bare live part, with a body part or hand-held rigid objects (for example, a fishing rod, flag or selfie stick). The outer limit of the air clearance distance is also defined as the boundary of the danger zone described in Section 11 which, for electrical safety purposes, is considered to be the same as direct contact with a hazardous-live-part (see Section 5.10.2);
(b) touching a live part, with a body part or where applicable with hand-held rigid objects, by reaching along, across or over the obstacle;
(c) touching a live part, with a body part or where applicable with hand-held rigid objects, by penetrating an opening in the obstacle with a body part or with hand-held rigid objects (for example, a scaffolding pole or maintenance object); and
(d) otherwise creating a conductive pathway between live parts and persons (for example, use of liquid jets or discharging rainwater drainage directly onto live parts).

An example of a common type of obstacle for humans is a bridge parapet. EN 50122-1 provides information on how the bridge parapet is to be dimensioned and the material forms that are permissible. It states that a parapet should be 1.8 m high measured from the standing surface on the bridge and the parapet top is designed to prevent persons sitting or standing on it. Where bridge parapets cannot be modified to provide the 1.8 m height, perhaps due to architectural restrictions, EN 50122-1 states that it should be at least 1.0 m high and a horizontal obstruction provided on the bridge face below the parapet such that the straight-line distance between the top of the parapet and the live parts at the edge of the horizontal obstruction is not less than 2.25 m. Alternatively, a horizontal obstruction may be fitted to the top of the parapet such that it reaches 1.5 m beyond the parapet and the position of its outer edge is 1.5 m measured vertically from the standing surface. Horizontal obstacles shall not provide a standing surface that could be used by trespassers or vandals and therefore tend to be inclined upwards or downwards to achieve this. Their design shall not direct liquids, including rain and melting snow, onto the live parts. While EN 50122-1 describes these protective provisions, individual railway administrations may vary these to comply with national safety regulations and address specific safety risks to persons and animals.

6.2.5   Overhead contact line system structures

OCLS structures are required to hold the conductors in their correct three-dimensional position. This is particularly important for the contact and catenary (messenger) wires whose position is essential to achieve the reliable and consistent transfer of electricity to a moving train pantograph. The structures therefore have to be capable of carrying the static (or dead) loads of the OCLS equipment and tensions, and live loads from accidental conditions such as dewirement and environmental loads from wind, snow and ice. The structure also provides a protective conductor to carry fault current back to the traction return circuit in the event of an insulator failure or a flashover following a lightning strike. In general, the size of an OCLS structure, together with its bolted or welded connections, as determined by the conductor and other physical loads, is more than adequate to carry any fault current. However, there are occasions where OCLS structures require a flexible connection, such as a pin or hinged connection, for structural reasons at the base or the connection between mast and boom. At these types of connections, an assessment is needed to determine whether such a connection can carry the fault current over the life of the structure or whether a flexible bond is required across the connection. Such a bond would be sized using the equations described in Section 4.5.5 and takes account of the bond material insulation and initial temperature based on environmental considerations and any traction current that the bond would carry under normal conditions.

Some OCLS structures use a head span arrangement rather than a conventional boom arrangement. This comprises three tensioned wires that support the individual overhead catenary system assemblies for each track. In this arrangement, the span wires need to be rated to carry traction fault current back to the traction return circuit. EN 50119 describes how the wires can be sized. If the span wires include tensioning arrangements such as turnbuckles, then these will have to be bonded across with cable that is rated the same as the span wire.

6.2.6   The OCLS connection to traction return

The connection between the OCLS and the traction return is via the OCLS structure. The OCLS structure can be either directly bonded to the traction return rail or more usually in recent years, to an uninsulated aerial earth conductor (AEC) that is mechanically clamped to the OCLS mast. The AEC is connected with cable to the rails and other conductors that form part of the traction return circuit at traction cross-bonds. The advantage of an AEC over direct structure to rail bonds is the significantly reduced occurrence of bonds being damaged through track maintenance and other lineside works and means the overall traction return circuit is more robust.

The sizing of the uninsulated AEC is undertaken in accordance with EN 50119. In many electric traction systems, the AEC is both a protective conductor and an integral part of the traction return circuit. In this situation, the AEC sizing has to take account of the normal traction current that it is carrying, as well as the ambient temperature, solar gain and wind speed, in order to establish the correct initial conductor temperature when the fault occurs. This is important because unlike insulated cables, the limiting temperature is determined by the need for it to remain mechanically strong when carrying fault current and also by the allowable sag that maintains the conductor height within permissible limits. For example, the maximum temperature for an aluminium cable with PVC insulation is 160 °C, whereas for a bare aluminium alloy aerial and tensioned conductor, it is 130 °C (EN 50119:2020 Table 1).

6.2.7   Equipment mounted on OCLS structures

Various types of equipment are mounted on OCLS structures, the majority of which are for isolating electrical sections within the OCLS or reconfiguring the electrical feeding arrangements. Other equipment includes transformers to provide low voltage (LV) supplies, surge arrestors, booster transformers and earthing devices. Many of these can be effectively bonded to the traction return via their mechanical fixings to the OCLS structures. Others, and particularly transformers and surge arrestors, will require to be specifically bonded to the OCLS structure as their construction is such that their mechanical attachment to the OCLS structure is not sufficient to ensure a robust electrical connection. Traction bonds should be sized using the adiabatic equations and take account of the bond material insulation and initial temperature.

Should equipment require a separate LV supply, the LV supply earth is typically separated from the OCLS structure. This is discussed further in Section 7.

6.3   25 kV cables

6.3.1   Cable construction

Cables consist of five main components:

1. conductor;
2. insulation;
3. screen;
4. armour where installation conditions require; and
5. sheath.

Conductor: This uses either stranded or solid copper or aluminium conductors although the overall makeup of each conductor varies and is dependent on the application.

Insulation: In older installations, oil was used an insulation medium. Modern insulation, however, uses either a thermosetting or thermoplastics solid material that provides both basic insulation and also prevents persons or animals from coming into direct contact with live parts, and solid insulation used in cables provides basic protection:

(a) thermoplastic material such as polyvinyl chloride (PVC) has an operating temperature between 30 °C and 70 °C, and up to 170 °C for short-term operation; and
(b) thermosetting material such as cross-linked polyethylene (XLPE) has an operating temperature between 30 °C and 90 °C, and up to 250 °C for short-term operation.

Screen: Railways are electrically noisy environments. The electrical noise is either radiated, induced or conducted as electromagnetic interference (EMI). Control cables need a screen to protect them from the noisy environment and this screen provides a shield based on the Faraday cage. The screen also minimizes EMI from a conductor and reduces high-frequency induction into other electrical systems.

Armour: Where fitted, this provides mechanical protection and tensile strength. This is in the form of steel or aluminium wires spiralled around the insulation, or a corrugated tape that is wrapped around the insulation. The armour acts as a protective conductor that is bonded to earth. It is designed to carry a short-duration fault current, but it is not typically intended to carry load current during normal operation.

Sheath: This provides additional protection from the surrounding environment. The outer protection for cable of the voltage rating found in railways is typically made from PVC or medium-density polyethylene (MDPE). In older cable types, lead was used as a sheath, sometimes with a hessian-type outer covering.

Fault protection is provided by the cable armour or metallic sheath (where one is provided, for example, in lead-sheathed cables), or sometimes the cable screen which has been increased in cross section to carry fault current when the cable insulation fails. These provide the function of a circuit protective conductor (cpc) and are connected to Earth. The resultant fault current will flow into the cable cpc and activate the protection devices, and then disconnect the supply.

6.3.2   Cable terminations

Cables are terminated using outdoor-type cable terminations (sealing ends) or within enclosed cable termination boxes that form part of a transformer or switchboard assembly. Cable terminations:

(a) must provide a mechanically robust means of holding the various parts of a cable (conductor, insulation, sheath, screen and armour wires);

(b) control the electric stress at the end of the cable such that cable insulation does not break down;

(c) provide an adequate creepage path across its surface appropriate to the impulse voltage, climatic conditions and environmental pollution; and

(d) provide a means of connecting the cable cpc to earth.

Cable terminations are selected according to where they are to be installed (for example, on an OCLS structure to which the bare live OCLS conductors are connected or within a cable termination box mounted on a circuit-breaker or transformer). Cable termination boxes provide basic and fault protection because they are totally enclosed and earthed as part of the circuit-breaker or transformer. Outdoor terminations provide basic insulation but, in order to do this, they need to be placed out of reach to provide protection against direct contact. Fault protection for outdoor terminations is provided by equipotential bonding of the termination to earth.

6.3.3   Cable sheath/armour bonding

The electric traction system is essentially a single-phase system and, therefore, all cables are single-core type with a protective conductor formed of the sheath or armour, and sometimes the screen (see Section 6.3.1). The cable protective conductor is designed to carry short-term fault current but is not expected to carry current under normal conditions. Because the cables are single-core, this has an implication on how the cable protective conductor is bonded to earth.

Section 6 – Electrification system assets excluding the traction return

When single-core cables carry current, a voltage is induced in the cable protective conductor by mutual coupling with the cable conductor. If the cable protective conductor is bonded to earth at both ends of the cable (often described as solid bonding), then a current circulates through the cpc. If the cable is bonded only at one end or at its mid-point (often described as single-point bonding), no current will circulate but a voltage gradient will be created along the cable protective conductor, with the maximum value occurring at the end furthest from where the cable protective conductor is earthed. If the cable length is excessive, the induced voltage may be sufficiently high to break down the cable sheath and pose an electric shock hazard if the cable has exposed-conductive-parts accessible to persons. Typically, under normal conditions the voltage at the unearthed end of the sheath should not exceed 60 V as per the limiting value for long-term conditions defined in EN 50122-1. Under fault conditions the short-term limiting values are normally determined by the insulation capability of the cable oversheath. Care is required at the cable termination at the unearthed screen/armour to ensure that direct contact is not possible as these voltages may be in excess of the safety limits defined in EN 50122-1.

25 kV cables that are used to connect circuit-breakers at traction switching stations to the overhead contact wire tend to run parallel to the live overhead conductors for short distances. Voltages in the cable sheath are dominated by those induced by the current in the cable conductor. For lengths of up to 300 m, the induced sheath voltage at the unearthed end of the sheath is not sufficiently high to cause insulation breakdown of the oversheath or joints, even when there is a cable fault. The screens of these cables are therefore usually single-point bonded at the circuit-breaker termination to avoid traction current flowing through the cable sheath. Appropriate insulation is still required at the cable termination with the unearthed cable sheath to prevent direct contact by persons.

For longer lengths, a careful assessment of the sheath/armour voltage is required to ascertain whether single-point bonding of the cable sheath/armour is practicable or whether special arrangements, such as sheath voltage limiters, are required at the unearthed end. It is worth mentioning that terminology for this is not consistent and often the term 'cable sheath bonding' is used, irrespective of whether the sheath is conductive and is inclusive of the armour and screen as appropriate.

The negative phase autotransformer feeder used in autotransformer systems often uses cable rather than a bare conductor in tunnels or other locations where the surrounding infrastructure provides insufficient space to obtain adequate air clearance insulation around a bare conductor. These cables tend to run parallel to the live overhead conductors and therefore the induced voltage in the cable sheath needs to take account of the mutual induction from the contact wire as well as that from the current within the cable conductor for both normal conditions and where there is a fault within the cable, and also on the parallel live conductors. Multi-conductor modelling (see Section 3), is required to represent the various conductors and their spacing. A simple example is provided in Figure 6.1 for a 12 kA fault on the overhead contact wire near an autotransformer station and employs calculations for mutual impedances between the OCLS conductor and the autotransformer feeder cables, based on the equations described in Section 3. The fault current is distributed 6 kA in the contact wire and 6 kA in the feeder cable, but in the opposite direction.

Section 6 – Electrification system assets excluding the traction return

Figure 6.1   Cable sheath induced voltage calculation

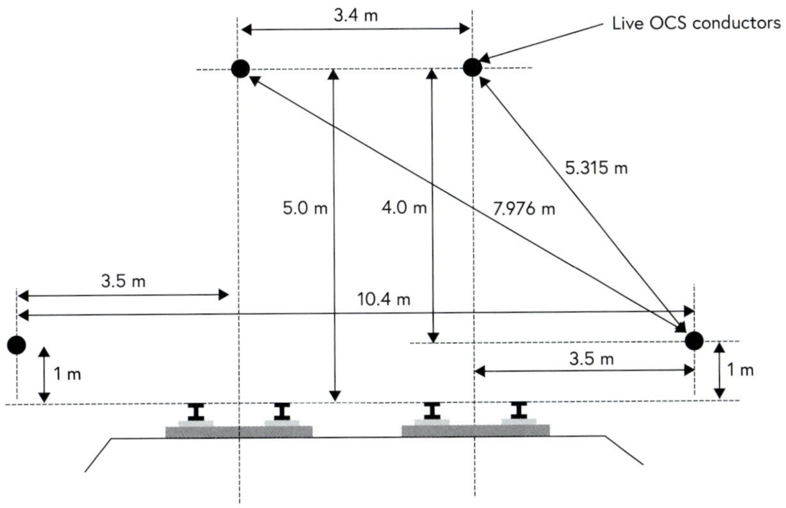

25 kV Autotransformer feeder cable details

Core cross section area	400 mm²
Core diameter	23.7 mm
Core AC resistance hot	0.061 ohm/km
Insulation thickness round core	12 mm XLPE
Area of copper screen	83 mm²
Diameter of screen	50 mm
Screen resistance cold	0.228 ohm/km
Outer insulation	3.1 mm MDPE
Overall diameter of cable	63.5 mm

Calculated values based on 3.5 m horizontal distance from nearest rail and cable. Calculated reactance values include the effect of earth currents. Mutual reactance between cable core and screen $M_{cc-cs} -= 0.639$ ohms/km. Mutual reactance between cable screen and nearest contact wire $M_{cs-cw} = 0.291$ ohms/km. Induced voltage Vs in cable screen per km = (cable core current) M_{cc-cs} − (contact wire current) M_{cs-cw} $V_s = (6,000)(0.639) − (6,000)(0.291) = 2.088$ kV/km. The negative sign between the two terms reflects that the currents in the cable and the contact wires are flowing in opposite direction. If the cable screen oversheath has an voltage withstand limit of 7 kV, then the cable length must not exceed 3.3 km if single-end bonded.

Table 6.2 shows how the induced voltage is affected by moving the cable further away from the track and the contact wire. In this example, the effect of the oppositely flowing current in the contact wire in reducing the induced voltage weakens as the distance increases and it is the voltage induced by the cable core that becomes dominant. The example illustrates how, when assessing induced cable sheath voltages for cables parallel to the overhead contact wire, the interaction of all current-carrying conductors require consideration.

Where single-point bonding is not possible, there are two options available:

1. Solid bond the sheath at bond ends and rate the cable sheath to carry current under normal conditions as well as in fault conditions. Because the sheath is carrying current as well as the cable conductor, the cable design has to take account of the additional heat source and will result in larger conductor sizes for the same cable rating, so as not to exceed the cable insulation limiting temperature (70 °C for PVC insulation and 90 °C for XLPE insulation). Concentric cables fulfil this by matching the sheath current rating to that of the cable conductor.

Table 6.2   Variation in cable screen voltage by horizontal separation

Horizontal separation distance to cable (m)	Distance contact wire to ATF cable (m)	Mutual impedance M_{cs-cw} (ohms/km)	Screen voltage induced from cable conductor 6,000 (M_{cc-cs}) V/km	Screen voltage induced from contact wire 6,000 (M_{cs-cw}) V/km	Total induced cable screen voltage V/km	Maximum cable length with single-point bonding km
3.5	5.315	0.291	3,834	1,746	2,088	3.35
10	10.77	0.247	3,834	1,482	2,352	2.98
15	15.524	0.225	3,834	1,350	2,484	2.82
20	20.396	0.208	3,834	1,248	2,586	2.71
50	50.16	0.154	3,834	924	2,910	2.41

2. Gap the cable sheath at intervals to divide them into discrete sections. One end of each section is solidly bonded to earth via the traction return circuit and the other end is bonded to earth via sheath voltage limiters and the traction return circuit. In this arrangement, the sheath no longer forms a continuous path for fault current, so the connection to earth has to be made to the traction return circuit, which will provide the continuous path back to the source for fault current.

Sheath voltage limiters are non-linear resistors whose function is to present a high resistance to the normal voltage induced on the cable sheath and enable the cable sheath to be single-point bonded. When the induced voltage rises due to fault or other transient conditions, the sheath voltage limiter has a low resistance, allowing the associated transient current to flow through the cable sheath to earth and enable the cable oversheath or cable joints to be protected from transient overvoltage. The form and selection of sheath voltage limiters is similar to surge arrestors and their specification is covered by the same standard IEC 60099. Sheath voltage limiters are selected using the following parameters:

(a) Minimum continuous operating voltage: this should be selected so as to exceed the maximum sheath voltage that will occur under normal conditions and under through fault conditions. A margin of 5 % or greater is typically added to take account of variations in sheath voltages that occur in practice.

(b) Rated voltage: this is determined by the sheath voltage limiter but usually falls between 1.15 and 1.25 times the maximum continuous operating voltage.

(c) Nominal discharge current: when a sheath voltage limiter operates it discharges a lot of energy very quickly. The nominal discharge current represents a peak value of lightning current impulse as defined in IEC 60099-4. Standard nominal discharge current ratings are 2.5 kA, 5 kA, 10 kA and 20 kA with values less than 10 kA usually appropriate for 25 kV cables.

(d) Peak residual voltage: this is peak value of voltage that appears between the terminals of an arrester during the passage of discharge current. This should be selected to be less than the peak impulse withstand voltage to earth of the cable oversheath and joints.

6.4 Switching stations

6.4.1 Basic protection

Basic protection is provided by basic insulation and equipotential bonding.

Most electrical equipment is metal-clad with air or solid or gas insulation that provides basic insulation and prevents direct contact with hazardous-live-parts. However, many conductors are often bare, with the insulation provided by air. For these conductors, protection against direct contact is provided by placing out of reach and the provision of obstacles.

The minimum dimensions for protection by placing out of reach and providing obstacles tend to follow utility practice for installations that are fenced from the public and where access is limited to trained and authorized staff. The clearance dimensions and requirements for obstacles described in IEC 61936 would typically apply rather than the EN 50122-1 clearances.

Because of the interconnection to the traction return conductors along the railway, equipotential bonding mitigates the touch and step voltage hazards exported from the rails into the switching station.

Sometimes gas-insulated switchgear is used in locations with limited space because it is physically smaller than other types of switchgear. However, the close coupling with phase conductors can result in steelwork and adjoining equipment forming closed loops through which induced currents flow under normal conditions. Gas insulated switchgear may also create surge voltages on the switchgear enclosure during normal switching or other transient condition that present a touch voltage hazard. Gas insulated switchgear requires particular earthing arrangements to be incorporated into the switching station earthing system design and these tend to be developed in close collaboration with the equipment manufacturer.

6.4.2   Fault protection

Fault protection is provided by equipotential bonding and earthing, and automatic disconnection of the supply.

Equipment normally equipotentially bonded and connected to earth include:

(a) OCLS structures;
(b) 25 kV cable sheaths and armours;
(c) transformer tanks;
(d) autotransformer centre point connection;
(e) return current busbar (or similar) where the connections from the return current circuit enter the traction switching station;
(f) metal-clad switchgear assemblies and cases, isolators and earth switch bases;
(g) metallic building structures including steel frames (bonded at each corner), rebar and piles;
(h) miscellaneous metalwork associated with oil and air tanks, screens, steel structures of all kinds;
(i) panels, cubicles, kiosks, LV AC equipment, lighting, and security masts;
(j) telecommunication cable screens unless advised otherwise by the telecommunications operator; and
(k) switching station fencing.

6.4.3   Earthing and equipotential bonding

6.4.3.1   Objectives

Earthing and equipotential bonding need to satisfy the following:

(a) It must carry current into the earth under normal and fault conditions without exceeding any operating and equipment limits or adversely affecting continuity of service. Typically, the earthing system is thermally rated and mechanically robust to accommodate the earth fault current with backup fault clearance times that can vary between 200 ms to as much as 3 seconds (EN 50388 Annex A). The actual values will be defined for a specific electric traction system and also the nature of the back protection provided by the electricity utility at the substation.
(b) It must prevent a person both within and in the near vicinity of the switching station from being exposed to touch and step voltages in excess of permissible safety limits that are discussed in Section 5. This is usually based on normal operating times of protection relays and circuit-breakers which can vary between 100 ms and 300 ms (EN 50388 Annex A). Normal fault clearance times, rather than backup fault clearance times, are justified through experience of the unlikely chance of a person being in direct contact with an exposed-conductive-part when the fault occurs and when the normal protection fails.
(c) The earthing system shall maintain its integrity over the installation lifetime, taking account of corrosion effects and environmental conditions.

It is worthwhile noting that an earthed object is not necessarily an object that is safe to touch. As discussed in Section 5, there is a body current that must not be exceeded to avoid death, and this can be related to a corresponding body voltage and therefore to limiting values of touch and step voltage. The resistance of an earth electrode is determined by the ground resistivity, and the potential rise of an earthed system is the product of current flowing to earth and earth resistance. A low earth resistance but high current may result in touch voltages higher than a similar system with high earth resistance but low earth current. Accordingly, the design of the earthing system requires knowledge of the fault current flowing to earth, the fault duration and the ground resistivity.

Earthing system design requires a methodical approach, and design process diagrams and guidance are provided in standards such as the IEEE Standard 80, AS Standard 2067 and EN 50522. In general they tend to follow a similar process:

Section 6 – Electrification system assets excluding the traction return

1. Obtain the substation or switching station equipment layout and overall site footprint.
2. Determine the ground resistivity.
3. Determine the maximum earth fault current and fault clearance times.
4. Determine safety limits for touch voltage.
5. Lay out the earth electrode system to accommodate the equipment, fencing, etc., and whether structural foundations are to be used as earth electrodes.
6. Calculate the earthing conductor sizes from maximum earth fault current and backup fault clearance time.
7. Calculate the earth electrode resistance.
8. Calculate the earth fault current flowing to earth taking account of reduction factors and normal fault clearance time.
9. Calculate the earth potential rise (EPR).
10. Compare with touch voltage safety limits. If the EPR is less than or equal to the touch voltage safety limit, then the design is considered acceptable.
11. If EPR exceeds the touch voltage safety limits, check that transfer voltages and touch voltage at fence lines and within the substation (mesh voltage) are within touch voltage safety limits. If true, then the design is considered acceptable. Note that European standards such as EN 50522 and EN 50122-1 apply the 'a' factor described in Section 5.11.1 as an acceptable threshold limit for EPR, whereas IEEE Std 80 does not and requires calculations to be undertaken.
12. If touch voltages exceed safety limits and touch voltage limits are exceeded, revise the earth electrode design until acceptable values are achieved. This may involve measures to reduce the earth fault current, such as neutral earthing resistors in the HV system, as well as means to reduce the earth electrode resistance, such as soil treatment and additional earth electrodes. Other recognized measures described in EN 50522, include:
    - the provision of crushed stone or similar within the installation to improve the insulation of the standing surfaces, insulated fencing and gates at the boundary of the substation or switching station;
    - the provision of external (grading) electrode 1 m outside the external boundary fence and connected to the earth electrode system; and
    - extending the buried earth grid and/or provide earthed standing platforms where equipment operating handles are located.

6.4.3.2 Ground resistivity

Ground resistivity is the key parameter in determining the resistance of an earth electrode. Equations suitable for manual calculation have been developed for various earth electrode arrangements, involving combinations of rods, mesh, and horizontal electrodes. These assume a homogeneous soil structure and therefore uniform resistivity and can be used to undertake an initial design. In practice, the ground is often made up of layers with different resistivities. For two layers, the standard equations may be enhanced using the approach described in IEEE Standard 80 but, beyond that, more sophisticated methods of calculation involving computer modelling are required. Although most countries have geological maps that indicate the typical geology and therefore the typical ground resistivity, these tend not to reflect local variations and different ground layers at the traction switching station site. For these reasons, it is important that the ground resistivity is measured using an established method such as the Wenner array technique, and which are described in Section 8.

6.4.3.3 Earth potential rise

When fault current flows into the ground, the voltage of the earth electrode rises in relation to the ground and is called the earth potential rise (EPR).

Because the earth electrode is bonded directly to the rails, the EPR at the electrode raises the rail potential in the vicinity of the switching station. In many non-traction environments, the EPR is not necessarily a safety issue per se, providing that touch and step voltages within and at the perimeter of the installation do not exceed permissible safety limits. For traction switching stations it is desirable that the EPR is limited by the maximum values of rail potential stated in EN 50122-1.

In order to establish the EPR, the fault current to earth within the switching station needs to be determined (see Section 2). Normally electrical traction systems are defined in terms of their nominal peak fault level (for example, 6 kA, 12 kA or 15 kA for 25 kV systems). This value usually represents the steady state short-circuit current and, assuming that the supply from a traction substation is far from a generator, can be taken as the initial short-circuit current used for earthing system design. Where generation is near, the initial fault current will be higher than the steady state fault current and this higher value should be used. Not all the fault current will flow to earth via the switching station earth electrode. The mutual induction from the catenary system and, for autotransformer switching stations, the direct connection of the autotransformer mid-point connection to the earth electrode encourages the current to flow into the rails from where it dissipates to earth, and the negative feeder in the autotransformer systems. These can significantly reduce the proportion of fault current that flows to earth by as much as 90 %. This amount by which the earth fault current is reduced is often described as a reduction factor.

6.4.3.4 Earthing conductors

Exposed-conductive-parts of equipment must be connected to the earth electrode using earthing conductors, which are typically made of bare copper, steel or aluminium and are sized using the adiabatic equation 6.1 for the maximum earth fault current and such that a limiting maximum temperature is not exceeded. This equation is from IEC 60949 but can be found in slightly different forms in various national standards, such as IEEE Standard 80 and BS 7454.

$$A = \frac{I_{sc}}{K} \sqrt[2]{\frac{t_f}{\ln\left(\frac{\theta_f + \beta}{\theta_i + \beta}\right)}} \tag{6.1}$$

where:

I_{sc} is the short-circuit current flowing to earth (A)θ_f is the final temperature (°C)
θ_i is the initial temperature (°C)
β is the reciprocal of the temperature coefficient of resistance of the current-carrying component (earthing conductor) at 0 °C (example values are shown in Table 6.3)
A is the geometrical cross-sectional area of the current-carrying component (earthing conductor) (mm^2)
t_f is the duration of the short-circuit current (seconds)
K is the constant depending on the material of the current-carrying component (earthing conductor) (A/mm^2) and is calculated using the adiabatic equation (example values are shown in Table 6.3).

Table 6.3 Material constants for use in the adiabatic equation

Material	β (°C)	K (A/mm^2)
Copper	234.5	226
Aluminium	228	148
Steel	202	78

Section 6 – Electrification system assets excluding the traction return

When specifying earthing conductors, the following shall be considered:

(a) The fault duration time shall be the time for the backup protection to operate.

(b) Earthing conductors tend to be bare and therefore their limiting temperature is not constrained by insulation but by the temperature at which the material melts (or fuses). However, there are other considerations such as mechanical strength, fixing arrangements, means of jointing earthing conductors, and proximity to flammable materials and therefore it is not unusual for a much lower final temperatures to be selected. For example, within the UK, final short-term temperatures for earthing conductors that do not have bolted connections tend to be limited to 400 °C for copper and 320 °C for aluminium.

6.4.3.5 Earth electrodes

Earth electrodes at traction switching stations are usually formed of buried horizontal electrodes that connect all plant items to earth, resulting in a grid configuration. The placement of the horizontal earth electrodes is typically at least 1 m away from the connected plant to control touch voltage hazards at arm's length. Additional buried horizontal connections are added to provide multiple current paths to earth and therefore improve resilience. Although the current in any conductor is discharged fairly uniformly along its length, the larger proportion of current flows to earth at the periphery of the grid rather than at the centre. This leads to higher touch and step voltages at the outer meshes of the grid and especially at the corners. Vertical earth electrodes (ground rods) can make a significant reduction to the earth electrode resistance and also can reduce the touch and step voltages, particularly where they can reach ground layers of lower resistivity located below the grid. Vertical earth electrodes can discharge more current to earth than horizontal earth electrodes and are usually positioned at the edge of the grid where the current is highest. In urban environments, or where space is limited, the structural foundations of the switching station building may be used to provide the earth electrodes. Standards such as IEEE Standard 80, IEEE Standard 142, BS 7430, AS Standard 2067 and EN 50522 provide guidance and techniques that will provide an arrangement appropriate for the installation.

It is desirable that the target earth electrode resistance limits the EPR to within the touch and step voltage safety limits described in Section 5 for the largest earth fault current and normal fault clearance time. An EPR above these limits requires the layout of the earth electrode to be carefully assessed so that step and touch voltages within the switching station, and immediately outside the boundary fencing, remain within safety limits. A grid is not a uniform conductive surface because it is formed as a mesh arrangement, with typically varying mesh sizes to reflect the equipment layout and space availability. The ground potential at the horizontal electrodes that form the edge of a mesh is greater than the ground potential at the centre of the mesh bounded by these electrodes. It can be visualized in terms of peaks and troughs and is described as the mesh voltage within IEEE Standard 80. It may result in a touch voltage greater than that experienced under the 1 m distance scenario. Because the EPR is higher at the periphery of a grid electrode than at the centre of the grid, the earth electrode seeks to limit the potential difference between the electrodes and the centre for a mesh at the periphery of the grid to within touch voltage safety limits.

The earth electrode resistance is determined by all the horizontal and vertical earth electrodes that are installed in the ground for a given ground resistivity. Earth electrode resistance equations have been developed for typical combinations of vertical and horizontal electrodes and can be found in various international standards, such as EN 50522 and particularly IEEE Standard 80, as well as national standards such as BS7430 in the UK. Where the arrangement is more complex and there are multilayer ground resistivities, computer modelling is employed on account of the increased complexity.

Section 6 – Electrification system assets excluding the traction return

Most switching stations and feeder switching stations tend to have a broadly rectangular or square footprint. While computer-based tools exist to support earth electrode design, equations exist to undertake manual calculations, which can be useful for initial design and for checking. The following example based on the Schwarz equations described in IEEE Standard 80, shows the approach for a broadly rectangular or square shaped earth electrode formed of a buried grid with vertical electrodes.

The earth resistance R_g of the combined grid and rod electrode is calculated from equations 6.2 and 6.3:

$$R_g = \frac{R_1 R_2 - R_m^2}{R_1 + R_2 - 2R_m} \, (\Omega) \tag{6.2}$$

where:

R_1 is the earth resistance of the buried grid (Ω)
R_2 is the earth resistance of all the vertical rods (Ω)
R_m is the mutual earth resistance between the rods and grid and reflects the change in overall earth
    electrode resistance due to the proximity of the rods and the grid (Ω).

Taking each term separately, the earth resistance R_1 of the buried grid can be found from:

$$R_1 = \frac{\rho}{\pi L_c} \left[\ln\left(\frac{2L_c}{a^2}\right) + \frac{k_1 L_c}{\sqrt{A}} - k_2 \right] (\Omega) \tag{6.3}$$

where:

ρ is the soil resistivity (Ω.m)
L_c is the total length of all connected grid conductors (m)
$a' = \sqrt{(2ha)}$ for grid conductors buried at depth h (m) below the ground surface, or equals a for grid
    conductor just buried under the earth's surface (m)
a is the radius of the grid conductor (m). If there is variation in grid conductor cross section, the most
    representative value of radius should be used.
A is the area covered by grid (m^2)
k_1, k_2 are the Schwarz coefficients provided in the graphs in Figure 24(a) and (b) of
    IEEE Standard 80:2013.

The earth resistance R_2 of the earth electrodes, taking account of their grouping is given by equation 6.4:

$$R_2 = \frac{\rho}{2\pi n_r L_r} \left[\ln\left(\frac{4L_r}{b}\right) - 1 + \frac{2k_1 L_r}{\sqrt{A}} \left(\sqrt{n_r} - 1\right)^2 \right] \tag{6.4}$$

where:

L_r is the length of each rod (m). In general, the rods will be about the same length, but should there
    be significant variation in length, the modal length should be chosen rather than the average to
    avoid the impact of very long or very short rods
b is the radius of rod (m)
n_r is the number of rods placed in area A

Section 6 – Electrification system assets excluding the traction return

It is worth noting that the first part of this expression for R_2 is equation 6.5 for a single rod R_{rod} when n_r is 1:

$$R_{rod} = \frac{\rho}{2\pi L_r}\left[\ln\left(\frac{4L_r}{b}\right) - 1\right] \tag{6.5}$$

If the rod is backfilled in a low resistivity material such as conducting concrete or is formed of reinforcement within concrete piles, then the resistance equation for these rod electrode forms can be used in the equation for R_2. The equation for a concrete backfilled earth rod is given in Section 4. Where reinforcement within concrete piles is used, the arrangement of the reinforcement within the piles influences the earth resistance and therefore the resistance equation takes account of the geometric mean between the reinforcement rods as described in the BS 7430.

The mutual earth resistance between the grid and the rods is R_m and is calculated using equation 6.6:

$$R_m = \frac{\rho}{\pi L_c}\left[\ln\frac{2L_c}{L_r} + \frac{k_1 L_c}{\sqrt{A}} - k_2 + 1\right] \tag{6.6}$$

Typically, a switching station earth electrode resistance of around 0.5 Ω is achieved, as this, together with the rail-to-earth impedance, should ensure the earth potential rise is limited to well below the limits defined within EN 50122-1.

Individual horizontal and vertical (rod) conductors are sized using the same adiabatic equation as for earthing conductors (see Section 6.4.3.4). Where a grid configuration is proposed, the earth fault current will be distributed through the various branches of the grid. A rule of thumb is to size the conductor on 60 % of the fault current where there are two or more robust and permanent earth conductors connected to where the fault current enters the earth electrode.

The earth electrode must be capable of dissipating the electrical energy into the ground surrounding it. Failure of an earth electrode when carrying fault current is due to an excessive temperature rise at the electrode surface that is caused by the current density, duration and the properties of the ground. Typically, ground has a negative temperature coefficient of resistance. This results in an initial fall in resistance, but as the moisture is driven away from the soil/electrode interface by the heating effect of the fault current, the resistance coefficient becomes positive, and the resistance increases. There is limited experimental work on this effect, but for short-term conditions the ground temperature at which a complete failure of the earth electrode occurs is generally accepted to be approximately 100 °C. The limiting value of current density is given in BS 7430 by equation 6.7:

$$J = 10^3\sqrt{\frac{57.7}{\rho t}} \quad (A/m^2) \tag{6.7}$$

where:

ρ is resistivity of the ground (Ω.m)
t is earth fault duration (seconds)

The surface area of all the buried horizontal and vertical electrodes are added together to obtain the total surface area of the earth electrode. The earth fault current is divided by the total surface area to obtain the current density which should be less than the limiting value. This assumes that the fault current is equally distributed throughout the earth electrode. This is not strictly correct but is enough for a first assessment.

IEEE Standard 80 does not provide a formula but recommends that the current density at the earth electrode should not exceed 200 A/m^2, irrespective of the ground resistance. EN 50522 explicitly does not provide guidance because for the type and configuration of earthing systems used in HV systems and the typical earth fault currents in HV utility substations, it notes that experience has shown that this is not a significant issue. For traction switching stations where space for an earth electrode is limited or ground resistivity is high, this could be an issue because the long-term integrity of the earth electrode is essential to fulfil its protective function in holding rail potential to within safe limits under normal and fault conditions.

6.4.4   Switching station fences

The fence provides a boundary between the switching station in which protective measures against electric shock can be provided and maintained in good condition. Granting access only to authorized staff strengthens the measures against electric shock through the application of strict procedures. Outside the fence, the railway infrastructure manager responsible for the switching station has no control of people or animals that can approach and touch the fence. The design and installation of the earthing system must therefore consider touch (humans) and step voltage (particularly for four-legged animals) hazards outside the fence and install an earthing system that achieves safe touch and step voltages.

Two arrangements are typically considered. The first is the provision of an independently earthed fence that is segregated from the switching station earthing system, and the second is to bond the fence to the earthing system.

6.4.4.1   Independently earthed fence

There are key requirements for an independently earthed fence. A 2 m separation is introduced such that the EPR at the fence is lower than at the periphery of the earth electrode. The EPR gradient is less steep and therefore the touch voltage at the fence is below the safety limit. It must be stressed that the 2 m separation is a distance that works for most ground resistivities. Should there be high ground resistivity, this distance may need to be increased and therefore an assessment of the touch voltage at the fence must be undertaken as part of the earthing design.

This arrangement depends on the 2 m separation from the periphery of the switching station earth electrode and a prohibition of the installation of any item within this zone that could compromise this distance throughout the life of the installation.

6.4.4.2   Fence bonded to the switching station earthing system

The independently bonded fence arrangement has the advantage that the measures to prevent electric shock to persons and animals do not extend beyond the fence into areas that could be in different ownership. However, it does require a larger area for the switching station, which may not be possible. In those situations, the fence may have to be bonded to the switching station earth electrode.

In this arrangement, the key safety provision is an outer electrode (or grading electrode) installed 1 m on the outside of the fence. Bonding the fence to the switching station earth electrode means that the maximum EPR gradient is at the fence. The outer earth electrode reduces that significantly so that the touch voltage at the fence remains within safe limits.

6.4.4.3 Fences that are not part of the switching station fences

Metallic fences that do not form part of the switching station fence should not be joined to the switching station as that would create a transfer potential risk along the fence from the switching station. Similarly, where such fences do not connect to the traction switching station fence but are within 2 m, there is a touch voltage hazard through simultaneous touching of the two fences.

If such situations exist, there are various measures that could be adopted to mitigate the electric shock hazards. These include replacing the fence with one made from a non-conductive material, the provision of insulated fence sections, or relocating adjacent fences such that the separation is greater than 2 m.

6.4.5 Externally provided utility and electrical supplies

Externally provided conductive utilities such as gas, water or drainage should have insulated sections provided to prevent transfer potential from within the switching station and prevent such utilities introducing an extraneous potential from outside the switching station.

The EPR will export a voltage to remote locations along incoming telecommunication cables with conductive screens or armour that are terminated within the switching station. This may create a touch voltage hazard at the remote location if its magnitude exceeds the safety limits for the telecommunications operator. Typically, telecommunication safety limits defined on a national basis but usually are based on the International Telecommunication Union (ITU-T) directives. Wherever practicable, the earth potential should be less than the safety limits; but, if that is not practicable, the acceptable mitigation measures have to be agreed with the telecommunications provider.

A switching station requires a LV supply for the various ancillary functions. If that LV supply is brought into the switching station from a source external to the switching station, its earth shall be separated from the earthing system of the switching station to avoid traction current flowing to earth via the LV protective conductor. This can be achieved by use of either a two-winding isolation transformer at the switching station or a TT type LV supply.

6.5 Feeder switching stations and substations

6.5.1 Configuration

Descriptions of the feeder switching station and a substation for single-phase and two-phase autotransformer electric traction systems are provided in Section 2. The feeder switching station is typically owned, operated and maintained by the railway infrastructure owner, and the substation is owned, operated and maintained by the electricity utility from which electric power for the electric traction system is obtained. The operational interface between the railway and the electricity utility is typically provided at a utility-owned disconnector compound located adjacent to the feeder switching station. The compound contains switching used to disconnect and isolate the incoming supplies from the substation and the return circuit to the substation by either the utility owner or the railway infrastructure manager.

6.5.1.1 Earthing arrangements at the feeder switching station

The earthing and bonding arrangements at the feeder switching station are similar to those for a traction switching station described in Section 6.4. The significant difference is the connection to the substation from where traction power is obtained.

Section 6 – Electrification system assets excluding the traction return

The electricity utility disconnector compound and the feeder switching station have their own earth electrode, but because of their proximity they are normally interconnected for equipotential bonding to avoid hazardous touch and step voltages. The two earth electrodes are normally designed such that each is not dependent on the other to meet acceptable touch and step voltages.

Within the feeder switching station, the conductors of the traction return circuit are connected to a return current busbar or similar. The return current circuit to the substation is taken from the return current busbar via the disconnector compound. The return current busbar is connected to the feeder switching station earth electrode in order to control the rail potential within the feeder switching station to within safe limits under normal and fault conditions and provide a current path back to the substation for earth faults within the feeder switching station.

6.5.1.2   Earthing arrangements at the substation

The design of the earthing systems within the substation must be such as to prevent:

(a) danger to personnel and plant due to the possibility of transferred potentials between the feeder switching station and the substation during faults on the electric traction system voltage or higher voltage systems of the electricity utility; and
(b) damage to equipment or devices due to stress voltages between earthed parts during earth faults on the electric traction system voltage or higher voltage systems of the electricity utility.

Where the substation electric traction system voltage and higher voltage earthing systems exist in proximity to each other within the substation, part of the EPR from the higher voltage system can be impressed on the electric traction system (and vice versa).

Two arrangements are presently used:

1. Interconnection of the electric traction system voltage and higher voltage earthing systems within the substation. Under this arrangement a neutral earth resistor is sometimes provided between the transformer electric traction system voltage neutral terminal and the substation earth.
2. Segregation of the electric traction system voltage and higher voltage earthing systems within the substation.

Figure 6.2 illustrates the two arrangements for a 25 kV electric traction system. For an interconnected system the earth connection of the 25 kV railway supply is made within and is marked as A. For a segregated arrangement, the earth connection is made at the 25 kV disconnector compound adjacent to the feeder switching station and is marked as B.

In order to determine which arrangement is most appropriate, the following points should be considered:

(a) The earthing of the substation shall satisfy its own requirements for maximum permissible touch and step voltages.
(b) The earthing of the feeder switching station shall satisfy its own requirements for maximum permissible touch and step voltages.
(c) Typically, the electricity utility and railway infrastructure owners have standards that define safety limits, but each make different assumptions about body impedance and footwear resistance, which results in dissimilar permissible touch voltages (see Section 5).
(d) Where the earthing of the substation is such that the permissible touch voltages at the feeder switching station satisfies the touch and step voltage limits of the railway infrastructure owner, then the earth electrodes of the traction earthing system and higher voltage earthing systems within the electricity utility substation can be interconnected. Where that is not possible, then the

Section 6 – Electrification system assets excluding the traction return

Figure 6.2   The two arrangements for a 25 kV electric traction system

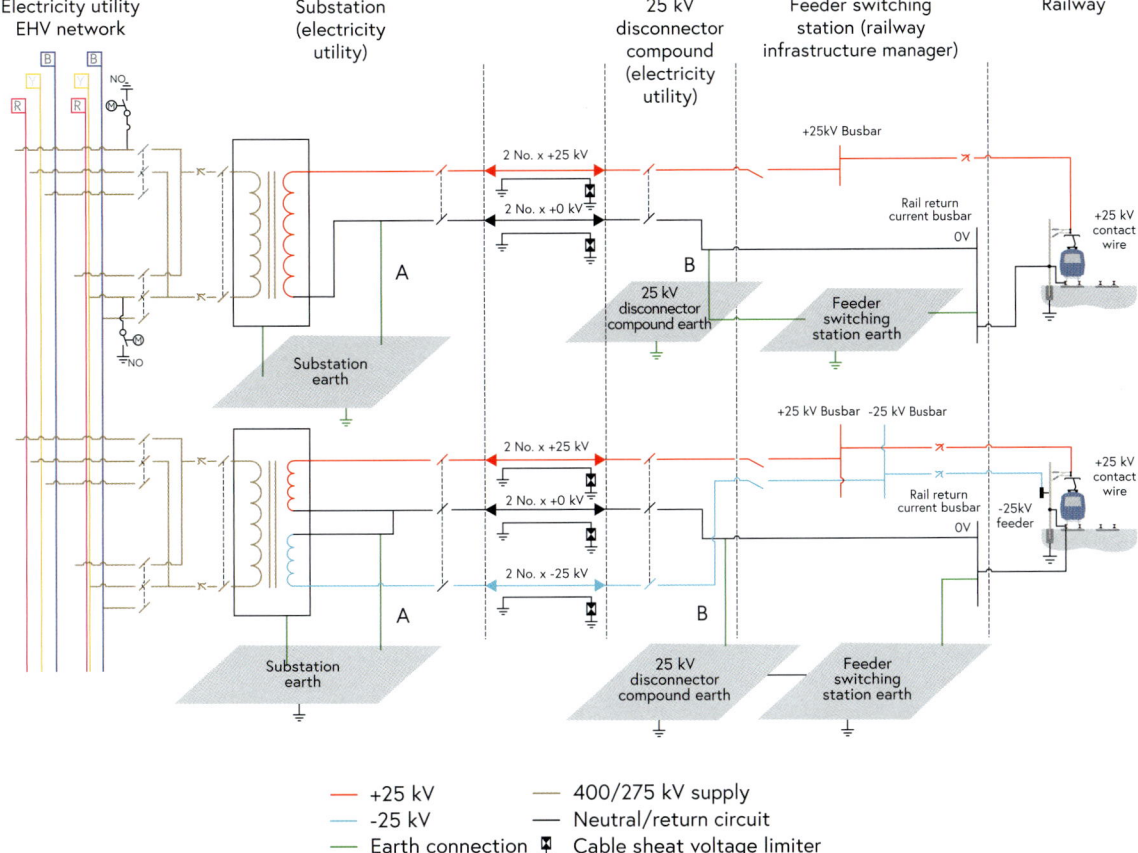

—— +25 kV	—— 400/275 kV supply	
—— -25 kV	—— Neutral/return circuit	
—— Earth connection	⊟ Cable sheat voltage limiter	

earth electrodes of the traction earthing system and higher voltage earthing systems within the electricity utility substation should be segregated in order not to export high 'transfer potential' to the railway infrastructure owners feeder switching station. In the latter arrangement, the traction earthing system is earthing at the electricity utility's disconnector compound at the feeder switching station.

In the interconnected arrangement, it is possible for traction return current to return to the substation via the electricity utility earth. This is undesirable as the continuous flow of traction current may damage the earth electrode. To improve this problem, the electric traction feeder cable and the neutral return current cable, and the autotransformer feeder cable in an autotransformer system, are laid together as closely as possible. This close coupling means that the positive traction current induces a voltage within the neutral return current cable which encourages the traction to flow through the neutral return current cable rather than through earth. In an autotransformer electric traction system, it is the difference in current between the feeder cable and the autotransformer feeder cable that induces the voltage in the neutral return cable.

Under both arrangements, cable sheaths are single-point bonded to prevent traction return current flowing through these. However, in order to prevent damage to the sheaths under fault conditions, sheath voltage limiters are often provided at the unearthed end (see Section 6.3).

 **Section 7**

Assets not part of the electric traction system

There are assets along the railway that can be endangered by an electric traction system. These are not part of the electric traction system and include civil and structural assets (such as stations, bridges and tunnels) and electrical assets (such as low and high voltage distribution systems). The threats from the electric traction arise under both normal and fault conditions. Traction fault currents can flow in assets that have not been designed to carry such large currents following insulation breakdown or breakage of live overhead conductors. Under normal conditions, inductive coupling from the overhead contact line system (OCLS) and galvanic coupling means that traction current can flow through structural reinforcements and protective conductors. This section describes the protective provisions that are typically applied to protect these assets and systems.

7.1 Civil and structural assets

7.1.1 Nature of the threat

Civil and structural assets are exposed to the threats from the electric traction system shown in Table 7.1.

Table 7.1 Electric traction system threats to civil and structural assets

	Threat description	Source mechanism
1	Traction current flowing through structure reinforcement.	Galvanic connection to the traction return circuit.
2	Hazardous transfer potentials from the traction return circuit to third-party installation through structure reinforcement.	Galvanic connection to the traction return circuit.
3	Hazardous transfer potentials from the traction return circuit to third-party installation via utility services or electrical installations located on or within the structure.	Galvanic connection to the traction return circuit.
4	Induced current and voltage onto long metallic assets located parallel to the railway.	Mutual coupling from the overhead catenary system.
5	Hazardous voltages and traction fault current transferred on structures located within the overhead contact line zone (OCLZ) or current collector zone (CCZ). (Section 5 describes the OCLZ and CCZ.)	Broken overhead contact line or pantograph.

7.1.2 Tunnels

Tunnels have structural forms that are determined by many factors, including their dimensional arrangement, ground conditions and method of construction. For example, tunnels formed using tunnel-boring machines typically have concrete segments that are interlocked with keys, and the space between the segments and the ground filled with concrete grout or, in the case where there is significant ground water, a waterproof membrane. Where excavation is undertaken without boring machines, a form of sprayed concrete lining or, in older tunnels, brick lining may be used. Where tunnels are constructed through rock, the bare rock may be left exposed. Tunnels can also be constructed using a form of cut and cover in which the ground is excavated from above, a concrete or similar tunnel lining is constructed and the ground is backfilled around the lining. Tunnel linings may also be steel or a similar material as an alternative to concrete. Tunnels may be provided with cross-passages when separate bores are provided for different tracks. For longer tunnels, shafts will be provided at intervals to enable evacuation and intervention by emergency services in accordance with regulations such as the EU Technical Standard on Interoperability for tunnels.

Tunnels are often provided with tunnel systems that do not form part of the electric traction system but are required for the operation of the railway. Tunnel systems include firefighting water pipes, ventilation and smoke extract plant, lighting, communications equipment and data network cabling, and train command and control systems. Tunnel systems are installed in parallel along the full length of the tunnel and can either be fixed directly onto the tunnel lining or installed on cable management systems that are attached to the tunnel lining.

Faults can occur between the live OCLS conductors and the tunnel lining and by failure of the insulator at OCLS assemblies where the OCLS is attached to the tunnel lining. The former depends on the electrical clearance using air as per EN 50119 to achieve basic insulation. For economic reasons, the tunnel diameter is often minimized as far as is practicable and therefore these OCLS electrical clearances need to be considered when dimensioning the tunnel.

Although the air clearance dimension defined in EN 50119 does not provide a fault protection provision, the likelihood of the air clearance between live conductors and the tunnel lining being compromised within the tunnel is considered low at locations other than at the OCLS tunnel support assemblies, unless there is water ingress, or polluting rail traffic such as steam locomotives, or from carrying goods such as coal. Between the OCLS support assemblies, the natural fall of the messenger or catenary conductor typically provides electrical clearances to the tunnel more than the minimum defined in EN 50119 and greater than at the OCLS support assemblies. Water ingress would be mitigated through design for new tunnels; the problem would be known for existing tunnels provided with an overhead contact electric traction system and appropriate measures implemented to control the water ingress. Polluting rail traffic would have to be resolved by increasing the electrical clearance and installing OCLS insulators with enhanced surface creepage distances appropriate for high-polluting environments; although the better solution would be to prohibit such traffic wherever practicable. For auxiliary conductors, such as the autotransformer feeder conductor used in an autotransformer electric traction system, an insulated cable that would provide both basic and fault protective provision can be installed if the electrical clearances using air could not be provided.

The fault current from a fault to the tunnel lining at a location other than at an OCLS support assembly would find its way back to the traction return circuit through the tunnel structure to the nearest OCLS supports where it would flow into the aerial earth conductor (AEC) via the OCLS support assembly tunnel fixings. It is likely that damage to the tunnel lining will occur at the fault location and possibly at other places depending on the route taken by the fault current and the current density. Options to protect concrete or brick or exposed rock linings with no steel reinforcement are very limited and usually difficult to justify against the low probability of this type of fault occurring.

A fault at an OCLS support assembly is straightforward to mitigate because the OCLS support assembly is connected by an AEC to the traction return circuit at frequent intervals. Fault current from an insulator failure would flow directly into the AEC with very little flowing into the tunnel lining. For faults occurring close to an OCLS support assembly, most of the fault current in the lining would leave through the assembly, which significantly reduces the extent of tunnel lining exposed to the damaging fault current. If the concrete tunnel lining contains reinforcement, then this typically would be bonded to the OCLS support assemblies and at traction current return circuit conductor cross-bond locations. A fault to the tunnel lining would damage the concrete at the fault locations but the bonded reinforcement would effectively carry the fault current back to the OCLS AEC and the other return circuit conductors with minimal damage to the rest of the tunnel lining. The reinforcement should be rated to carry traction fault current without damage to the concrete and standards such as IEEE Standard 80 and 142, EN 50522

Section 7 – Assets not part of the electric traction system

and BS 7430 provide information and calculation methodology. If the reinforcement is continuous, it has to be rated to carry traction current under normal conditions and will require the multi-conductor analysis described in Section 4 to determine this current. If steel or other metallic tunnel lining is used, then the tunnel lining is directly bonded to the traction return at cross-bond locations as well as via the OCLS support assembly tunnel conductive fixings. The metallic tunnel lining will ensure that traction fault current flows back to the return circuit via the shortest routes without damage to the lining.

Where the OCLS uses tensioned conductors, some railway administrations fit external, longitudinal copper or steel strips on tunnel walls and concrete walkways, and which are connected to the traction return circuit as a mitigation against structure damage from a falling live overhead conductor should there be a breakage in the OCLS. A broken conductor will strike the tunnel at multiple points, including the rails, and therefore judicious positioning of these strips increases the likelihood of the live conductor striking a bonded part and reduces the opportunity for significant fault current to flow through the structure. Where extensive longitudinal tunnel services are installed on cable trays that are bonded to the traction return circuit, these can fulfil the function of the longitudinal strips, but they need to be designed and installed such that they can carry the traction fault current. Where a rigid overhead contact line system is used, it is unlikely that a breakage will result in it striking multiple points within the tunnel, and therefore this tends to remove the need for longitudinal strips.

For an electric traction system, tunnels provide resistances between the traction return circuit to earth significantly higher than for the open route. OCLS structures fixed to a concrete lining segment can have an earth resistance of approximately 24 Ω compared to a single OCLS structure earth resistance of 12 Ω: a factor of two higher. This means that the rail potential under normal and fault conditions can exceed the limiting values described in Section 5, which in turn present a touch voltage hazard within the tunnel and a transfer voltage hazard beyond the confines of the tunnel. As modern electric trains require greater power and electric traction fault levels are increased to accommodate this, the management of rail potential, touch voltage and transfer voltage becomes more critical. Depending on the tunnel's construction, there are various approaches that can be used, often in combination, but they all require close design and construction coordination with the tunnel designer and contractor. These may include:

(a) Provide cross-bonding between all traction return conductors at regular intervals. This is straightforward for single-bore tunnels but where separate bores are provided, such track-to-track cross-bonding is typically determined by the location of cross-passages and intervention shafts. Where cross-passage intervals are large, bonding the traction conductors together within each separate bore, in addition to the track-to-track bonding at cross-passages, will distribute traction current between the various traction return conductors at more frequent intervals and help reduce the rail potential.

(b) Utilize the reinforcement within the tunnel lining. The OCLS support fixing arrangements can be electrically connected to the reinforcement to provide a reduced earth resistance. The reinforcement is effectively a horizontal earth electrode. In general, this, of all the approaches described, has the most significant effect on reducing rail potential. If the tunnel reinforcement is electrically continuous, it will, by mutual coupling with the live overhead contact line, carry some current under normal conditions. This needs to be determined using the methods mentioned in Section 4, and the reinforcement electrical rated accordingly to avoid damage to the tunnel lining. If an AEC is used, it is not necessary to ensure that the reinforcement is electrically continuous as the longitudinal continuity is achieved by the AEC. The reinforcement also has to be rated to carry traction fault current such that the thermal forces on the concrete from the reinforcement would avoid damage to the concrete. The tunnel reinforcement is connected to the rails and other traction return circuit at cross-bond locations. The use of tunnel reinforcement may not be practicable if the tunnel requires an outer membrane for waterproofing. Electrically, earth electrodes should be installed on the outside of the membrane and a connection taken through it. However, maintaining the integrity of the waterproofing would be a major concern for the tunnel contractor and the infrastructure owner and it may not be possible to achieve this in practice.

(c) Provide earth electrodes at track-to-track cross-bonding. This arrangement is practicable for a tunnel with cross-passages where space is available to install earth electrodes through the cross-passage walls. It would only be appropriate if there was no main running tunnel reinforcement or the tunnel reinforcement could not be utilized. Cross-passage earth electrodes reduce the maximum rail potential because they lower the rail potential at the cross-bond location, but the ratio of minimum to maximum rail potential between adjacent cross-bonds does not change. Unlike use of the tunnel reinforcement, it does not reduce the OCLS support assembly earth resistance. The practicality of installing an earth electrode through the cross-passage lining is similar to those for the main tunnel lining in respect of maintaining the integrity of any waterproofing.

(d) Provide earth electrodes at shafts and at tunnel portals. Generally, tunnel shafts and portal structures are provided with electrical and mechanical plant required to ensure the tunnel can operate safely and, in the event of accidents, passengers can be evacuated with emergency services able to safely access the tunnel. Shafts and portals are normally provided with earth electrodes for this plant that are connected to the traction return circuit. These earth electrodes tend to be typically less than $1\ \Omega$ and their connection to the traction return make a significant contribution to holding the rail potential to well below the safety limits of EN 50122-1 (see Section 5). Providing an earth electrode also tends to eliminate transfer voltage risks as both the traction return circuit and the shaft and portal electrical systems are bonded together, and bonded to a common earth electrode.

(e) Increase the number of longitudinal traction return conductors. Providing more traction return conductors reduces the voltage drop between track-to-track cross-bond locations and therefore the maximum rail potential. In some tunnels the non-traction longitudinal systems, such as firefighting water pipes and cable management systems, are bonded to the traction return circuit as parallel conductors. It should be noted that these longitudinal systems with their joints and connections must now be rated to carry traction current under normal conditions. Because of their role in controlling rail potential they have an enhanced safety purpose, and their on-going maintenance needs to ensure their integrity with the same importance as all traction return circuit conductors.

A feature of tunnels is that all conductive surfaces of plant and equipment generally sit within the overhead contact line zone (OCLZ) or current collector zone (CCZ) where the overhead contact line system uses tensioned conductors. If a rigid OCLS is used, the OCLZ is much reduced, and tunnel conductive parts may fall outside the OCLZ. However, in practice, given the limited space typically found within tunnels, all accessible conductive parts tend to be equipotentially bonded to the traction return so as to provide a path for fault current and ensure that touch voltages are kept below safety limits. Because of the equipotential bonding, the actual rail potential in itself may not cause a safety hazard provided that the rail potential is not transferred outside the confines of the tunnel. This requires attention to detail and close working with the tunnel contractor and the tunnel plant and equipment installers to make sure that the traction return system is effectively earthed where access to the tunnel is permitted, such as shafts and portals, and that no extraneous-conductive-paths (for example, metallic pipes, telecommunication cable sheaths and fences) are installed that may result in hazardous transfer voltages.

7.1.3 Overline structures

Most of these types of structures are bridges that span over the railway and carry roads, railways, canals and pathways, as well as utilities such as water, gas, and electricity. Low voltage (LV) electrical installations such as street lighting are provided, especially in urban environments. Other overline structures include transverse structural props in retained cuttings and roof slabs that span over subsurface stations that are used to support the station buildings, including ground-level station access and egress; these may also be used for urban realm or oversight property development. Overline structures are often used to support the overhead catenary and typically represent locations where the

Section 7 – Assets not part of the electric traction system

air gap distances described in Section 6 between live parts and the structure, which is nominally at earth potential, are at their smallest. At these locations there is a greater risk of failure of the air insulation and flashover from the live parts to the structure.

The bonding of overline structures requires close cooperation with the structure designer, contractor and the structure owner because the electrical bonding requirements cannot compromise the structural integrity. More often it creates additional complexity that can be difficult to address. Bonding discussions should start at the concept or outline structural design stage and should seek to address the following bonding principles:

(a) exposed-conductive-parts are bonded to the traction return circuit for protection against electric shock;
(b) prevention of damage and elimination of hazards to utilities, services and electrical installations;
(c) prevention of traction return current flowing through the structure under normal electrical conditions, typically by providing a single-point connection to the traction return circuit for the whole structure; and
(d) endeavour to eliminate features that could provide opportunities for birds, bats, etc., to roost or nest on the underside of overline structures at locations close to live, uninsulated OCLS conductors.

For overline bridges the soffit is usually electrically connected to the structure abutments on both sides of the outer tracks and then to the foundations of the structure to provide a connection to earth. In a multi-span structure with piers and abutments, the piers closest to the track, rather than the abutments, may be electrically connected to the soffit and the pier foundations will provide the connection to earth. For simplicity, the connection from the soffit to the traction return circuit is usually made to the AEC associated with the OCLS. If no AEC exists, the connection would need to be made to the rail or any other suitable traction return conductor.

Steel-type structures are normally straightforward to bond. In general, the welded or bolted or riveted connections between the parts of the structure are adequate to ensure that there is electrical continuity through the structure. Where non- or partly conductive parts are provided (for example, for expansion or for movement on bridge bearings), appropriately rated flexible bonding cables should be provided to give electrical continuity across these parts.

Concrete or masonry is not considered to be an insulator at the nominal voltage of an AC electric traction system because of its hygroscopic nature. A flashover to this type of structure will cause current to find its way back to the source via the structure, resulting in damage and creating a touch voltage safety hazard to persons or animals on the structure. EN 50122-1 recognizes this safety hazard and provides protective provisions as mitigation for structures formed of these types of material. Where reinforcement is used, the preferred approach in EN 50122-1 is to bond the reinforcement within the structure to the traction return. No information is provided within EN 50122-1 as to how this is to be achieved, and therefore the requirements need to be drawn from other standards such as IEEE Standard 80 and 142, EN 50522, BS 7430 and EN 62305. Each railway administration will develop its own standards and practices, but Table 7.2 provides an example to illustrate how this could be addressed.

The preferred EN 50122-1 approach of bonding the structural reinforcement is suitable for new structures that contain reinforcement, but where there are structures on existing routes that are being electrified for the first time, this approach is often impracticable. Accordingly, alternative protective provisions are used, all of which involve attaching an electrically conductive obstacle or a bare conductive part to the outside of the structure and connecting these to the traction return circuit. Both provisions are similar, but the dimensional requirements of an obstacle are stricter than those for a bare conductive part. An obstacle has to span the full width of the CCZ and protrude beyond each face of the bridge by 0.5 m. The weight of an obstacle can be significant, and the structure design needs to accommodate this additional load.

Section 7 – Assets not part of the electric traction system

Table 7.2   Example of structural reinforcement bonding requirements

Item	Requirement
1	The reinforcement in the layer of structural reinforcement of the surfaces closest to the OCLS (including the autotransformer feeder conductor where provided) shall be electrically interconnected to form an earthing conductor mesh. This shall be electrically connected to a point at which the connection from the AEC shall be attached. The reinforcement shall be considered as an earthing (or grounding) conductor as defined in earthing standards, such as IEEE Standard 80, EN 50522 and BS 7430, and shall be suitable to carry the traction fault without damage to the structure.
2	The reinforcement bars used as earthing conductors shall be welded or clamped together at intervals to form the earthing conductor mesh. Normal reinforcement wire wrapping is not suitable.
3	The interconnection between reinforcement bars can be made to suit the structural design. For overline structures, interconnection should be provided at the following positions: (a) at the centre line of each track measured vertically from the centre line of the two rails; (b) at the position of the connection point; and (c) where the soffit reinforcement is connected to the reinforcement within the pier or abutment that is used as an earthing conductor to the earth electrode provided by the foundations.
4	Reinforcement within piers, abutments, retaining walls and similar that are used as earthing conductors shall be rated for the traction fault current. Where multiple reinforcement bars are utilized, a minimum of two electrical connections that interconnect all these reinforcements bars shall be provided. One of these should be close to where the connection is made to the earth electrodes in the foundations.
5	Reinforcement in structural foundations that will perform the function of earth electrodes need to be specifically identified.
6	Reinforcement that is not being used as an earthing conductor is bonded together to form an equipotential surface within the structure. The normal structural reinforcement wrapping will be sufficient to provide this function, but checks need to be taken to ensure the reinforcement design provides the electrical continuity required.
7	When used an earthing conductor mesh, or an earthing conductor, or as earth electrodes in foundations, the reinforcement and its associated connections is a hidden component and therefore needs to satisfy the lifetime requirement for critical inaccessible elements specified by the railway infrastructure owner.

A bare conductive part has no dimensional requirements specified in EN 50122-1. They therefore tend to be formed of copper or steel strips attached to the structure face, and on the bridge soffit for all or part of the structure length. Bare conductive parts attached to the structure soffit may also be installed near the OCLS along the length of the structure, but their position needs to be carefully coordinated with the OCLS design so as not to reduce the air gap insulation clearance to live parts. Bare conductive parts are lighter than obstacles and therefore easier to install. However, they do not provide the same effectiveness as an obstacle, so their position and size need to cover those parts of the structure that are most likely to experience flashover from live parts.

Both electrically conductive obstacles and bare conductive parts need to be rated to carry the traction fault current for a time corresponding to the fault clearance time of the electric traction system backup protection.

7.1.4   Underline structures

The structural form of an underline structure tends to be such that it is physically well separated from the live parts of the OCLS and therefore there is no risk of a flashover due to air insulation failure. The protective provision is required to address the hazard arising from a broken OCLS live conductor and to avoid hazardous touch potentials between accessible conductive parts.

Section 7 – Assets not part of the electric traction system

Underline structures that are formed of steel or similar material are bonded directly to the traction return circuit in the same manner as metallic overbridges. Because they are beneath the track, the connection is made directly to the rails or other traction return conductor.

For concrete structures with reinforcement, the preferred approach from EN 50122-1 is to bond the structural reinforcement to the traction return in a similar manner to that described for overline structures of this type. Where existing structures contain no or some structural reinforcement and form part of an existing route that is being electrified, this approach is not practicable. One approach would be to provide bare conductive parts in the form of copper or steel strips on the surface of the structure for their full length and at a position most likely to be struck by a broken live OCLS conductor. The bare conductive parts would be connected to the rails or other traction return conductor, such as a buried earth conductor. Unlike an overline structure, the protective provision is only addressing the hazard arising from a broken OCLS live conductor, which will touch many points including the rails that are part of the traction return circuit. Another approach is to rely solely on the rails, on the assumption that broken live conductors will always fall onto the rails, ensuring that the traction supply is disconnected quickly. Essentially, both approaches involve a balance of risk and consequence against the cost of providing the protective provision. Accordingly, railway infrastructure owners have differing approaches to the bonding of the types of structures that take account of structure length and form, and the position of the rail.

For all types of underline structures, all exposed-conductive-parts, such as handrails, walkways, parapets, etc., are bonded to the traction return circuit. This ensures that fault current from a broken OCLS live conductor gets back to the traction return and ensures disconnection of the supply, and also eliminates hazardous touch voltages that may occur between exposed-conductive-parts.

Viaducts are long underline structures for which the bonding requirements described above equally apply. Viaducts, however, are more susceptible to lightning strikes and, therefore, appropriate lightning protection systems are required. In many situations the structural reinforcement of the viaduct can be used as lightning earth electrodes and down conductors, and these can be used as part of the traction return circuit bonded reinforcement preferred by EN 50122-1. These must be bonded to the traction return circuit to provide an equipotential zone in the event of a lightning strike. Section 9 provides more information on lightning protections systems that are applied to railways.

There are many situations, particularly in urban environments, where third parties attach LV electrical installations, such as street lighting, CCTV or illuminated advertising panels, to these structures. Traction bonding of the underline structure may result in a hazardous transfer potential imposed upon these installations. Protective provisions for these electrical installations are described in Section 7.1.6.

7.1.5   Railways crossing other railways

Particular care should be taken where one electrified railway crosses another electrified railway that is part of different electric traction system. For AC traction systems it is often satisfactory to bond the structure to the return circuit of both electric traction systems. Bonding conductors would need to be rated for the highest fault current and longest fault clearance times of the two systems. If the AC systems are of different frequency (say, 50/60 Hz and $16^2/_3$ Hz), inter-bonding for electrical safety may be appropriate, but the different frequencies may adversely affect train detection systems. Should one railway be a DC electric traction system, then direct bonding between the electric traction systems is not permitted due to the galvanic corrosion effect on the structure from DC traction currents and possibly adverse interference with train detection systems. In these cases, the bonding of the structure is typically undertaken via a device of the type described in Section 10. Such devices provide a non-permanent bond between the AC and DC systems.

7.1.6 Low voltage electrical installations on a structure

Electrical installations without a protective earth do not require bonding, even though they may have metallic covers. For LV equipment this is typically Class 2 insulated equipment in accordance with IEC 61140.

Where LV electrical installations do not comply with Class 2 requirements, and it is not practicable to relocate them from the structure, it is necessary to provide protective provisions that prevent the transfer of traction voltages into the LV installation via either their protective earthed neutral (PEN) conductor or protective earth (PE) conductor, which may present a touch or step voltage hazard at the LV supply source and at other installations connected to the same LV distribution feeder cable. Table 7.3 illustrates some typical protective provisions that can be used individually or in combination, depending on the structure and electrical installation.

Table 7.3 Examples of protective provisions for electrical installations on structures

Item	Protective provision
1a	A minimum straight-line distance shall be provided between exposed-conductive-parts of the LV installations and of the structure that are bonded to the traction return circuit such that it not possible to touch both simultaneously with outstretched arms.
1b	Electrical installations shall be installed in non-conducting containment or insulated pipes.
1c	Electrical installation shall not be mounted on the sides of the structure where it is over or adjacent to the railway. The installation could be within the OCLZ and the CCZ where it is likely to be at risk from flashover due to insulation failure or direct contact from conductor or pantograph mechanical breakage.
Should some of the requirements 1a, 1b and 1c not be practicable, the following provisions should be considered:	
2	Fit non-conductive screening or sheath to accessible conductive parts of the installation.
3	Fit non-conductive material to the exposed-conductive-parts of the structure.
4	Convert the earthing arrangements of the LV installation to a TT supply (as per BS 7671) for the electrical service, and bond the installation to the structure.
5	Modify the LV electrical installation, so as to comply with the requirements for Class 2 insulation stated in IEC 61140.

7.1.7 Utilities installed on or within a structure

Utilities which use metallic pipes or ducts or have electrically exposed-conductive-parts should be installed within the structure in non-conducting containment or electrically insulated pipes, such as those made from high-density polyethylene.

Utilities should not be installed on the outside of a structure because they will usually lie within the OCLZ or CCZ, and therefore are at risk of being made live from broken pantographs or live conductors as well as from flashover due to insulation failure.

On many existing railways that are to be provided with an electric traction system, utilities may well exist on the outside of a structure and within the OCLZ or CCZ.

Section 7 – Assets not part of the electric traction system

If it is not practicable to relocate them, the following protective provisions can be considered:

(a) Provide an insulated section in the utility on both sides of the railway and bond the part of the utility that crosses the railway to the traction return circuit. In this arrangement, the utility is exposed to traction return voltage and current under both normal and fault conditions. The insulated sections prevent this voltage and current from being transferred along the utility to locations remote from the railway.

(b) Bond the utility to the traction return circuit via a voltage limiting device (VLD). In this arrangement, the utility is not normally bonded to the traction return circuit. In the event of a short circuit from the OCLS, the rise in voltage on the utility will cause the VLD to operate and connect the utility to the traction return. Once the fault is cleared by the electric traction system protection devices, the VLD will open and disconnect the utility from the traction return circuit. There should be no galvanic connection between the structure and the utility if the structure is bonded to the traction return circuit.

(c) Install a conductive obstacle between the live parts of the OCLS and the utility. The conductive obstacle should extend the full width of the railway and be bonded to the traction return circuit. There should be no galvanic connection between the obstacle and the utility and between the structure and the utility if the structure is bonded to the traction return circuit.

In all cases, it is important to discuss and agree the protective provisions with the utility owner.

7.1.8 Fencing

Metallic fencing and its associated gates that are situated within the OCLZ or CCZ are bonded to the traction return circuit to ensure that traction fault current is returned to its source effectively and to enable automatic disconnection of the supply in the event of being struck by a broken live overhead line conductor. Therefore, the metallic fence construction has to carry the fault current for a sufficient time, albeit suffering considerable damage as a consequence. Not all metallic fence types are suitable and it may be necessary to replace fencing with more suitable types. Strained wire and chain link fencing are examples of unsuitable fence types.

Outside the OCLZ and CCZ, there is no requirement to bond the metallic fence to the traction return circuit, but measures are required to prevent the fence from acquiring hazardous voltages from capacitive and inductive coupling with the electric traction system. To mitigate these voltages, the following measures have typically been used:

(a) fence panel support posts should be metallic and driven at least 1 m deep into the ground to earth the fence.

(b) where fences are mounted on concrete they shall be solidly bolted onto the concrete without resilient pads or plastic type washers, packers or similar that may introduce electrical insulation. Should these be needed for the fence installation, a separate earth electrode shall be driven into the ground, to which the fence shall be bonded to limit voltages from capacitive or inductive coupling.

An insulated section can be installed within the fence every 1,000 m to limit the voltage induced along the fence by mutual coupling from the OCLS. Typical arrangements that have been used within the rail and electricity supply industry include the following:

(a) installing a 2.5 m (or longer) insulated fence panel made wholly of insulating material.

(b) installing a 2.5 m (or longer) metal fence panel mounted on insulated supports. The insulation needs a voltage withstand capability of 1,000 V AC. The fence panel mounted on insulated supports is a floating section and no conductive asset shall be fixed to it to prevent it from being electrically bypassed.

These protectives measures can readily be assessed and implemented for all fencing within the responsibility of the railway infrastructure owner, but there may be instances where long third-party metallic fences that are close to the railway could acquire hazardous touch voltages from capacitive and inductive coupling. There are no absolute dimensional limits between the fence and the live conductors of the electric traction system, above which no safety issues exist, as the distance is influenced by the length of fence that runs parallel to the live conductors; the traction fault current; the screening effect of the return circuit (for example, rail return, booster transformer return or autotransformer system); the ground resistivity; and how the fence is galvanically connected to earth. A rule of thumb that has been used to trigger an assessment of hazardous touch voltages is where the third-party fence is located 100 m or less from the nearest live conductor and runs roughly parallel to the railway for 1 km or more.

7.1.9 Segregation from third-party electrical installations

Where there is a third-party electrical installation within simultaneous touching distance of a fence that is bonded to the traction return circuit (see Section 5.3), means of mitigating the hazard of transferring traction potentials to the third party need to be considered. The following are possible mitigations:

(a) insert an insulated fence panel that is longer than the simultaneous touching distance;
(b) attach an insulated barrier to the bonded fence;
(c) use a TT type electrical supply (as defined in IEC 60364-1 or the CENELEC harmonized document HD 60364-1) and bond the installation to the fence;
(d) relocate the third-party installation to more than the simultaneous touching distance from the fence; and
(e) provide insulation on the third-party installation.

7.1.10 Stations

Stations can vary from the simple type, where there are no buildings and the facilities have limited platform lighting and passenger information, to large and complex stations typically found in cities and major rail interchange locations. Stations may also be shared with tracks that use DC electric traction systems or AC electric traction systems of different voltage and frequency. Accordingly, there is no single solution but there are a number of key principles that are applicable to any type of station:

Conductive structural and building elements within the OCLZ or CCZ: All conductive assets within these zones should be bonded to the traction return circuit. Care is required in the bonding arrangement to avoid paths to circulate within the station structure. Typically, platforms are bonded individually in a radial configuration and connected together at a common connection point (normally the station main earth terminal (MET)), from which a single-bond connection to the traction return is made. Where there are structural components, bonding cables are not normally required across the structural connections unless reliable electrical continuity cannot be provided. Bonding cables would be required for smaller components or where components are electrically separate, and also for the connection to the common connection point. Bonding cables would be sized using the following approach:

(a) the bonding cable does not normally carry current and therefore the initial conductor temperature can be the ambient temperature;
(b) the temperature rise is calculated for the maximum traction fault current for the normal fault clearance time using the adiabatic equation described in IEC 60949 or in the slightly different forms in various national standards, such as IEEE Standard 80 and BS 7454, and checked against the maximum permissible for the type of insulation;
(c) if the maximum temperature exceeds the permissible limit, the protective conductor size shall be increased until the temperature rise is acceptable; and
(d) a voltage drop check, as described in Section 7.2.2, is carried out to ensure that the voltage drop along the protective conductor does not exceed safety voltage limits.

Section 7 – Assets not part of the electric traction system

The steel reinforcement within concrete structures, particularly in platforms, may also be bonded to the traction return via the common connection point, but that is not universal practice where the reinforcement is not exposed and accessible. Bonded reinforcement is considered as an earthing conductor and should satisfy the requirements of Table 7.2.

Electrical clearances: The air gap electrical clearance to the live overhead contact system conductors should not be less than the basic dimensions provided in EN 50119. It is preferable that these clearances should always be considered as minimum values and, wherever possible, they should be as large as practicable. Station canopies and footbridges are those parts of the station which are at greatest risk of infringing these clearances.

Station LV electrical installations: The MET of the LV electrical installation should be connected to the traction return circuit at a single location. Bonding to the traction return circuit provides a path for traction fault current should a traction fault occur within the station, and also controls the touch voltages arising. The single-point connection reduces the likelihood that traction current may circulate within the LV protective conductors.

Electrical equipment within the OCLZ and CCZ require protective conductors capable of carrying traction fault current. Either the LV protective conductor to this equipment is increased in size or a separate bonding conductor is provided. For those parts of the station electrical installation outside the OCLZ or CCZ, no additional requirements for traction fault current are required and the design of these parts should satisfy the fault protection requirements of IEC 60364-4-41 or CENELEC HD 60364-4-41. It can be seen that it is desirable that the LV installation should, wherever practicable, provide separate circuits for equipment within the OCLZ and CCZ, and for equipment outside these zones.

There are circumstances where LV circuits do not require provisions against traction fault currents, even though they may be in the OCLZ or CCZ. For example, if the circuits run beneath a station canopy and that canopy is bonded to traction return, the canopy provides an electrically connected obstacle as defined in EN 50122-1. The LV circuit contained within the canopy therefore requires no protective provision for traction fault current.

Platform screen doors (PSDs): These are increasingly being used at stations where large numbers of passengers wait on the platform. The key safety objective is to limit the potential difference between the PSD and the traction return system under peak traffic conditions, such that the passengers are not exposed to current in excess of the threshold of perception value (see Section 5). The application of the threshold of perception value is used because the consequential risk of injury from the startle reaction, leading to trips and falls while passengers board and alight from the train, is more significant than that of receiving an electric shock.

The preferred approach is described below:

(a) The PSD shall be made electrically continuous for the purpose of direct connection to the traction return circuit. A through-bonding conductor rated for the traction fault current and normal fault clearance time shall interconnect all conductive parts of the PSD and each door.
(b) A connection is made from the bonding conductor at each PSD and the platforms and at the mid-platform location to the traction return circuit. This would provide a total of three connections to traction earth for the entire PSD assembly.

Interface with other electric traction systems: Where other electric traction systems exist at a station, care needs to be taken to ensure that the bonding of the station takes account of the largest fault current and longest fault clearance times. Providing electrical separation between platforms that have different types of electric traction systems is very often impossible to achieve and therefore the electric traction systems are often interconnected. For AC electric traction systems, this may not be a problem, but if they operate at different frequencies there will probably be significant safety risks to the train detection arrangements that need to be addressed. For DC traction systems, the station is not bonded to the rails because of the long-term corrosive effects of DC traction current. The connection between the station and the DC traction return rails is usually made via a VLD. This makes the connection between station and the DC traction return circuit only when the voltage difference is more than the safety voltage limits defined in EN 50122-1. Section 10 provides information about the earthing and bonding at the interface between AC and DC electric traction systems.

7.2 Non-traction power systems

The railway infrastructure requires electrical power for a variety of purposes in addition to the electric traction system. The non-traction power can be supplied at both high and low voltages and is distributed along the railway corridor to supply equipment associated with signalling and telecommunication systems, trackside lighting, switch heating, pumps and tunnel ventilation.

Supplies will be obtained either directly from an electricity utility or, less typically, from generating stations owned by the railway infrastructure owner. Supplies may be obtained at either high voltage (HV) (>1,000 V AC) or at low voltage (LV) ($\leq$ 1,000 V AC). Where supplies are obtained from an electricity utility, they need to be agreed and formalized through an appropriate connection agreement.

7.2.1 Threats from the electric traction system

The design and installation of these non-traction power systems will be in accordance with the standards and codes of practice appropriate to their nominal voltage and configuration. They also have to coexist with an electric traction system which creates the threats shown in Table 7.4.

Table 7.4 Electric traction system threats to non-traction power systems

	Threat description	Source mechanism
1	Traction current flowing through protective conductors within non-traction power system.	Galvanic connection to the traction return circuit.
2	Traction current flowing through earth electrodes and protective conductors of the electricity supplier.	Galvanic connection to traction return circuit.
3	Hazardous transfer potentials from the traction return circuit to the electricity supplier's installation.	Galvanic connection to the traction return circuit.
4	Hazardous touch voltage between traction return circuit (or equipment directly connected) and exposed-conductive-parts connected to LV installation protective earth conductors.	Proximity allows simultaneous touching.
5	Induced current and voltage within protective conductors.	Mutual coupling from the overhead catenary system.
6	Induced current and voltage within cable sheaths.	Mutual coupling from the overhead catenary system.
7	Hazardous voltages and traction fault current transferred on assets located within the overhead contact line zone (OCLZ) or current collector zone (CCZ). (Section 5 describes the OCLZ and CCZ.)	Broken overhead contact line or pantograph.

Section 7 – Assets not part of the electric traction system

EN 50122-1 categorizes these threats into two groups:

1. threats from the traction return system that endanger the non-traction power systems. These tend to occur through galvanic connection or mutual coupling (items 1 to 5 in Table 7.4).
2. threats arising from their location within the OCLZ or CCZ (item 7 in Table 7.4)

The threat of hazardous voltages and currents from the mutual coupling of the OCLS (items 5 and 6 in Table 7.4) is not covered in EN 50122-1 but requires assessment where the non-traction distribution system extends over long distances and is parallel to the railway. The assessment involves multi-conductor modelling described in Section 4.

Additional protective provisions are required to address these threats that are not normally addressed in the standards and codes of practice for HV and LV non-traction power systems.

7.2.2 Low voltage non-traction power systems

7.2.2.1 Supply arrangements

The electric traction system imposes threats 2 and 3 in Table 7.4 on the electricity supplier. In the case of HV supplies, faults on the electricity utility installation could result in a hazardous transfer potential appearing at the railway non-traction supply intake substation. The normal protective provision employed, which is described in EN 50122-1, is to separate the railway non-traction power system earth from the earth (or combined earth and neutral) conductors of the supplier.

The supply arrangements vary depending on whether the supply is provided at HV and transformed down to LV, or whether the supply arrangement directly provides power at the correct low voltage. These arrangements and the protective measures typically adopted to mitigate the effects of the threats 2 and 3 of Table 7.4 are described in Table 7.5.

The above arrangements, in which there is no protective earth (PE) connection between the railway system and the electricity supplier's system, provide the protective measures most commonly used and makes it easier to obtain connection agreements. It is, however, possible with the agreement of the electricity supplier for the supplier's earth connection to be used by the railway in a similar manner to that provided for most commercial and domestic users. Typically, the electricity supplier will want to be satisfied that the rail potential at the point where the LV supply intake is bonded to the traction return circuit is considerably lower under normal conditions than the long-term rail voltage limits defined in EN 50122-1. For example, in the UK the rail potential shall not exceed 25 V under normal electric traction system conditions, and under fault conditions the rail potential shall not exceed 430 V for fault clearance times greater than 0.2 s and 650 V for fault clearance times of 0.2 s or less. The railway infrastructure owner also has to be satisfied that a fault within the electricity supplier's system should not create a hazard to the railway system. A broken combined PEN conductor is typically the most significant hazard.

7.2.2.2 Distribution arrangements

The LV distribution has to be provided with basic and fault protection in accordance with IEC 60364-4-41 or CENELEC HD 60364-4-41. There is a mismatch between the long-term safety voltage limit of 60 V used for the electric traction system and the 50 V limit that is used in IEC 60364-4-41 or CENELEC HD 60364-4-41, as explained in Section 5. Touch voltages above 50 V are not permitted under IEC 60364-4-41 or CENELEC HD 60364-4-41. If the traction rail potential exceeds 50 V, final circuits within the LV distribution should have their live and neutral connections disconnected under fault conditions. This can be achieved using a two-pole circuit-breaker, or by a two-pole residual current circuit-breaker (without integral overcurrent protection)

Table 7.5   Low voltage non-traction power supply arrangements

Supply arrangement		Railway non-traction LV power system arrangement	
Supply voltage	Earthing connection to consumer	System earthing type (IEC 60364-1/ CENELEC HD 60364-1)	Protective measure
Low	Protective earth (PE) or combined protective earthed neutral (PEN)	TT	Supply PE conductor or (PEN) conductor not connected (i.e gapped) to the railway earth at the railway intake cubicle. Earth electrode provided for the railway at LV main earthing terminal (MET). Equipotential bonding of railway intake cubicle to earth electrode. Automatic disconnection of supply in accordance with IEC 60364-4-41 or CENELEC HD 60364-4-41.
Low	PE or PEN	TN	Supply PE conductor or PEN conductor not connected (i.e gapped) to the railway earth at railway intake cubicle. Earth electrode provided at railway intake cubicle. Two-winding (typically 1:1 ratio) isolation transformer. The neutral (star point) of a three-phase secondary winding or one end of a single-phase secondary winding is connected to an earth electrode to provide a TN-S LV supply. Equipotential bonding of railway intake cubicle and transformer tank to earth electrode. Automatic disconnection of supply in accordance with IEC 60364-4-41 or CENELEC HD 60364-4-41.
Low	PE or PEN	IT	Supply PE conductor or PEN conductor not connected (i.e gapped) to the railway earth at railway intake cubicle. Earth electrode provided at railway intake cubicle and connected to LV MET. Two-winding (typically 1:1 ratio) isolation transformer with NO connection from secondary winding to earth electrode or LV main earthing terminal. Equipotential bonding of railway intake cubicle and transformer tank to LV MET.Automatic disconnection of supply in accordance with IEC 60364-4-41 or CENELEC HD 60364-4-41.
High	PE	TN	Step down two-winding transformer located on railway infrastructure. Supply PE conductor not connected (i.e gapped) to the railway earth at railway intake.Earth electrode provided at railway intake. Transformer tank and equipment etc. at railway intake equipotentially bonded to railway earth electrode. The neutral (star point) of a three-phase secondary winding is connected to an earth electrode.LV MET connected to traction return circuit. Automatic disconnection of supply in accordance with IEC 60364-4-41 or CENELEC HD 60364-4-41.
High	PE	TT	Step down transformer not located on railway infrastructure. HV earth electrode provided at transformer location. Transformer tank, transformer secondary winding and HV equipment etc. equipotentially bonded to HV earth electrode. Separate LV distribution earth electrode installed at LV distribution cubicle. Transformer secondary earth not connected to separate LV earth electrode.

(Continues)

Section 7 – Assets not part of the electric traction system

Table 7.5 *cont.*

Supply arrangement		Railway non-traction LV power system arrangement	
Supply voltage	**Earthing connection to consumer**	**System earthing type (IEC 60364-1/ CENELEC HD 60364-1)**	**Protective measure**
			Separate LV earth electrode located sufficiently far from HV earth electrode to limit earth potential rise (EPR) at LV earth electrode from HV earth faults.LV distribution cubicle equipotentially bonded to separate LV earth electrode. Cable sheaths/armour of LV cable between transformer and LV distribution cubicle not connected (i.e. gapped) at LV distribution cubicle.Automatic disconnection of supply in accordance with IEC 60364-4-41 or CENELEC HD 60364-4-41.
High	PE	IT	Step down transformer located on railway infrastructure. Supply PE conductor not connected (i.e gapped) to the railway earth at railway intake. Earth electrode provided at railway intake and connected to LV MET. Transformer tank and equipment etc. at railway intake equipotentially bonded to LV MET. No connection from transformer secondary winding to earth electrode or LV main earthing terminal. Automatic disconnection of supply in accordance with IEC 60364-4-41 or CENELEC HD 60364-4-41.

(RCCB) and an overcurrent protective device in the live conductor, or by a two-pole residual current circuit-breaker (with integral overcurrent protection) (RCBO).

The non-traction power system and the electric traction systems are in close proximity because of the limited space available in a typical narrow railway corridor and its connection to equipment that is close to the rails. It is not practicable to electrically separate the two systems and therefore a bond is provided between the traction return circuit and the LV system. Typically, this is made at the MET of the LV installation where the protective conductors and the earth electrode of the LV installation are connected. This bond mitigates against hazardous touch voltages (threat 5 in Table 7.4) between the LV installation and the traction return circuit conductors (and also any other equipment with exposed-conductive-parts that may be connected to these conductors). It also provides a connection for traction fault current that may flow in the event of a broken live OCLS conductor or pantograph that may strike the LV installation where parts of the installation within the OCLZ or CCZ are indirectly bonded (threat 7 in Table 7.4).

Exposed-conductive-parts of the LV installation that are within the OCLZ or CCZ are endangered from live OCLS conductors or from the pantograph that may break and fall onto the LV installation (threat 7 in Table 7.4). The exposed-conductive-parts need to be bonded to the traction earth so that traction fault current returns to the source to limit the touch voltage imposed on the LV installation to within the safety limits defined in Section 5 and minimizes the damage caused. Two approaches to bonding that are not mutually exclusive are typically adopted and are described in EN 50122-2:

1. Utilize the LV protective conductor that is connected to the LV installation METs, which is bonded to the traction return circuit. In this approach, the LV protective conductor needs to be rated to carry the maximum traction fault current for the fault clearance time as well as fulfilling the fault protection provision of IEC 60364-4-41 or CENELEC HD 60364-4-41.

Section 7 – Assets not part of the electric traction system

2. Bond the exposed-conductive-part of the LV installation directly to the traction return circuit and do not connect the LV protective conductor to the exposed-conductive-part. In this approach, the LV protective conductor needs only to be rated to satisfy the LV fault protection provision of IEC 60364-4-41 or CENELEC HD 60364-4-41. The direct bond to the traction return circuit will carry the traction fault current.

EN 50122-1 does not provide a name for these two arrangements, but within the UK the description indirect bonding is used for the first approach and direct bonding for the second. These terms are used as a convenient shorthand in the discussion that follows. The arrangements will also vary between LV equipment with Class 1 and Class 2 protection as defined in IEC 61140. It should be noted that Class 2 protection relates only to the nominal voltage of the LV installation. Class 2 protection against the nominal voltage of the electric traction system is not typically practicable for economic reasons.

Indirect bonding – Class 1 protected LV equipment

LV equipment is not bonded locally to the traction return. Normal traction load current does not flow in the LV protective conductor, but traction fault current will in the event of a traction fault to the LV equipment.

The protective conductor of the LV distribution system shall be rated for the maximum prospective traction fault current and fault clearance time as follows:

(a) The protective conductor is sized to meet the fault protection requirements of IEC 60364-4-41 or CENELEC HD 60364-4-41.
(b) The temperature rise is calculated for the maximum traction fault current for the normal fault clearance time using the adiabatic equation described IEC 60949 (or in slightly different forms in various national standards, such as IEEE Standard 80 and BS 7454), and checked against the maximum permissible for the type of insulation.
(c) If the maximum temperature exceeds the permissible limit the protective conductor size shall be increased until the temperature rise is acceptable.
(d) A voltage drop check is carried out to ensure that the voltage drop along the protective conductor does not exceed the safety voltage limits (see Section 5). Equation 7.1 can be used. It should be noted that this does not consider the contribution due to the cable inductance but for many situations it is sufficient.

$$V_\mathrm{d} = I_\mathrm{SC} \times R'\,(\mathrm{V/m}) \tag{7.1}$$

where:
  R' is $R_{20}(1 + \alpha_{20}(\theta_\mathrm{f} - 20))$ Ω/m (from IEC 60287-1-1 section 2.1.1) and represents the resistance per metre of the protective conductor (Ω/m)
  V_d is the voltage differential along the length of the cable per metre (V/m)
  I_SC is the short-circuit magnitude (A)
  R' is the DC resistance at the final short-circuit temperature (Ω/m)
  R_{20} is the DC resistance per unit length at 20 °C (Ω/m)
  α_{20} is the constant mass temperature coefficient at 20 (°C/k) (IEC 60287-1-1 Table 1)
  θ_f is the conductor temperature after the short-circuit event (°C)

Direct bonding – Class 1 protected LV equipment

The PE and cable armour shall not be connected (i.e.gapped) at equipment that lies within the OCLZ or CCZ so that no traction current will flow through the LV protective conductor. The protective conductors of the LV distribution system shall be rated for the prospective LV fault current and fault clearance time for LV faults up to the gap at the equipment. The path for LV fault current from LV faults at the directly bonded equipment shall be via the traction return system. The current path for HV traction faults is via the bond to the traction return circuit and not via the LV protective conductor.

Section 7 – Assets not part of the electric traction system

It is important that the LV protection is not compromised by this method of bonding. If adequate protection is not achievable and it is not possible to avoid bonding the equipment to the traction return, then it may be necessary to connect the LV protective conductor, which results in the protective conductor being connected to the traction return at two points. This arrangement will result in fault and load currents from the traction system flowing via this protective conductor. These traction currents can cause overheating and loss of conductor integrity, and the current flow can also give rise to a fire hazard if conductors are not suitably dimensioned.

Bonding of Class 2 LV insulated equipment

No LV protective conductor is provided. LV Class 2 insulated equipment is installed within a metallic enclosure that is bonded to the traction return circuit.

7.2.3   High voltage non-traction distribution systems

Power distribution to the various equipment and systems is required at low voltage. This requires many LV supply intake points, as there is a limit to how far an LV distribution system can extend along the railway before the voltage drop becomes unacceptable and earth fault loop impedances become so high that discrimination between fault and load currents is very difficult. An alternative is for the railway infrastructure owner to provide its own HV distribution system. Substations would be provided as required along the railway with transformers and switchgear to provide LV supplies to a localized area. Fewer connections from the electricity utilities are therefore required, albeit the required power capacity of these will be much greater. Figure 7.1 illustrates a combined HV and LV system.

Figure 7.1   Combined HV and LV non-traction power system

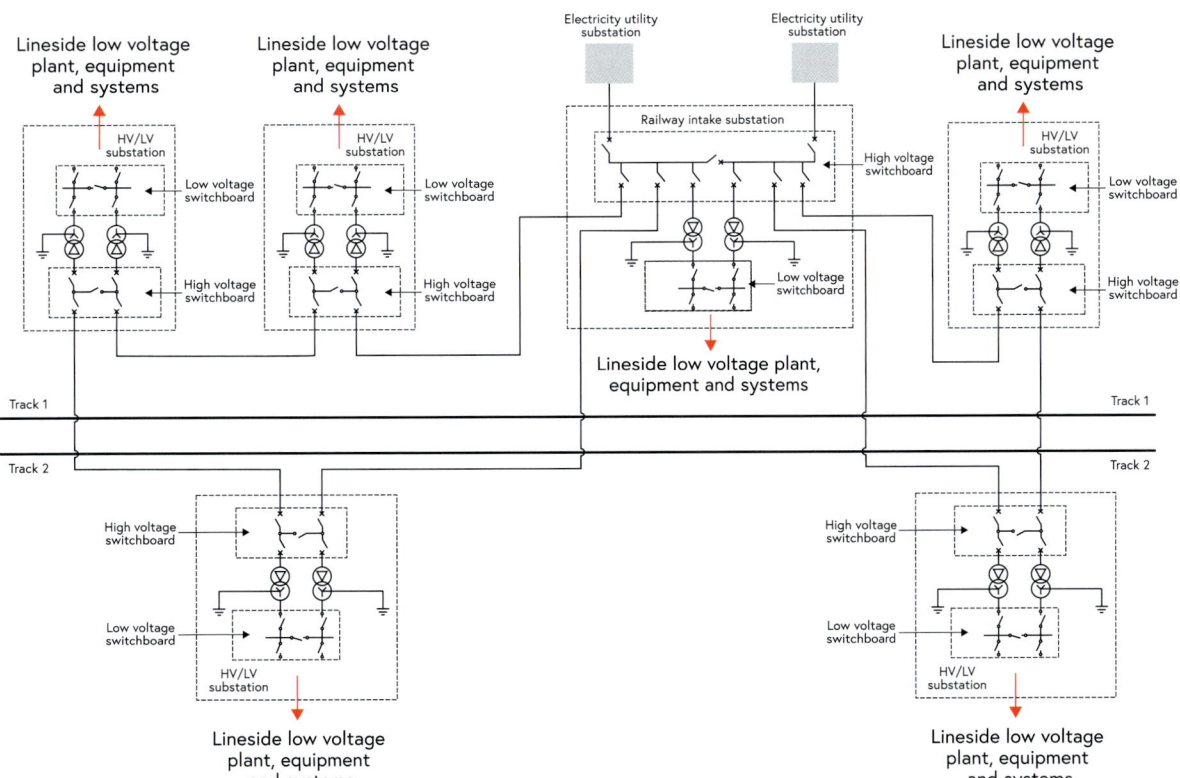

Section 7 – Assets not part of the electric traction system

7.2.3.1 Supply arrangements

Threats 2 and 3 in Table 7.4 are applicable to the HV supplies provided by the electricity utility.

The earthing systems at the electricity supplier substation and at the railway infrastructure owner's intake HV substations should each be designed, as far as practicable, to be safe without the electricity supplier being dependent on the railway infrastructure owners' earthing system, and vice versa (i.e. the earthing systems are independent). This independent arrangement enables the isolation and earthing arrangements of equipment to be much simpler and avoids the long-term asset maintenance obligations on both parties to be satisfied that the combined earthing system remains fit for purpose.

Although the two earthing systems are independent, it is preferable for the electricity supplier earth and the railway infrastructure owner's earth to be interconnected through the sheaths and armour of the interconnecting cables. This continuity provides a significant reduction, through the mutual coupling between cable conductors and sheath/armour, in fault current flowing to earth during faults within the railway infrastructure owner's HV installation. The earth potential rise (EPR) is reduced and consequently controls touch voltage to within the safe limits described in EN 50122-1. Should interconnection not be practicable, location-specific earthing arrangements shall be jointly developed by the electricity supplier and the railway infrastructure owner, and appropriate operational and maintenance processes agreed to avoid the occurrence of dangerous touch and step voltages.

If the electricity supplier's substation is separate to the railway infrastructure owner's intake substation, the earthing and bonding arrangements at that substation shall meet the electricity supplier's requirements for permissible touch and step voltage. The railway intake substation earthing and bonding arrangements shall meet the railway infrastructure owner's requirements for permissible touch and step voltages, and generally with those defined in EN 50122-1. For the interconnection of the two earthing systems via cable sheaths and armour:

(a) a fault on the electricity supplier's system must not create a transfer voltage at the railway infrastructure owner's intake substation that exceeds the permissible touch voltage limits defined in EN 50122-1; and

(b) a fault on the railway infrastructure owner's HV system must not create a transfer voltage in excess of the electricity supplier's permissible touch voltage limits.

7.2.3.2 Low voltage distribution arrangements

The LV distribution arrangements are essentially the same as described in Section 2.2.2, but the design of the HV and LV earthing at the railway HV/LV substations requires particular attention to limit the EPR on the LV system arising from faults on the HV system. The hazard is the transfer potential along the LV protective conductor that presents a touch voltage hazard at LV equipment remote from the HV/LV substation. The consequence of the interconnection of the earths differs according to whether the LV installation is indirectly or directly bonded.

Indirectly bonded LV systems. These are earthed at one point, normally at the HV/LV substation, where the star point or neutral connection on the transformer is also earthed. Their connection to the traction return circuit is usually at the same point. The following needs to be considered to enable the HV and LV earthing systems to be interconnected:

(a) the HV earth system shall limit the EPR such that the stress voltage between earth parts and insulated parts of the LV system do not exceed the limits, such as those defined in EN 50522 and IEC 61936-1.

(b) The value of the EPR shall ensure that the transfer potential shall not create a touch voltage that exceeds the values defined by EN 50122-1 for normal disconnection times.

Section 7 – Assets not part of the electric traction system

Directly bonded LV systems. The LV protective conductors are earthed at the source but gapped at the LV equipment, which is bonded to the traction return. The EPR from an HV earth fault will create a voltage across the gapped protective conductor, and the insulation of the protective earth gapping shall be rated for the EPR and prevent direct contact in order that the HV and LV earthing systems may be interconnected.

The alternative to interconnecting the HV and LV earths is to install the LV earth sufficiently far away from the HV earth electrode such that the EPR at the LV earth electrode is reduced and the transfer potential hazard mitigated. This arrangement is common practice in the utility sector, but requires more space than is often available within the railway corridor. As the LV distribution panel would be bonded to the LV electrode, considerable care is needed in the design and construction of the HV/LV substation to provide the separation between the LV and HV electrodes.

 # Section 8

System measurements

8.1 The need for system measurements

Electric traction systems are inherently complex. The earthing of the traction return circuit to achieve human safety also results in traction currents flowing through the earth, which in turn create current loops that are mutually coupled with lineside cables and other long and parallel conductive assets.

There are multiple and variable loads from trains that progress along the railway, which means that voltage and current flowing through the electric traction system conductors are never constant. The electric traction system imposes unbalanced load on the electricity supplier and the risk of unacceptable transfer voltages under normal and fault electric traction system conditions.

World-wide electric traction experience over many years has enabled standards and codes of practices to be developed and, together with the advances in sophisticated computer modelling techniques, has enabled many of these complexities to be addressed during the design with a high degree of confidence. However, modelling, standards and codes of practice can never fully represent the alignment, ground conditions, electrical topology and spatial constraints that are particular to each railway and which, in turn, result in the installed and integrated electric traction system differing from the more idealized representation used in the design. Furthermore, the application of modelling, standards and codes of practice will lead to a design which is then revised in many small ways to address the specific difficulties that always emerge during installation.

Therefore, system measurements remain essential to provide inputs to system modelling and to validate that the complete and integrated electric traction system will meet its safety and performance requirements. They should never be considered as a task that can be defined and planned at the end of the project, and need to be scoped and developed during design and planned for implementation during installation.

8.2 Limitations of system measurements

System measurements are expensive and require extensive organization given that electrified railways run for many hundreds of kilometres. There has to be an assessment of the extent of testing to which any particular system should be subjected. It should consider the degree of complexity and uniformity of the electric traction system configuration, the location of the highest electrically loaded sections, the complexity of the interface with the electricity supplier at the traction feeder station and the locations where touch step and transfer voltages are significant. It should be determined by the individual characteristics of the system in question and the degree of confidence that can be apportioned to the results of design and special studies. Accordingly, it is essential that system measurements are taken by competent personnel who have the experience and knowledge to define, set up and safely implement the measurements in an efficient manner, while comprehensively documenting the measurement arrangements and results.

System measurements are useful, but it should be understood that they have limitations, including:

(a) they are undertaken when the electric traction system is configured in particular states. It cannot absolutely demonstrate compliance with safety and performance requirements for all of the defined normal and degraded states.

(b) they are location-specific and representative of the system configuration. As a type test they demonstrate that the system performs as intended. They cannot absolutely demonstrate that similar tests undertaken elsewhere will necessarily produce the same results.

(c) system measurements on a complete and integrated electric traction are typically not intended to provide verification for subsystems, although they may identify deficiencies.

(d) specific tests are required where there is intersection or parallel operation of AC and DC railways.

(e) they cannot take into consideration any changes to the system topology that may occur in the future.

8.3 Scope of this Section

Overall system measurements will seek to validate the safety and performance requirements of electric traction system subsystems and the complete and integrated electric traction system. Subequipment and components are measured using testing and measurement methods as typically defined in equipment and component standards, and then, once assembled, they are tested as a subsystem. In this way the electric traction system is measured in a structured manner based on the concept that subsystem measurements cannot proceed until equipment and component satisfactorily pass their own tests. Similarly, measurements on an integrated electric traction system cannot proceed until the subsystems have passed their testing.

In the context of this Guide, the system measurements discussed are those associated with earthing and bonding and are shown in Table 8.1.

Table 8.1 System measurements associated with earthing and bonding

Subsystem	System measurement	Project phase
Earth (ground) electrodes at switching stations, feeder switching stations and substations (including those for lightning protection systems)	Soil resistivity	Design
	Earth (ground) electrode resistance	Installation/periodic test
	Earth connection resistance/ground grid integrity	
	Earth conductor joint resistance	
Switching stations/feeder switching stations	Earth potential rise/ground potential rise	Integrated electric traction system testing post energization
	Touch, step and transfer voltage	
Overhead contact line system (OCLS) structures	Touch voltage	
Rails	Rail potential	
Stations, bridges, tunnels, viaducts and depots	Touch voltage	
Lineside parallel cables	Induced longitudinal and transverse voltage	

8.4 Description of measurements

8.4.1 Soil resistivity

8.4.1.1 Measurement methods

The resistance of an earth electrode is comprised of three factors:

1. the resistance of the (metal) electrode;
2. the contact resistance between the electrode and the soil; and

Section 8 – System measurements

3. the resistance of the soil surrounding the electrode through which current flows outwards from the electrode surface to the mass of Earth.

Of the three resistances, the first two can be made to be relatively small with respect to the third and can be neglected for all practical purposes. As the Earth is vast in size compared to the earth electrode, it has capacity to absorb a virtually unlimited supply of current which suggests that it has a very low resistance. However, the current flowing into the earth is transmitted across the very small metal electrode–earth soil interface. The soil resistance is therefore the dominant factor. IEEE standard 142-2007 – *IEEE Recommended Practice for Grounding of Industrial and Commercial Power Systems* 'Table 4-1' provides an example based on a single 3-metre-long rod electrode, which shows that 25 % of the resistance occurs within 0.03 m, 68 % within 0.3 m, and 94 % within 3.0 m of the electrode surface. A knowledge of soil resistivity and how it varies in depth as well as across a location is essential to the design of an earthing system and key design parameters, such as the earth (ground) grid resistance and earth (ground) potential rise, are strongly dependent on it. It is essential for the accuracy of the design that these measurements and their interpretation are obtained early in the design process.

Soil resistivity data can be obtained by conducting various tests, the most established of which is the Wenner method, as illustrated in Figure 8.1. A test current is circulated through the soil between the outer current electrodes C1 and C2 and the potential difference between the inner two potential electrodes P1 and P2 is measured. By varying the inter-electrode spacing, a profile of the change in soil resistivity with depth can be ascertained.

Figure 8.1 Wenner method schematic arrangement (Image reproduced with permission by BSI from Fig NC.10 of EN 50522:2010)

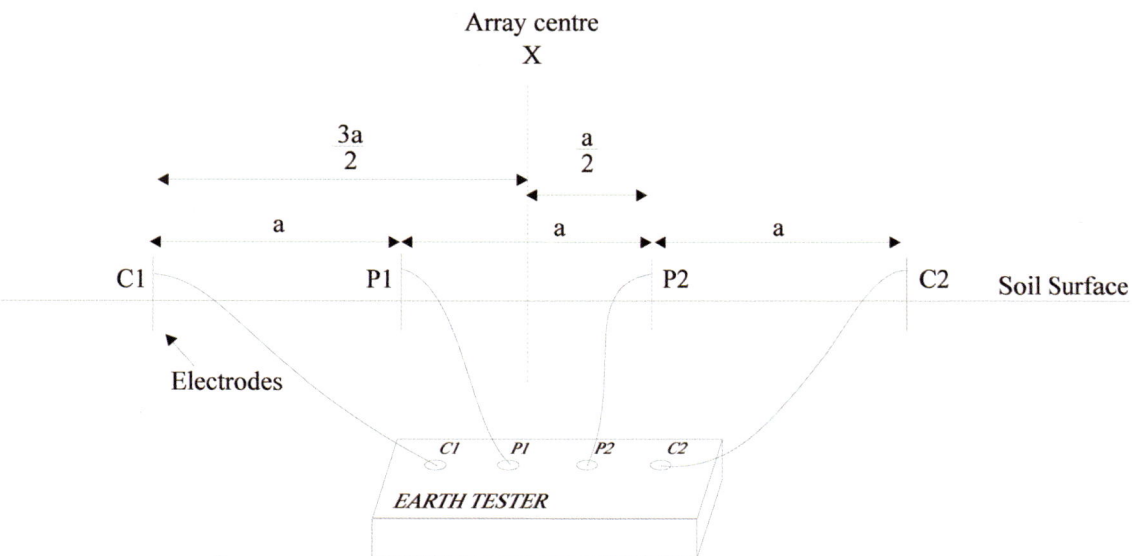

The ratio of the measured potential difference divided by the current circulated provides the measured apparent resistance. For equal spacing and a rod depth not exceeding 10 % of the spacing, the apparent resistivity is then calculated for each measurement location using equation 8.1:

$$\rho = 2\pi Ra \tag{8.1}$$

where:

ρ is the apparent resistivity (Ω.m)
R is the apparent resistance (Ω)
a is the Wenner spacing (m)

Section 8 – System measurements

A soil resistivity survey should determine the change of resistivity with depth. This is achieved by making a series of measurements over a range of Wenner spacings. The depth of soil having the most influence on the measurement is approximately equal to the Wenner spacing because the greater the spacing, the greater the depth to which the test current penetrates. Experience has shown that Wenner spacings of several hundred metres are desirable to achieve greater accuracy in subsequent calculations, even for relatively small earthing systems. However, sites that permit these large spacings will rarely be available and therefore a number of test locations centred around the site of interest are used. IEEE Standard 81-2012 – *IEEE Guide for Measuring Earth Resistivity, Ground Impedance, and Earth Surface Potentials of a Grounding System* provides additional information on the Wenner method and describes the Schlumberger–Palmer arrangement where it is not practicable to achieve equal Wenner spacings. In this arrangement, the spacings shown in Figure 8.1 between C1–P1 and between C2–P2 are larger than the P1–P2 distance. The P1–P2 distance remains fixed and the C1–C2 position varied. If the C1–P1 and C2–P2 distance is c and the P1–P2 distance is d, and c > 2d, the apparent resistivity ρ is calculated for each measurement location using equation 8.2:

$$\rho = \pi(c + d)\frac{R}{d} \tag{8.2}$$

The soil resistivity measurements from each of these tests illustrate how the apparent resistivity varies with electrode spacing and therefore the influence of the deeper soil layers. These values are used to construct a soil model which is then used in the design of the earth electrode.

There are locations, particularly in urban and built-up areas, where there is inadequate space to fully utilize the Wenner method. In such cases, a one-rod method, as described in BS 7430 (or variation of depth method as described in IEEE standard 80), may be used. A typical one-rod measurement arrangement is shown in Figure 8.2. The dimensions shown are examples based on UK-type distribution

Figure 8.2   One-rod soil resistivity measurement (Image reproduced with permission by BSI from Figure 18 of BS 7430:2011+A1:2015)

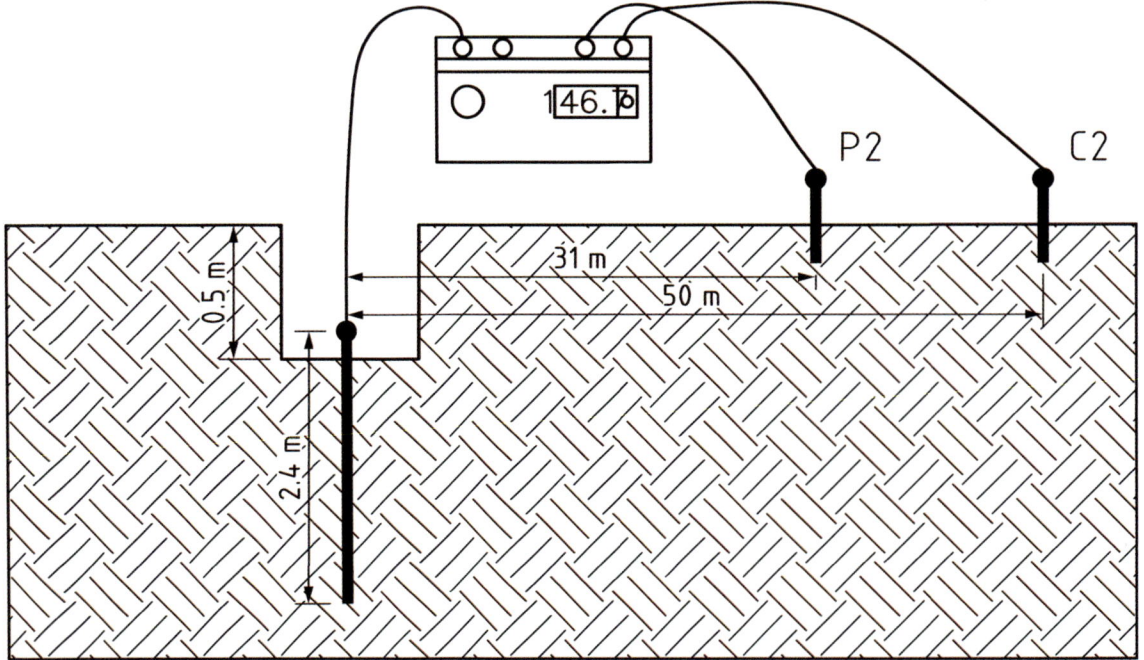

Section 8 – System measurements

substations but do give an idea of the greatly reduced space needed when compared with the Wenner method. By driving the test electrode rod to different depths, the variation of apparent resistivity can be measured. Unlike the Wenner method, this method only provides soil resistivity local to the test electrode rod.

The apparent resistivity can be calculated using equation 8.3:

$$\rho = \frac{2\pi R_f L}{\left[\log_e \left(\frac{8L}{d} \right) - 1 \right]} \tag{8.3}$$

where:

ρ is the apparent resistivity (Ω.m)
R_f is the resistance to earth of the test electrode rod (Ω)
L is the length of the test electrode rod (m)
d is the diameter of the test electrode rod (m)

8.4.1.2 Soil models

The basic objective is to derive a soil model that is a good approximation of the actual soil. A knowledge of the soil resistivity at different depths enables an effective earth electrode to be provided. Vertical earth electrodes (ground rods) are better suited to where the resistivity of the lower soil layers is less than that of the top layer, whereas an earth grid may be more effective where the top layer has the lower resistivity. Soil models can fall into one of three types:

1. **Uniform soil model**. As the name suggests, this assumes that the soil composition and structure is uniform, both vertically and horizontally. The soil resistivity is therefore the same throughout the site. A uniform soil model can be developed either manually using the equations provided in IEEE Standard 81 or by computer modelling. A uniform soil model should be used only when there is a moderate variation in apparent resistivity. If there is a large variation in measured apparent resistivity, the uniform soil model is unlikely to yield accurate results.
2. **1D multilayer soil model** (see Figure 8.3). This model takes account of the soil variability by dividing the soil into vertical layers, with the interlayer boundary at fixed depths. The interlayer boundary does not necessarily reflect the geological structural boundary but represents a position where the resistivity changes. Multilayer soil models provide a one-dimensional vertical view, and it is implicitly assumed that the same layering is constant in the horizontal plane. Manual calculations are provided in IEEE Standard 81 for a two-layer model but more layers than this require computer-based methods. For electric traction systems this type of model is typically adequate for an earth (ground) electrode design.
3. **2D and 3D multilayer soil models** (see Figure 8.4). This is an extension of multilayer soil models but provides resistivity variation in the horizontal plane as well as the depth (z) into the ground. In a 2D model, the resistivity values are allowed to vary along one horizontal direction (x) but are assumed to remain constant in the other horizontal distance (y). A 3D model in which resistivity values are allowed to change in all three directions (x, y and z) would be more realistic but would require more complex (and expensive) measurement and computation, which is probably unnecessary for most electric traction systems.

The type of soil model to be used determines the measurement methodology and instrumentation and therefore needs to be established at the outset.

Section 8 – System measurements

Figure 8.3   1D soil model principle and example of computer 1D model output

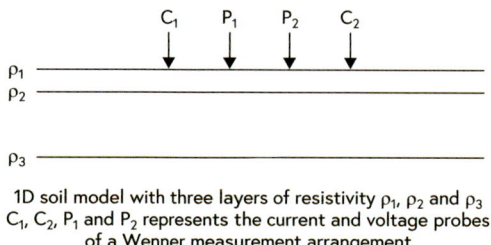

1D soil model with three layers of resistivity ρ_1, ρ_2 and ρ_3
C_1, C_2, P_1 and P_2 represents the current and voltage probes
of a Wenner measurement arrangement

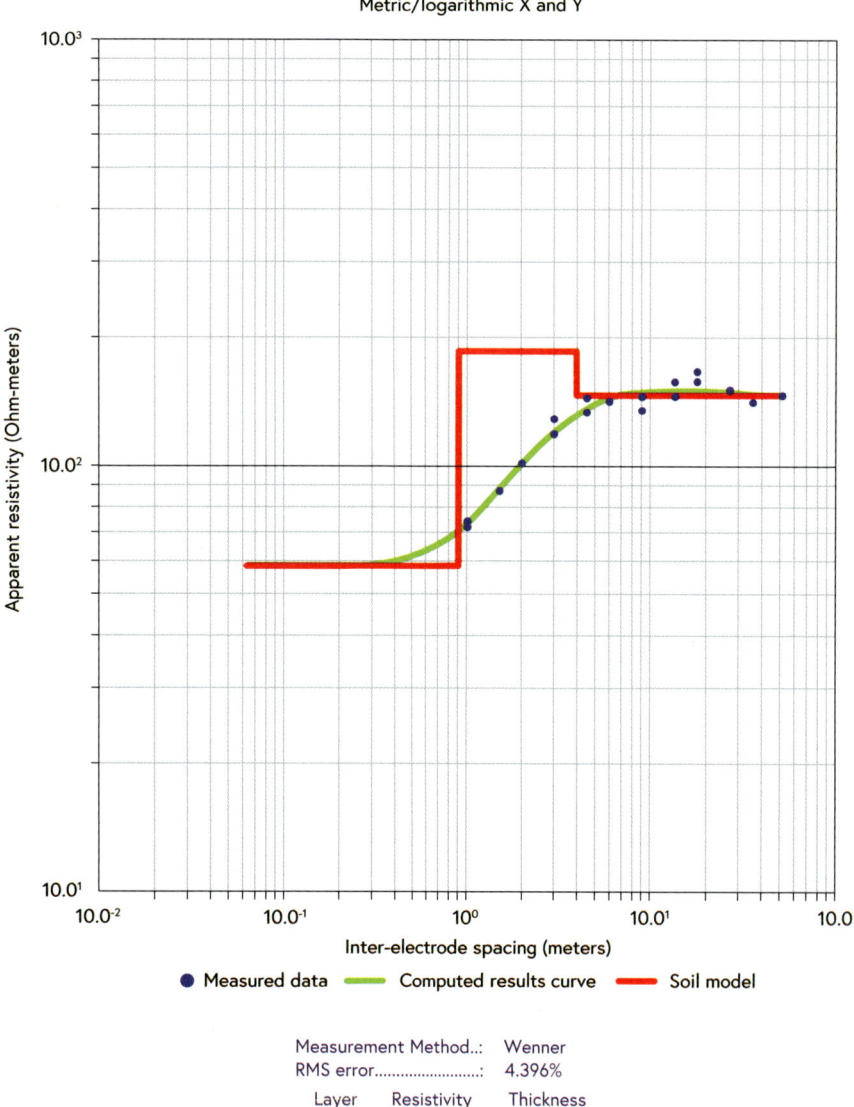

● Measured data   ━━ Computed results curve   ━━ Soil model

| Measurement Method..: | Wenner |
| RMS error.........................: | 4.396% |

Layer Number	Resistivity (Ohm-m)	Thickness (Meters)
Air	Infinite	Infinite
2	57.74248	0.9040243
3	184.6032	3.017681
4	145.7706	Infinite

8.4.2   Earth electrode resistance

Once an earth electrode system has been installed, its resistance to earth should be measured before further construction works progress. This is particularly important where the reinforcement in concrete

Section 8 – System measurements

Figure 8.4   Examples of 2D resistivity modelling

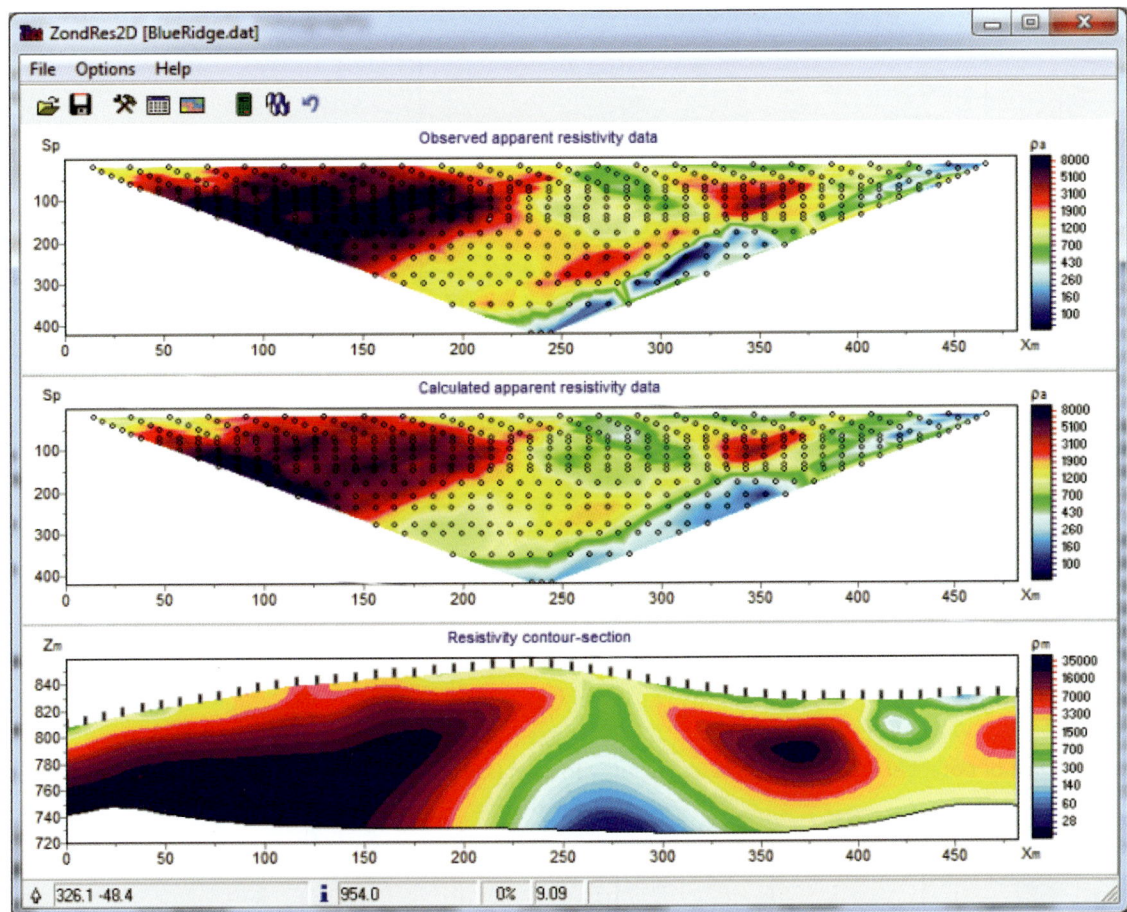

foundations is being used and where these will be contained within the overall structure, resulting in limited access for measurement. Typical locations are underground switching stations or stations, or where installations are located in urban environments and these can be the most challenging of which to undertake earth electrode measurements because of the lack of space or difficultly in driving earth measurement probes. This section describes the most common methods, but other methods or a combination of methods may be used to suit the location, and which can be found in standards such as IEEE Standard 81, EN 50522, BS 7430 and specialist measurement literature.

In general, the fall of potential method is considered to be the most thorough and reliable. A fall of potential measurement may be conducted using the arrangement shown in Figure 8.5, in which three points of contact are made with the soil. One is the connection to the electrode under test. The other two are probes, one connected to C for supplying the test current, and one connected to P for measuring the potential at a given position in the soil. The tester acts as a current source and circulates a current through the soil between the current probe and the electrode under test. The potential probe is used to make voltage measurements at multiple locations along a test route between the current probe and the electrode under test. The tester uses Ohm's Law to calculate and display the resistance at each voltage measurement location.

To minimize inter-electrode influences due to mutual resistances, the current probe is generally placed at a substantial distance D from the earth electrode under test. The potential probe P is typically placed in the same direction as the current probe C and the resistance measured at a number of positions along this axis. The results are plotted as shown in Figure 8.6. They show rising values and then a plateau

Section 8 – System measurements

Figure 8.5 Fall of potential measurement arrangement

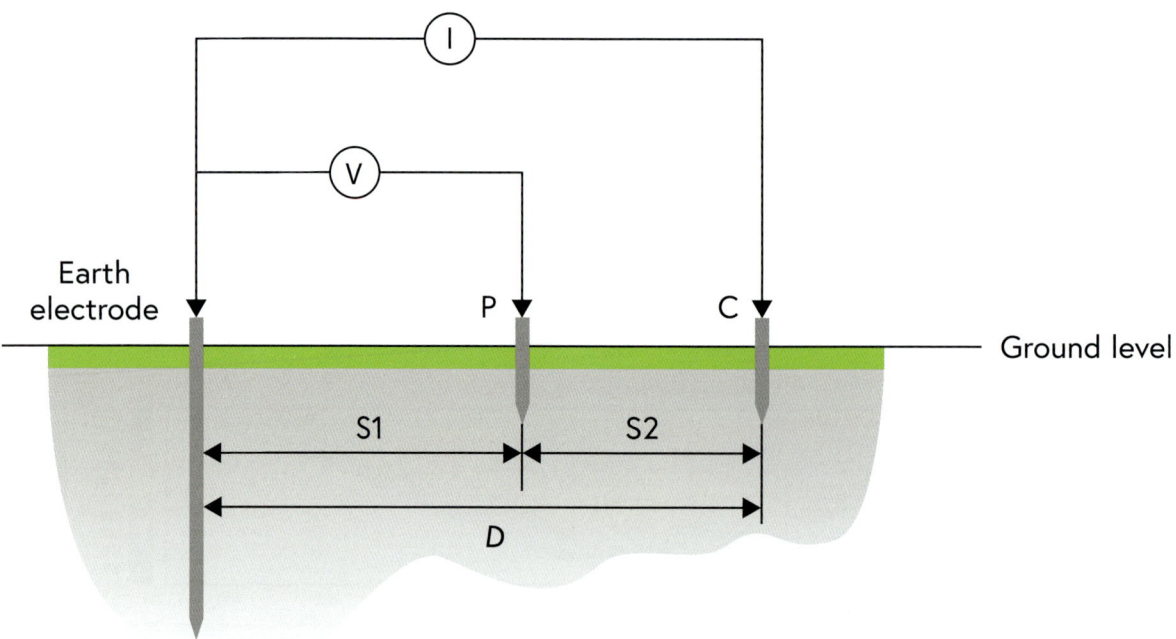

Figure 8.6 Fall of potential results (Image reproduced with permission from Figure NC.4 of EN 50522:2010)

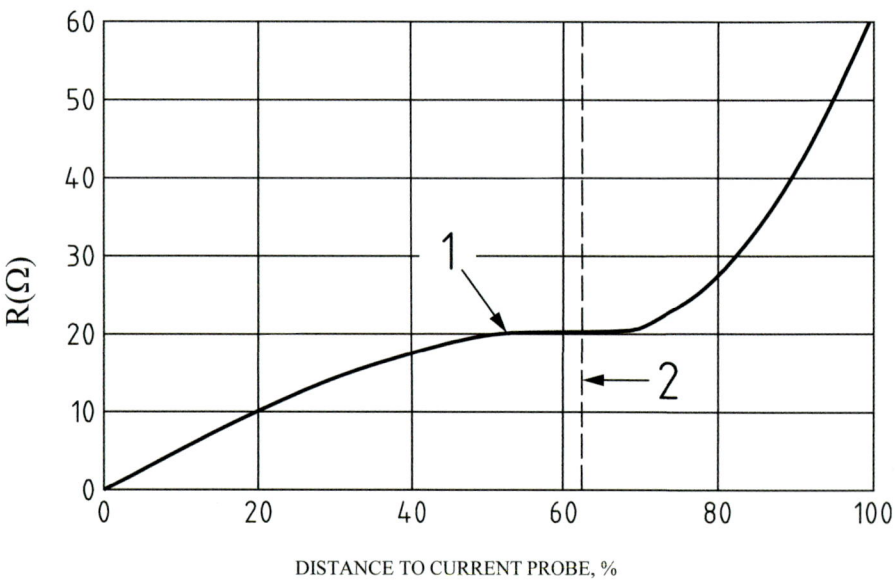

DISTANCE TO CURRENT PROBE, %

region before rising again as the current probe is approached. The value of resistance at the plateau is then taken as the earth electrode resistance.

Where space is limited for taking multiple measurements, one of the positions along S1, between the earth electrode and the potential probe P, is typically chosen to be 62 % (IEEE Standard 81) or 61.8 % (EN 50522) of the distance *D* between the earth electrode and the current probe when current and potential probes are in the same direction. This distance is based on the theoretically correct position

Section 8 – System measurements

for measuring the exact electrode impedance for a soil with uniform resistivity, and that the distance between the earth electrode under test and the test probes is sufficient to consider the test probes to be a hemisphere. Taking a single measurement at the 61.8 % position can only be considered a reasonable estimate and other earth resistance measurement techniques should be used to validate the result. Figure 8.6 shows the 61.8 % (62 %) line and illustrates its correspondence with the plateau region.

Where an earth electrode is large or there are difficulties in positioning the probes at sufficient distance, frequently it is not possible to detect the plateau region of Figure 8.6. A variation of the fall of potential method, called the Slope method described by Dr C.F. Tagg[14] is widely used and is described in IEEE standard 80 and EN 50522. It is based on the fall of potential where the current probe cannot be situated far enough away to produce the graph shown in Figure 8.6. Somewhere on the graph line is the point at which the test ground's field ends and the earth electrode resistance maximizes, but it is not visible from the graph because of the proximity of the current probe. The Slope method uses the same three probe test set-up and takes resistance readings R1, R2, and R3 at corresponding voltage probe positions S1 at 0.2, 0.4, 0.6 times the distance D shown in Figure 8.5.

A slope coefficient μ is calculated using equation 8.4:

$$\mu = \frac{R3 - R2}{R2 - R1} \tag{8.4}$$

The Slope method uses calculations to produce a slope coefficient table from which the ratio S1/D can be read. Knowing the position of the current probe D, the theoretical distance S1 for the voltage probe can be determined and earth electrode resistance measured at that point. Where the value of μ falls outside the slope coefficient table, the current probe should be moved further away if practical, or alternative techniques used.

A further variation of the fall of potential method developed by Tagg is the Intersecting Curves method[15]. It involves considerable work to set up and process the results but can be used when space is limited. The procedure involves connecting the earth tester at any convenient point on the earth electrode to produce a fall of potential curve that will not have the clearly defined plateau. The measurement equipment is repositioned along a different axis from the earth electrode to produce a second curve, and then repositioned again to produce a third curve. The correct value of resistance can be read at 61.8 % of the distance from the true electrical centre of the earth electrode to the current probe which is known. The true electrical centre is defined as distance x from the earth electrode connection. If the actual distances between the electrical centre and the current probe is D_c and the potential probe is D_p:

$$x + D_p = \frac{61.8}{100}(x + D_c) \tag{8.5}$$

$$D_p = \frac{61.8}{100}D_c - \frac{38.2}{100}x \tag{8.6}$$

For each of the three measurement axes and using equation 8.6 for D_p and a range of values for x that are likely to reflect the earth electrode shape, earth resistance values can be read from each of the fall of potential curves. The various values for x are plotted against their corresponding resistances on the same

---

[14]*Measurement of the resistance of physically large earth-electrode systems* – Proceedings of the IEE, Volume 117, Issue 11, 1970, pp. 2185–2190
[15]Proceedings of the IEE, Volume 116, Issue 3, 1969, pp. 475–479

graph, with one plot for each of the three measurement axes. Because of the variability in soil composition, the three curves will form a triangle, the centre of which represents the correct earth resistance value.

8.4.3   Earth conductor joint resistance measurement

At switching stations, feeder switching stations and substations, joints in the earth conductors and where earthing conductors are connected to plant and equipment are measured with a four-pole micro-ohmmeter. The measured resistance of a joint should be compared with the resistance of the same length of conductor but without a joint. The pass/fail threshold value is set appropriately, but typically any joint with a resistance that exceeds the resistance of the unjointed conductor by 50 % should be remade.

The joint resistance is very low and accordingly a four-pole measurement technique is used as illustrated in Figure 8.7. DC current connections to the joint are made separately to the outside of the voltage measurement connection location. This ensures that the resistance of the current probes is not included within the voltage measurement and that the current flow through the joint is fairly uniform.

Figure 8.7   Four-pole joint resistance measurement

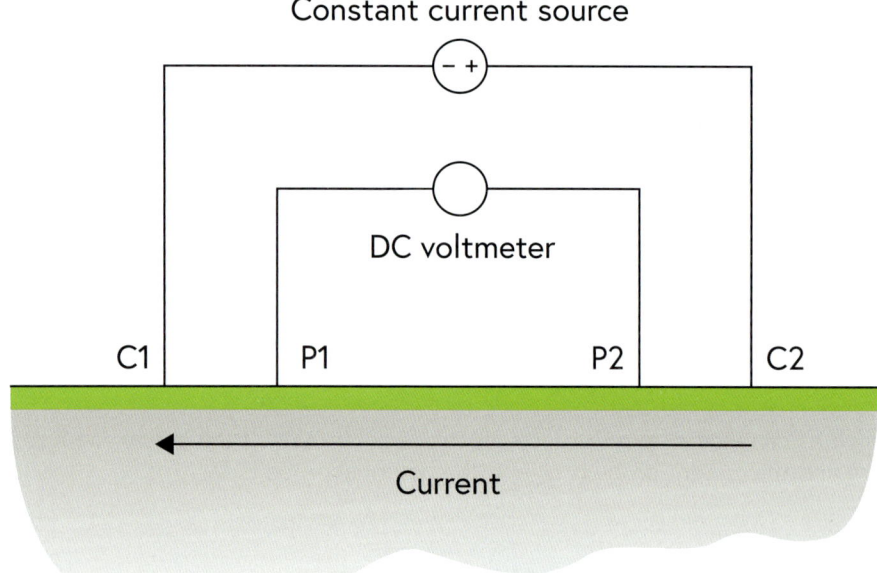

At other locations where bonds are made to the traction return and are expected to carry traction normal and fault currents rather than provide equipotential bonding, similar measurements should be made.

8.4.4   Earth electrode integrity

This measurement is usually required within switching stations, feeder switching stations and substations where the earth electrode system provides an equipotential surface that controls touch

Section 8 – System measurements

and step voltage as well as being capable of carrying traction current under fault conditions. The measurement methods described are based on the high current test and resistance test methods described in IEEE Standard 81. The high current test method is more expensive and time consuming and therefore the resistance test method may be sufficient at sites where simple and accessible earthing systems are provided and where assurance may also be provided through progressive visual and joint testing.

High current test method

A typical test set consists of a variable AC voltage source with voltage and current ranges typically 0 V to 35 V and 0 A to 300 A. A test lead T1 is connected to a reference connection point on the electrode and a second test lead T2 connected to the earth connection on a plant item as shown in Figure 8.8. A test current flows between the connections and the voltage drop across the electrode measured. The second test lead is moved around to another plant item and the voltage across the electrode measured. The current flowing into the earth electrode is measured at the test connection point T2 to assess the current split between test connection point and other connections to the earth that may be present, as in general at least half the test current needs to flow into the earth grid at the T2 connection. This is repeated for all the plant items and other bonded structures until the entire substation earth electrode is tested. The voltage drop for each measurement should broadly be similar and in accordance with the typical value of 1.5 V per 15 m separation stated in IEEE Standard 81 for copper earth grids. Where significantly different results occur, investigation is required to identify and rectify the defects.

Figure 8.8 High current earth (ground) integrity testing configuration

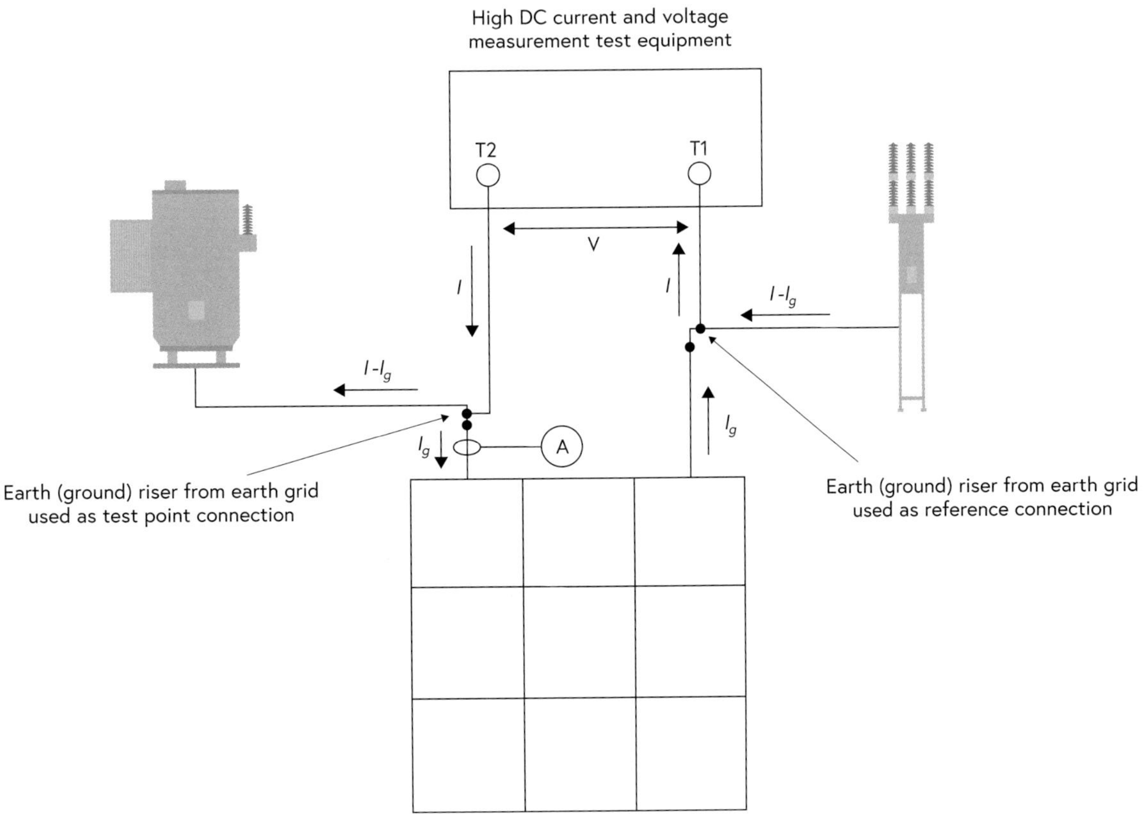

Section 8 – System measurements

Resistance test method

The resistance between plant items and the earth electrode is measured using a micro-ohmmeter (see Figure 8.9). The methodology is the same as for the high current test method, with a series of resistance measurements between plant and bonded structures. The results will vary but should be within a reasonably consistent band for the size of the earth electrodes at switching stations. Any values that are significantly different should be investigated and corrected as necessary.

Figure 8.9 Earth grid integrity test by resistance measurement

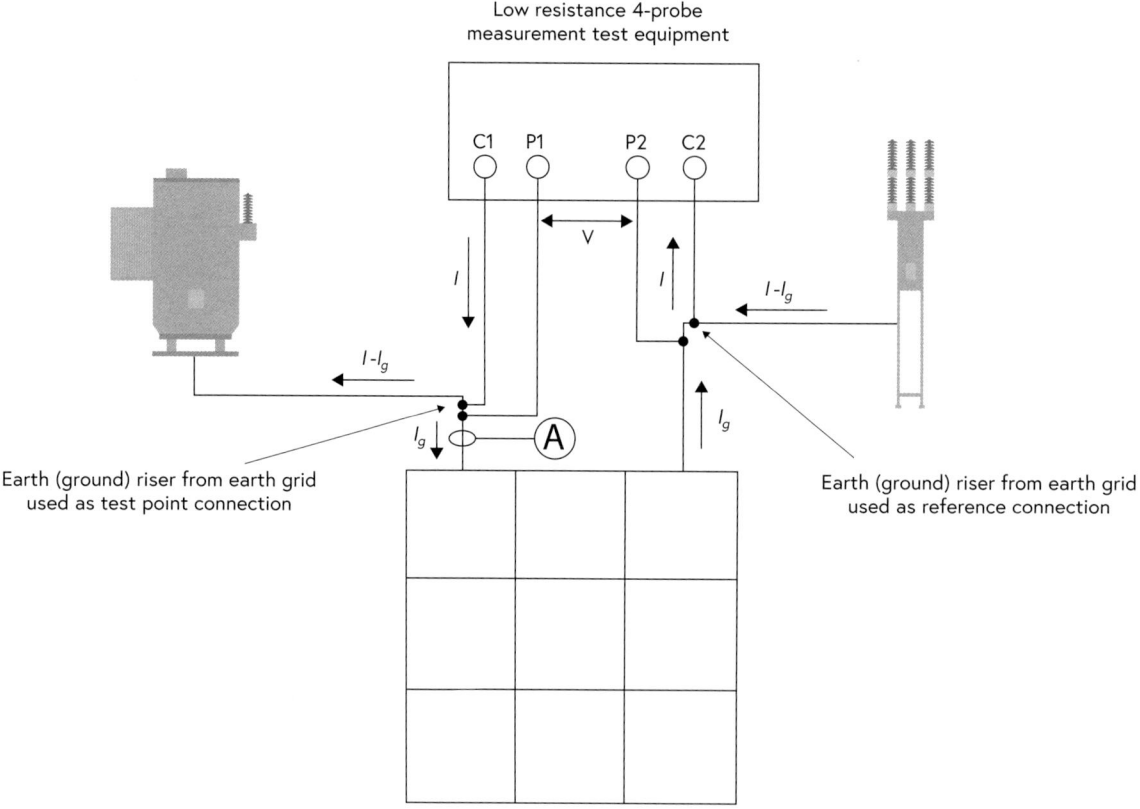

8.4.5 Rise of earth potential

In general, because the feeding arrangements for an AC electric traction system are radial (see Section 2), the highest values for rise of earth potential will occur at the substation and at the feeder switching station where the traction normal and fault currents are largest. These tests will generally therefore only be required at the substation and the feeder switching station. In most cases these measurements are taken as part of the testing prior to the feeder switching station being brought into service and will be undertaken as part of the short-circuit testing:

(a) Rise of earth potential may be measured by injecting a test current into the substation electrode and returning it through a remote electrode via a connecting conductor. Providing the remote electrode is located a large enough distance from the substation, relative to the size of the substation electrode, a rise of the earth potential profile will be set up around the substation proportional to that which would exist during fault conditions. The actual value can be found by extrapolating the measured voltage by the ratio of the fault current to test current.

Section 8 – System measurements

(b) Rise of earth potential can also be measured during a short-circuit type test. The location of the remote test probe for the rise of earth potential measurement needs to be established before the short-circuit test. Appropriate data acquisition instrumentation has to be set up and synchronized with the short-circuit test. It needs to be capable of high data sampling (in the order of up to thousands of samples per second) to fully record the voltages which, by nature of a short circuit, are essentially transient in nature.

8.4.6 Touch and step voltages

Touch and step voltage measurements are required to be conducted at locations where humans are most likely to be present on the railway, and this includes passenger stations, tunnels, viaducts, depots, bridges and overhead line masts. At passenger stations, touch voltage measurements should be taken at platform screen doors (PSDs) where these are provided. As described in Section 7, touch voltages at PSDs should limited to prevent persons form being exposed to current in excess of the perception threshold current. Touch voltage measurements should also be taken at the switching station fence where there is public and animal access, and at locations within the substation that calculations show have the highest touch voltages.

Short-circuit testing is therefore undertaken at various locations in addition to feeder switching stations. To implement the short-circuit test, a short-circuit device is usually located on an overhead contact line system (OCLS) structure which provides an easy and safe means of connecting the short-circuit device to the OCLS conductors. A typical measurement set-up is shown in Figure 8.10. The transducers that are used to record the voltages are connected to the data acquisition system or store their own data. If it is the latter, it has to be accurately time synchronized with the short-circuit test. Item 1 represents the hand-to-feet touch voltage where the feet are placed at 1 m from the exposed conductive part. Item 3 represents a hand-to-hand (metal-to-metal) touch voltage which could be hazardous if the exposed-conductive-part is not bonded to the traction return.

Figure 8.10   Typical test measurements at short-circuit site

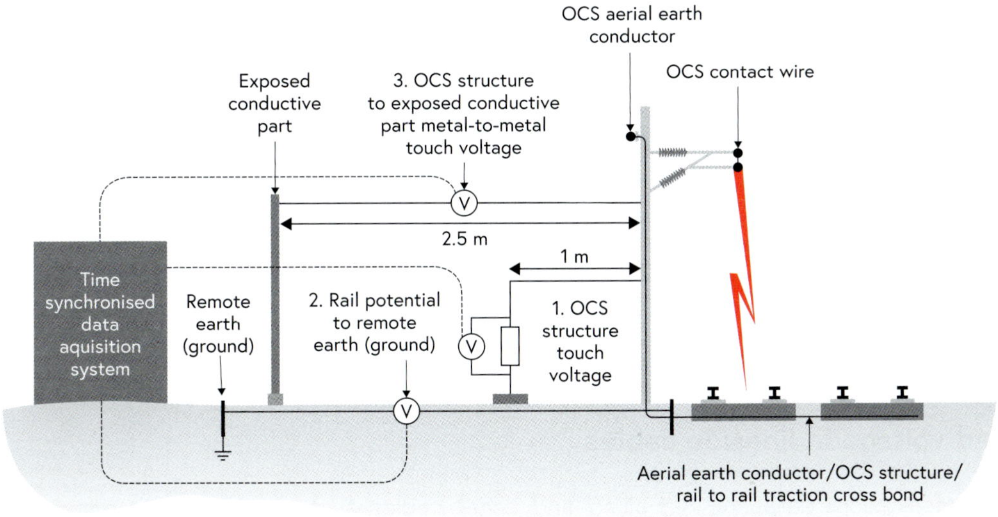

Touch voltage may be measured as a prospective touch voltage or as an effective touch voltage. Effective touch voltage is the voltage that the body experiences when touching an exposed-conductive-part (as described in Section 5) and it is this parameter that is used to specify safety limits. Measurement of effective touch voltage requires a representative body impedance to be placed between the exposed-conductive-part and the ground, as shown in Figure 8.11.

Section 8 – System measurements

Figure 8.11 Measurement of effective touch voltage

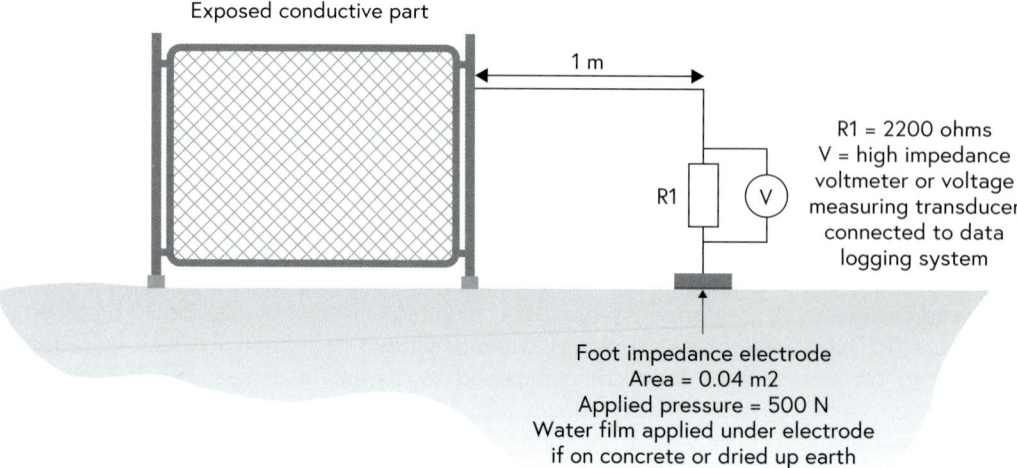

Exposed conductive part

1 m

R1

V

R1 = 2200 ohms
V = high impedance
voltmeter or voltage
measuring transducer
connected to data
logging system

Foot impedance electrode
Area = 0.04 m2
Applied pressure = 500 N
Water film applied under electrode
if on concrete or dried up earth

The value of body impedance R1 shown in Figure 8.11 is taken from EN 50122-1. If the safety limits are derived from IEEE Standard 80, then the value of R1 needs to be 1,000 Ω.

To measure prospective touch voltage, the resistor R1 is removed and replaced by an electrode driven 0.1 m into the soil, where practicable. If the value of prospective touch voltage is within the safety limiting value for the effective touch voltage, then it can be considered unnecessary to specifically measure effective touch voltage.

Step voltage is measured in a similar manner to touch voltage, with two electrodes (to represent each foot) placed 1 m apart. If an effective step voltage is to be measured, an equivalent body resistance representative of the hand-to-hand current path through the body would be connected between the two electrodes and the voltage across the resistance measured. Alternatively, the prospective step voltage could be measured between two electrodes driven approximately 0.1m into the ground and 1 m apart. In general, step voltages would only require measurement in switching stations.

8.4.7   Rail potential

Rail potential is measured in a similar manner to rise of earth potential. A connection is made to the rail typically at the short-circuit location and a remote earth electrode driven into the soil typically around 50 m perpendicular to the rails. Item 2 of Figure 8.10 illustrates the rail potential measurement.

8.4.8   Induced voltage in lineside cables

There are telecommunication, data and signalling and other cables that run over long distances, alongside the railway and parallel to the OCLS conductors. As described in Section 3, voltages are mutually induced by the live OCLS conductors into metallic cable conductors, screens, sheaths and armours. Where these are earthed at one end, a longitudinal voltage V_L is established along the cable with a maximum value at the unearthed end of the cable. In these types of cables there is also an imbalance between cable pairs with the mutual coupling of the OCLS conductors establishing a transverse voltage V_T between the paired conductors. Transverse voltage is not usually a safety hazard, but it has a significant adverse impact quality that can result in intelligibility problems in audio systems

Section 8 – System measurements

and in data corruption in data transmission systems. Mitigations against excessive longitudinal voltages, such as screening cables and limiting cable lengths, are applied but their effectiveness can only be proved through testing under traction short-circuit conditions and normal current conditions when train load is present. In many cases, the limiting safety values for V_L are those set by the ITU-T standards (see Section 5). However, for a common earth bonded traction return system, there is a mismatch in safety limits between ITU-T and EN 50122-1. In such arrangements, the safety values of EN 50122-1 are used for consistency.

Longitudinal voltage measurements are required to be controlled to ensure that touch potential does not exceed the safe operational levels under continuous load or fault conditions. At the remote end of the cable, the conductor pairs are connected to earth, whereas at the testing end, both conductors are joined and the voltage between the pair and earth is measured with a digital acquisition system.

When measuring the transverse voltage, both ends of the cable pair are terminated with a 600 Ω resistor which is typically the characteristic impedance of the cable. Because V_T affects the transmission quality, the transverse voltage is additionally passed through a psophometric filter before being measured by digital acquisition system. Psophometric weighting provides an indication of the audible disturbance caused by noise on speech circuits. Standard ITU-T O.41 defines a set of weighting coefficients over the frequency range 16.66 Hz to 6,000 Hz that matches the response of the human ear (see Figure 3.17). Alternatively, voltmeters with built-in psophometric filters and or 600 Ω terminating resistors can also be used.

Figure 8.12 illustrates a typical test configuration to measure longitudinal V_L and transverse V_T voltages.

When measuring the induced V_L and V_T of a railway system, it is important that the 25 kV and V_L and V_T measurements are synchronized, and the location is not affected by non-railway (electric power transmission line) interference.

Figure 8.12 V_L and V_T measurement configurations

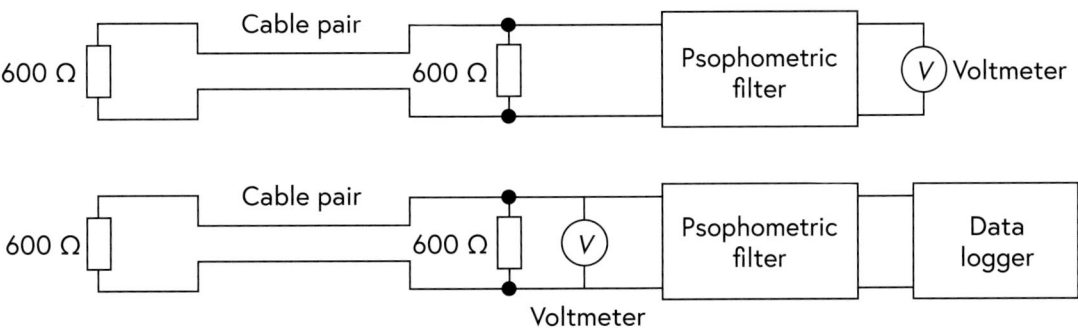

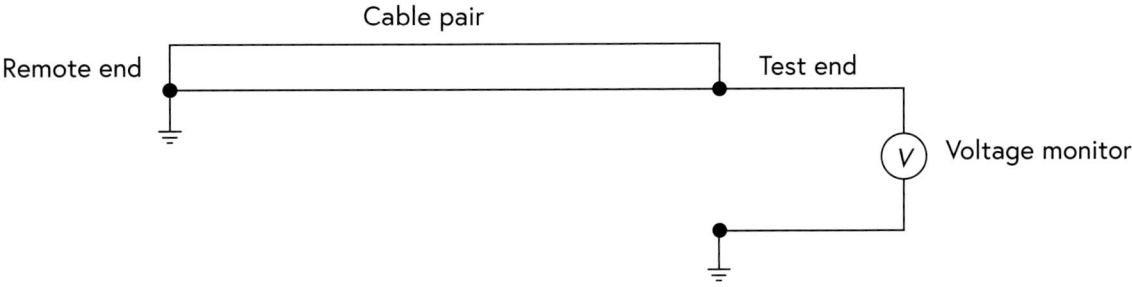

Section 8 – System measurements

The following system data must be recorded.

Short-circuit testing (V_L measurement):

(a) feeding arrangement of the overhead line;
(b) the configuration of the traction return system;
(c) unique identifier of the OCLS mast where the short circuit is applied;
(d) location (km) where the short circuit is applied (typically at a track-to-track cross-bond);
(e) location and length of the telecommunications/data pair;
(f) electrical feeding section where the short circuit is applied;
(g) the magnitude and shape of the short current; and
(h) time of the short circuit.

Initial dynamic testing – mobile load bank (V_L and V_T measurements):

(a) feeding arrangement of the overhead line;
(b) electrical feeding section of the train in section (no other trains should be in the electrical feeding section);
(c) location and length of the telecommunications/data pair;
(d) train measured data: time, position, speed and 25 kV traction current;
(e) telecommunications measured data: time, V_L and V_T; and
(f) train location and V_L and V_T measurements need to be time synchronized.

Longitudinal voltage is measured by a short-circuit test during the commissioning of the feeder station. Transverse and longitudinal voltage measurements are conducted under the defined normal operating conditions, and this requires a train service to operate, which is not possible until all the testing is complete and the required safety and functional performance is proven. As part of this testing, a load bank drawing a constant current is run along the railway. Both V_L and V_T are measured for this constant current and the measured voltage can then be extrapolated to predict the likely voltages when a full timetable is running. Typically, dynamic testing involves a gradual introduction in the number of trains in a very structured manner so further V_L and V_T measurements can be made to validate the initial results obtained from the load bank.

 **Section 9**

Lightning protection for railway structures

9.1 Lightning phenomena

Lightning is a naturally occurring electrostatic phenomenon during which a discharge occurs between two electrically charged regions in the atmosphere or ground. This causes an immediate release of as much as one gigajoule of energy. The discharge produces a wide range of electromagnetic radiation, including the visible light of the discharge itself. The lightning arc heats the air that it passes through so intensely that it creates shock waves, which are then heard as claps or rolls of thunder. The main kinds of lightning are distinguished by where they occur and include a strike within a thundercloud, a strike between two different clouds, or a strike between a cloud and the ground.

Most cloud-to-ground lightning strikes are negative, meaning that a negative charge of electrons travels downwards along the lightning channel. The reverse happens in a positive cloud-to-ground strike, where electrons travel upwards along the lightning path and a positive charge is transferred to the ground (see Figure 9.1).

Two basic types of flashes exist:

1. downward flashes initiated by a downward leader (-ve charge) from cloud to earth; and
2. upward flashes initiated by an upward leader from an earthed structure to a cloud.

Positive lightning strikes are less common than negative strikes, and on average make up less than 5 % of all lightning strikes

Mostly downward flashes occur in the flat territory, and to lower structures, whereas for exposed and/or higher structures upward flashes become common.

The negative lightning that strikes the railway infrastructure produces a current of typically 10–100 kA with a rise time of around 1 μs and can produce an overvoltage surge that travels in both directions along the overhead lines. In any electrical analysis of the system behaviour, the modelling/analysis should be at a frequency of the order of MHz (see Figure 9.2).

The current European international railway protection standards do not specifically address the threat of lightning to railway overhead electrification lines, including EN 50122-1 & IEC 62128-1 and Lightning Protection Standards (EN 62305); NFPA 780 (USA) and AS 1768. Some railway standards have addressed lightning protection, including AREMA (American Railway Engineering and Maintenance-of-Way Association) *Part 7 Traction Electrification System Grounding & Bonding* 'Section 7.7'; and *Western Australia Specification for Earthing and Bonding in the 25 kV AC Electrified Area*.

This Section will provide the necessary understanding of lightning phenomena and the additional design mitigations that are required to be addressed to close out risks to the railway infrastructure and hazards to passengers and maintainers.

9.2 Characteristics of a lightning strike

The electric current within a typical negative cloud-to-ground lightning discharge rises very quickly to its peak in 1–10 μs, and then decays more slowly over 50–200 μs.

Various lightning strike scenarios lead to surge currents being conducted into the ground. The height, shape, and isolation of structures are often the dominant factors in determining where lightning will directly strike.

Section 9 – Lightning protection for railway structures

Figure 9.1 Positive and negative charge lightning strikes

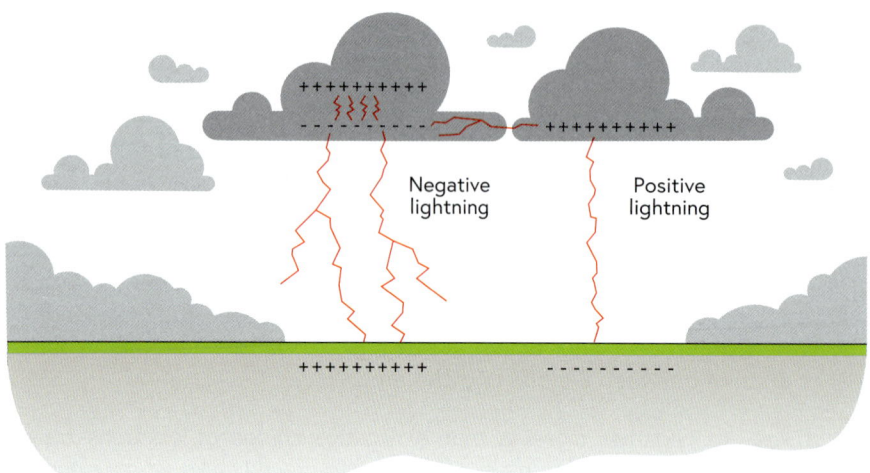

High-rise buildings often surround a railway, but they do not always protect the rail corridor, so direct and indirect lightning strikes to railway buildings and the railway overhead electrification lines are considered to be credible risks. It is important to understand the following types of lightning strikes and their associated behaviour.

The transient nature of the lightning strike current (see Figure 9.2) results in several phenomena that need to be managed in the effective protection of electrical installations and civil structures, for example, lightning surge, the rise of earth potential, electromagnetic radiation, fire and explosion.

Figure 9.2 Impulse current (Image reproduced with permission from EN 62305 Pt 1 and AS1768 Figure D1)

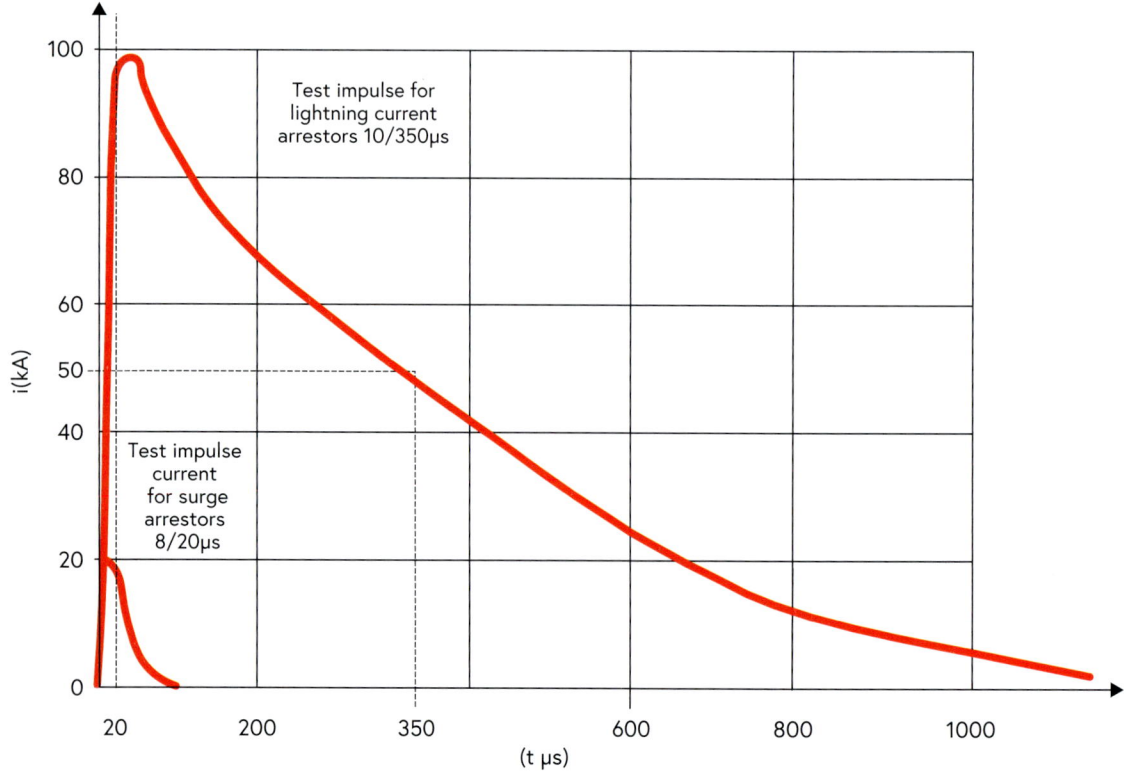

Section 9 – Lightning protection for railway structures

9.2.1 Lightning strikes

Direct lightning strikes: These occur when lightning strikes a structure or human directly, and accounts for 5 % of lightning-related injuries (see Figure 9.5). Where there is direct lightning, it is likely to have multiple injection points and can be several kilometres from the original strike point.

Indirect lightning strikes: These strike the ground remotely, and generally will only have a single injection point into the railway infrastructure (see Figures 9.6 and 9.7).

Indirect lightning strike – step potential: Where lightning current travels in the ground, it creates ground potential. The current creates step potentials through the ground, or near a person or an object. An unfortunate example recorded in 2016 occurred when the current in the ground from a nearby lightning strike killed more than 300 reindeer in Norway. Animals are particularly exposed as they have no protection for their feet during lightning storms.

Indirect lightning strike – side-flash: This occurs when lightning strikes an object directly, which then jumps through the air to strike another object or human. Humid air is often the cause of this phenomenon.

9.2.2 Rise of earth potential

Where the lightning current enters the earth, there is an earth potential rise (EPR), and its magnitude depends on the following factors: ground resistivity (at various depths), the magnitude of the surge current, and the distance from the injection point. The EPR affects the area surrounding the point of entry and significant ground voltages may be present several hundreds of metres away from the injection point. The severity of the EPR is made worse in poor ground such as granite or sandstone, due to high resistivity.

9.2.3 Lightning surge propagation

Surge impedance is the characteristic impedance of a lossless transmission line (see Figure 9.3). Since the line is assumed to be lossless, the series resistance and shunt conductance are assumed to be negligible. The voltage drop in the transmission line due to inductance is compensated by the capacitance. Where they are equal and opposite, this is called the surge impedance of the transmission line, and in this case, the voltage and current are in phase.

Figure 9.3   Transmission line model

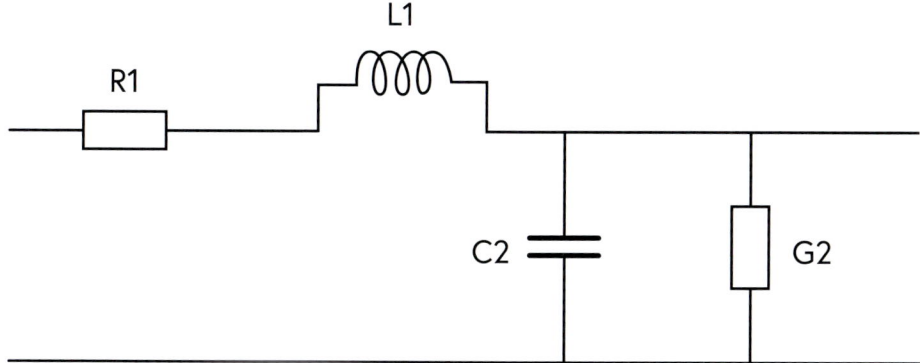

Section 9 – Lightning protection for railway structures

The characteristic impedance of a transmission line is given by equation 9.1:

$$Zc = Z/Y \qquad (9.1)$$

where:

R1 is the resistance per unit length, considering the two conductors to be in series
L1 is the inductance per unit length, considering the two conductors to be in series
G2 is the conductance of the dielectric per unit length
C2 is the capacitance per unit length
j is the complex operator
ω is the angular frequency.

The transmission line characteristic impedance Z_c is the series impedance per unit length per phase and Y is the shunt admittance per unit length, where R1 and G2 is negligible $\cong 0$ (see equations 9.2 and 9.3):

$$Z = R1 + j\omega L1 \ \Omega \qquad (9.2)$$

$$Y = G2 + j\omega C2 \ \Omega \qquad (9.3)$$

Capacitive volt-amperes (Var) (equation 9.4):

$$\frac{V^2}{XC} = V^2\omega C \qquad (9.4)$$

Inductive volt-amperes (Var) (equation 9.5):

$$I^2 XL = I^2 \omega L \qquad (9.5)$$

Where the transmission line is correctly terminated, the series inductance consumes the energy charged by the capacitance of the line and therefore the reactances are equal (equation 9.6):

$$V^2\omega C = I^2 \omega L \qquad (9.6)$$

The surge voltage is determined by the strike current and the surge impedance of the conductor.

Hence, according to definition of a lossless transmission line, the surge impedance Z_s (Ω) is defined by equations 9.7 and 9.8:

$$Z_s = \frac{V}{I} = \frac{\sqrt{L}}{\sqrt{C}} \qquad (9.7)$$

$$Z_s = \sqrt{\left(\frac{L1}{C2}\right)} \ \Omega \qquad (9.8)$$

Section 9 – Lightning protection for railway structures

Lightning surge travel (just below the speed of light) is dictated by the surge impedances of the various conductors and conductive structures. When a lightning strike injects current into an overhead aerial earth conductor (AEC), it generates a voltage surge in the conductor. The purpose of a lightning protection system (LPS) is to direct this lightning surge to earth without causing damage to humans or components. When this surge reaches a down conductor, it requires the shortest possible route to a lightning earth-termination system, to ensure that the impulse resistance and inductance in the arrestor path are kept to a minimum.

The surge impedance of the copper strip (down conductor) is between 300 Ω and 480 Ω, depending on the height above the local track-bed, whereas the surge impedance of the steel rail is over 8,000 Ω. The self-inductance of steel is significantly greater than the inductance of copper strip and so a major voltage drop will occur along steel conductors. It is best practice, therefore, for the lightning path to be directly connected to the earth-termination through a copper or aluminium down conductor. The conductive path and termination should ideally not include any steel rail component which has a relatively high surge impedance.

9.3   Lightning damage, danger, disturbance and disruption

Where railways are exposed, a lightning strike is likely to cause physical damage to civil structures, and the lightning impulse is responsible for danger to humans, and damage and disturbance to electrical systems, such as those for signalling, and data and communications. Lightning is subsequently responsible for the disruption to the operational railway and the loss of revenue.

Structural damage: Lightning strikes can cause structural damage to unprotected structures, and this may include physical damage, fire or explosion due to the hot lightning plasma arc.

Damage to services and systems: Lightning impulse current is likely to damage or disturb interconnected electrical supplies and electrical systems. Lightning current may overheat conductors, damage insulation, create fire and trigger explosions due to overvoltage resulting from the conductive or inductive coupling.

Lightning can affect the electrical services coming into the railway and cause damage to the physical line or pipe used to provide the service.

Danger to humans: During an electric storm, humans close to the civil structures that are inadequately bonded are liable to receive an electric shock. National and international standards and codes of practice require that structures are adequately bonded and that voltage differences are not able to cause danger to humans due to touch and step potential as a result of inductive coupling.

Disturbance and disruption to the railway: Direct and indirect lightning strikes can cause disturbance to the signalling and control systems due to inadequate lightning protection of railway structures, and the lack of zoning within buildings and equipment rooms. The lightning disturbance to signalling and communication systems is liable also to disrupt train movements and the railway's operational services.

The UK rail infrastructure reported[16] that lightning strikes damaged rail infrastructure on an average of 192 times each year between 2010 and 2013, with each strike leading to 361 minutes of delays. In addition, 58 trains a year were cancelled due to damage by lightning.

---

[16]National Rail Enquiries – Train Delays http://www.nationalrail.co.uk/service_disruptions/81158.aspx

9.4   Lightning strikes to railway electrification structures

The 25 kV overhead lines are in an exposed position and create a latticework of conductors that act as a lightning interceptor or air-termination system. The overhead line masts naturally conduct the lightning current to the base of the mast and act as a down conductor.

The base of the mast is connected with steel fixing bolts to reinforced concrete foundations. This performs the function of the earth-termination. The lightning current then splits between the track-to-track cross-bonding and the steel reinforcement of the mast foundations. Additionally, a fast transient copper earth rod could be installed at the base of the mast. As the current enters the mast foundations, this conduction produces a resistive and inductive voltage drop creating a high EPR, which creates a touch potential danger at the base of the mast for humans and livestock and may also damage any bonded equipment.

Where the lightning surge can propagate into the rails, it can travel along the rail (typically 25 m) and potentially disrupt any system that is electrically connected to the rails.

On open route sections of high speed railways, additional buried earth conductors (non-insulated copper) are bonded to the traction return circuit at the track-to-track cross-bonds. The primary function of the buried earth conductor is to provide functional and protective earth for the non-traction low voltage (LV) supply and trackside systems. Any lightning surge current that passes down the Overhead line masts to the base of the mast is then additionally able to discharge to earth into the buried earth conductor. Within metres of where the lightning current enters the buried earth conductor, the lightning can dissipate into the surrounding soil.

The buried earth conductor also acts as a fast transient earth and has a very low leakage to earth at 50 Hz of 0.5 Ω.km (ground resistivity 50 Ω.m from BS 7430:2011+A1:2015 Figure 14). Where track-connected systems and LV systems are bonded near mast down conductors, they are liable to experience a disturbance from the lightning impulse.

9.4.1   Lightning strike to AC conductors (direct strike)

When the lightning strikes the AC overhead line conductor, a lightning surge propagates along the conductor and eventually causes the AC insulator to flashover as the lightning overvoltage exceeds the basic insulation level (BIL) of the insulator. The lightning breakdown of an insulator (see Figure 9.4) will subsequently cause a flashover across the insulator. The fault current is then required to be cleared by the 50/60 Hz electrification protection system. Where lightning is striking the overhead lines in a particular location, it is not uncommon for there to be multiple trips of the 50/60 Hz power and disruption to the operational service.

Figure 9.4 depicts the three possible breakdowns that occur with a 25 kV insulator:

1. arc horn – breakover protection;
2. insulator flashover (breakdown of the insulation); and
3. insulation breakdown (structural damage to the insulator).

Section 9 – Lightning protection for railway structures

Figure 9.4   Insulator flashover, breakdown and break over

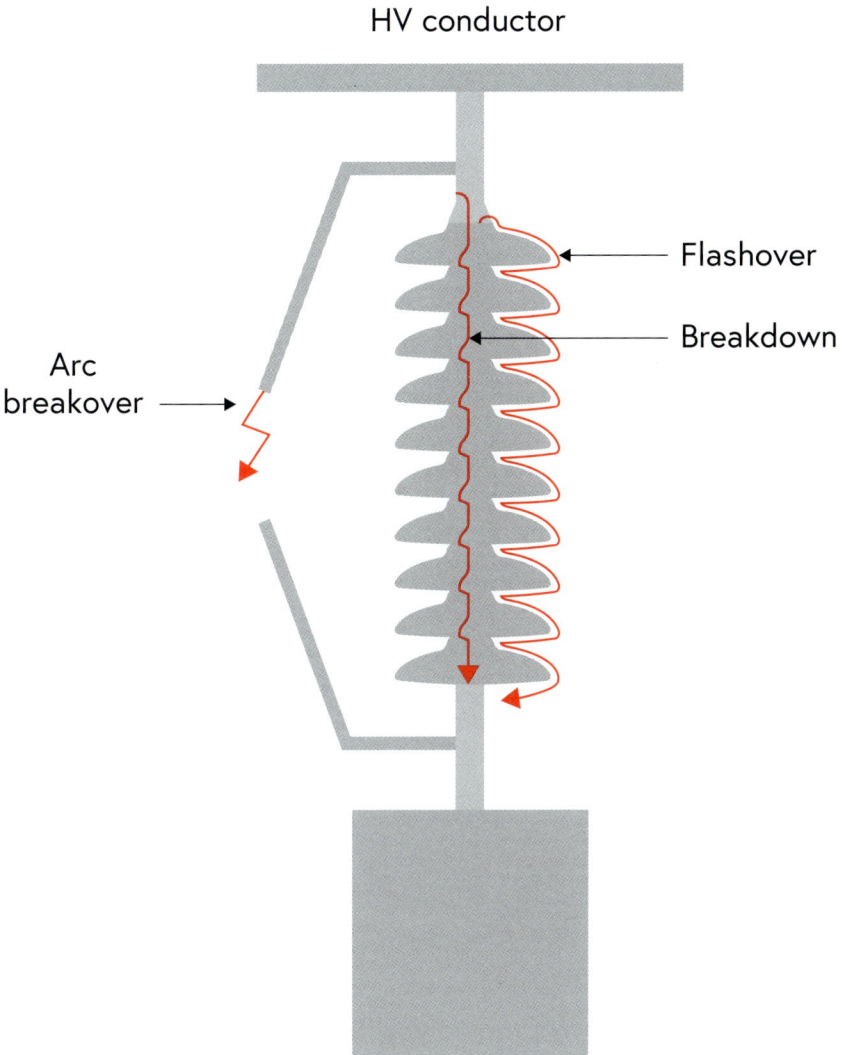

9.4.2   Lightning strike to overhead line structure (direct)

Air-termination: The lightning impulse that strikes the earthed overhead line structure creates a 1 MV lightning surge voltage, causing AC insulators to flashover or break down as the lightning overvoltage exceeds the BIL, and where the insulator breaks down there is a 50/60 Hz line to earth fault.

Down conductors – overhead line masts

The lightning current passes down the overhead line mast or the AEC bonding conductor (see Figure 9.5), which acts as the lightning down conductor. The current then distributes into the traction return circuit, splitting between the mast foundation, the interconnected earthed rails and the local earth.

To reduce the surge potential on the mast, it should be bonded to the reinforcement of the foundations or slab track. As the current conducts through the mast foundations and into the earth, it produces a resistive and inductive voltage drop, creating an EPR which, at the base of the mast, creates a touch potential danger for humans. The EPR is dependent on the magnitude of the current, the bonding of the reinforcements, the foundation resistance and the resistivity of the ground.

The effects of lightning strikes on railway overhead line portal infrastructure is shown diagrammatically in Figure 9.5.

Figure 9.5   Direct lightning strike to overhead line equipment

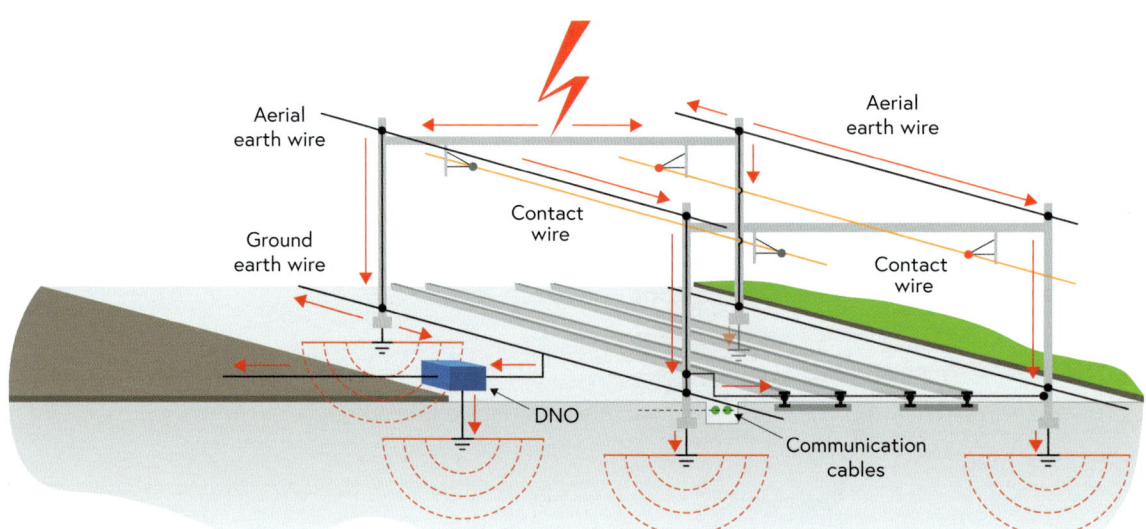

9.4.3   Lightning strike to adjacent high voltage tower line (indirect strike to railway)

A direct lightning strike to the live HV conductors creates a lightning surge voltage that causes the HV insulator to flashover or break down as the lightning overvoltage exceeds the BIL of the HV insulator. Other phases on the tower can then break down or back flashover, causing a 50/60 Hz phase-to-phase fault as well as a 50 Hz phase-to-earth fault in the power system. The 50/60 Hz fault current enters the ground and creates an EPR of up to 10 kV at the foundations of the tower. The EPR at the tower footing is dependent on the magnitude of both lightning (1 MHz) and 50/60 Hz current, the foundation resistance and the resistivity of the ground.

The consequence of lightning striking a tower line is that lightning current (1 MHz) distributes through the tower footings into the ground or any earth bonded system, creating an EPR. Where there is a local earth connected to the high voltage (HV) and LV systems, the lightning current can pass through the earthed path and may create a flashover to extraneous earth (an external earth to the railway) or cause a transient disturbance to the equipment earth port. Where extraneous earth enters the railway, they are required to have clearance to prevent touch voltages and flashover to the railway earth. The gapping of LV supplies to the railway infrastructure is addressed in detail in Section 7.

The effect of lightning strikes on a tower line is shown diagrammatically in Figure 9.6.

Section 9 – Lightning protection for railway structures

Figure 9.6   Lightning strike to an HV tower line close to the railway

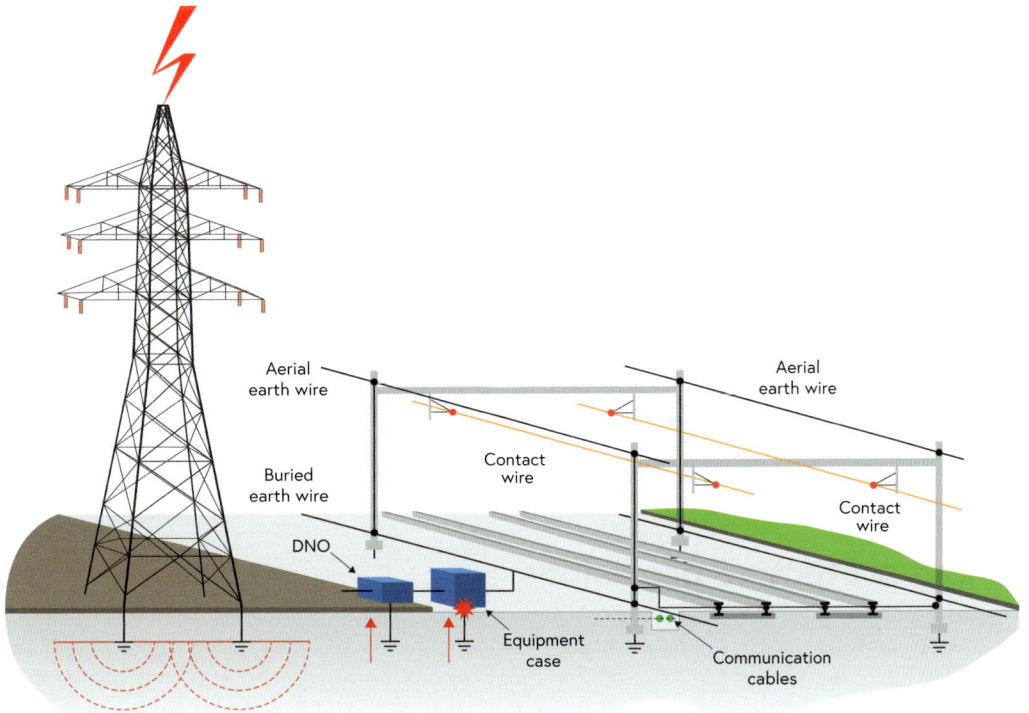

9.4.4   A lightning strike to the ground (indirect strike to railway)

Lightning strikes to the ground (see Figure 9.7) outside the area protected by the railway overhead lines are not uncommon as the railway passes through the open country as well as through urban and suburban areas. The lightning current enters the ground through non-railway structure foundations, trees or third-party earth mats, creating an EPR. The lightning current travels through the earth and buried utilities until it finally dissipates.

Where railways are designed with segregated earthing, lightning conduction through the ground creates voltage differences between the railway earth producing a transient voltage lasting for micro-seconds and possible disturbance to the control systems or damage to connected equipment.

Where the lightning current in the ground travels parallel to the railway, electromagnetic coupling (in MHz) can induce voltages that may cause flashovers in LV supplies, signalling and communications systems, or where lightning current can pass conductively through circuit protective conductors, screens or armours, it creates the possibility of disturbance to equipment earth ports.

Figure 9.7   Lightning striking the ground

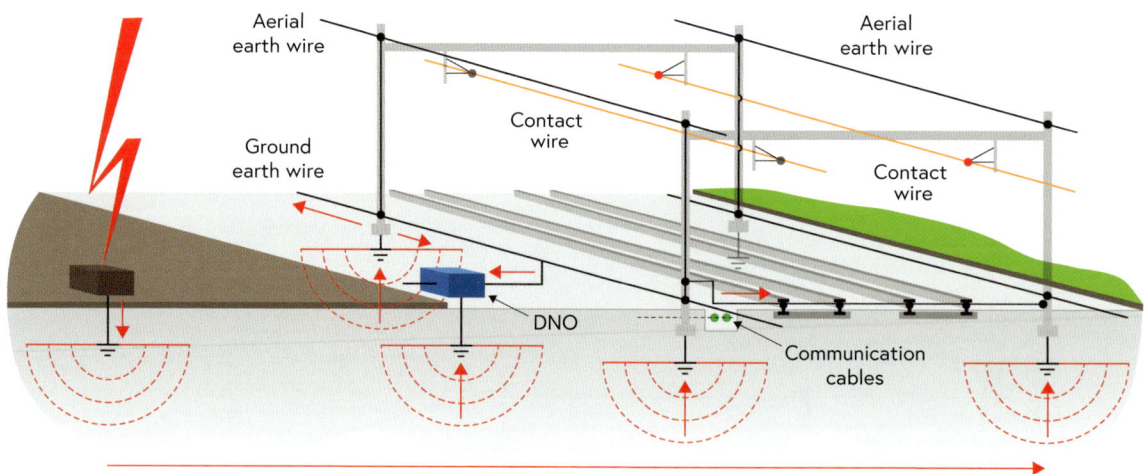

9.5   Lightning protective provisions

Sources of disturbance such as lightning are referred to generically as emitters, with humans and LV equipment being referred to as susceptors. A transfer function characterizes the coupling between a particular emitter and a particular susceptor. The main function of the LPS is to provide protective provision and reduce the transfer function between the lightning conduction path and the susceptors to a minimum.

To prevent danger, damage, disturbance and disruption (DDDD) to the railway network infrastructure, lightning protection is required on civil structures, 25 kV overhead lines, 25 kV traction units, electrification substations, passenger stations, and where equipment rooms contain LV systems.

On the AC traction networks, the overhead lines, earth wires, masts and rails act as a Faraday cage and thereby form an equivalent lightning protection zone (LPZ). Within this zone, equipotential bonding is required to protect humans from the danger of arcing, side-flash and disturbances to LV systems. The lightning current conducts down the masts and into the mast foundations, which act as an earth-termination.

With large multi-disciplinary projects, integrating the lightning protection requirements through the various engineering disciplines is a project interdisciplinary weakness. At the concept design stage, the architect may not specifically address the lightning requirements of building facades, and the interconnection to overhead lines and the railway traction return circuit.

At the design stages, the project systems integrator is required to integrate these requirements through interdisciplinary checks between civil electrification, signalling, station and LV designers.

9.5.1   Lightning protection system

High-rise buildings or protective structures do not always protect the rail corridor and, therefore, lightning strikes to railway buildings and the railway overhead electrification lines are considered to be a credible risk. If danger, damage, disturbance and disruption to service are to be averted, it requires design mitigations and controls.

The structures LPS comprises three principal elements: the air-termination, the down conductor and the earth-termination network. On civil structures and overhead electrified lines it is necessary to safely intercept a lightning flash with an air-termination, as this conducts the lightning current safely towards earth using a down conductor and then disperses the lightning current into the earth using an earth-termination network. The LPS is a passive means of preventing damage from the effects of the lightning strike and of discharging the electric charge produced by the cloud via a designated path into the ground.

The air-termination: This is the point of connection for a lightning strike. On a large flat or pitched station roof this can be a mesh conductors arrangement or vertical air rods on a prominent roof, including a railway overbridge or GSM-R mast (Global System for Mobile Communications-Railway mast). The overhead electrification system presents itself as a meshed network of steel structures, earth wires and AC overhead conductors.

The down conductors: These are the main element to conduct the lightning from the air-termination to the earth-termination network. The dominant characteristic of the down conductor is the inductive reactance (at 1 MHz) component. The down conductor route should therefore be kept short, and any curves should have a large radius. If these measures are not taken, then lightning current may arc over a resistive or reactive obstruction that it encounters in the conductor. This arc current may then damage the lightning conductor and can easily find another conductive path, such as the building wiring or plumbing, and cause fires or other damage.

Any voltage difference may be large enough to cause a dangerous side-flash (spark) between the two surfaces, creating damage to equipment or danger to humans. Equipotential bonding is the most effective way of eliminating any risk of a side-flash and ensure electrical continuity between structural conductive parts.

The earth-termination network: This is responsible for the dispersion of the lightning impulse current into the earth, without damage to the structure, through earth rods or structure (foundation) reinforcing bars. Where the ground has poor resistivity, or is very shallow, the earthing design can be augmented by additional earth rods, counterpoise (ground ring) conductors radiating away from the protected structure, or bonding to the building reinforcing bars, which can be used as a ground conductor.

9.5.2 Insulation coordination

The purpose of an insulation coordination study is to select the dielectric strength of equipment with the system voltages and transient overvoltages that can appear on the electrification system for which the equipment is intended. Many factors are taken into account, including the probability of potential surges, the environment, the insulation withstand characteristics and the arrester characteristics.

Figure 9.8 is derived from EN 50163 Figure A.1 and EN 50124-2 Annex A and displays the anticipated magnitude of overvoltages caused by a lightning impulse (μs), switching transients (ms) and temporary overvoltages (s).

The coordination study determines the dielectric withstand and the selection of equipment with the correct dielectric strength in relation to the voltages that can appear on the system, while taking into account the service, environment and characteristics of protection and control devices. Figure 9.8 is a key input to the specification of the surge arrestors voltage and the power supply voltage

Section 9 – Lightning protection for railway structures

Figure 9.8   Coordination of arrestor voltage and power system voltages

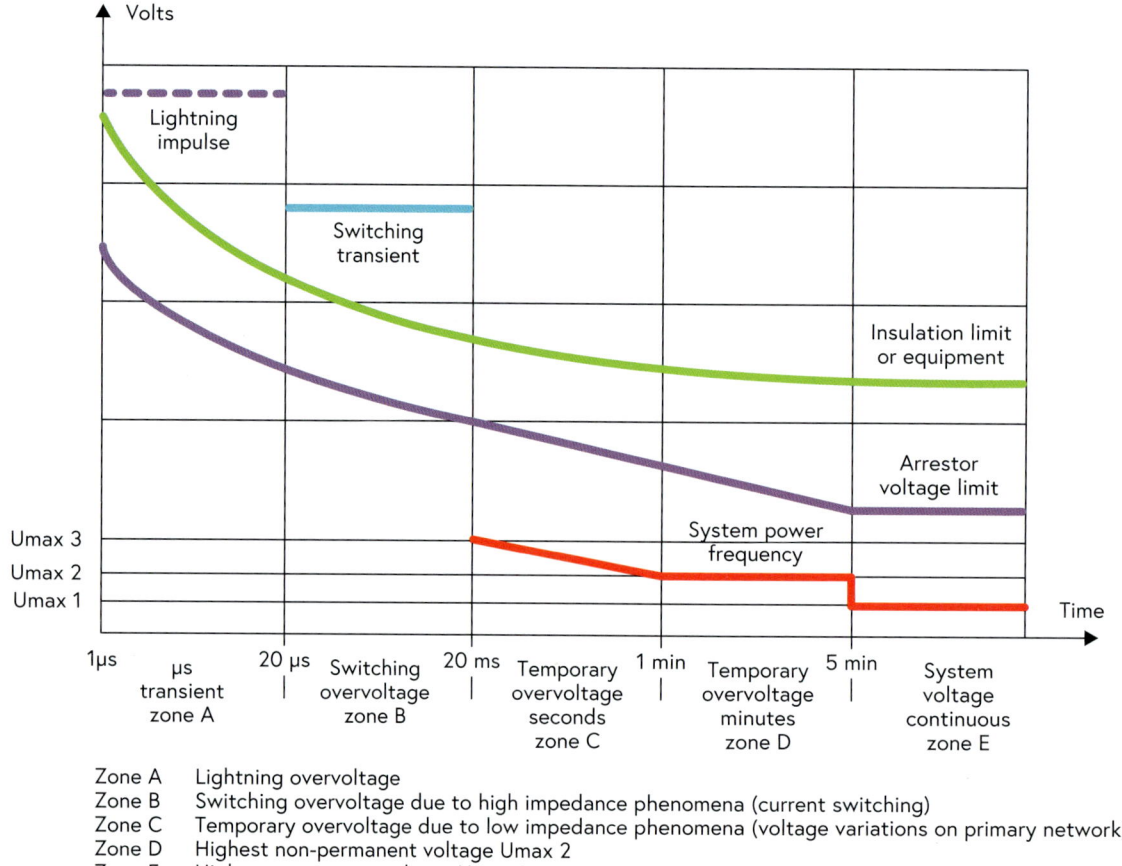

Zone A   Lightning overvoltage
Zone B   Switching overvoltage due to high impedance phenomena (current switching)
Zone C   Temporary overvoltage due to low impedance phenomena (voltage variations on primary network)
Zone D   Highest non-permanent voltage Umax 2
Zone E   Highest permanent voltage Umax 1

Insulation coordination studies are required where lightning is liable to strike AC bare conductors that pass close by to the undersides of bridges, station canopies or AC transformers, and to insulated cables. The study reduces the cost of disturbance caused by insulation failure to an economically and operationally acceptable level. Lightning design requires a detailed assessment of the overhead line 25 kV clearances from exposed-conductive-parts or partly conductive surfaces and the assessment is required to identify the level of insulation based on the exposure of the structure to humans and equipment, and the system reliability.

An insulation coordination study is required to assess the clearances of exposed live parts of the 25 kV overhead contact system from the earth and identify the correct dielectric withstand rating values for human and equipment protection. These values are coordinated with the values set out in Table A.3 of EN 50124-1:2017 (ref. Pollution degree PD4A & PD4B):

(a) Basic insulation: insulation required for human protection.

Rated impulse voltage (UNi) = 200 kV peak. Corresponding to an air clearance of 370 mm assuming the worst-case dielectric condition of the electrodes.

(b) Functional insulation: insulation required for equipment functionality.

Rated impulse voltage (UNi) $\geq$ 145 kV peak. Corresponding to an air clearance 270 mm assuming the worst-case dielectric condition of the electrodes.

(c) Reinforced insulation: insulation equivalent to double insulation.

Rated impulse voltage (UNi) $\geq$ 325 kV peak. Corresponding to an air clearance of 600 mm.

9.5.3   Rated impulse voltage

The rated impulse values are selected from EN 50124-1:2017, Table A.2, and are based on overvoltage category OV3 and OV4 and a rated insulation voltage (UNm) of 27.5 kV, which is equal to the highest permanent voltage present on the system (Umax1).

The necessary impulse earthing resistance for AC infrastructures based on a lightning current 60 kA with respect to the overvoltage category for AC railways is given in Table A.2 EN 50124-1.

9.5.4   The probability of lightning strike current

Cumulative frequency analysis is the analysis of the frequency of occurrence of values of a phenomenon less than a reference value. This phenomenon is time- and space-dependent and detailed in EN 62305-1 and IEEE 998.

9.5.4.1   The cumulative frequency distribution of lightning (EN 62305-1)

The lightning current parameters in EN 62305 are based on the results of the International Council on Large Electric Systems (CIGRE) data given in EN 62305, Table A.1.

9.5.4.2   IEEE 998 IEEE Guide for Direct Lightning Stroke Shielding of Substations

Since the strike current and striking distance are related, it is of interest to know the distribution of strike current magnitudes as explored in IEEE 998 – *IEEE Guide for Direct Lightning Stroke Shielding of Substations*. The median value of strikes to overhead earth conductors, structures and masts is usually taken to be 31 kA (see J.G. Anderson's *Transmission Line Reference Book,* 1987). This gives the probability that a certain peak current will be exceeded in any strike current, as specified in equation 9.9, giving the results in Table 9.1 below (reference IEEE 998 Section 3.4 'Determination of the probability function of lightning peak currents on flat ground').

$$P(I) = \frac{I}{\left(1 + \left(I/31\right)^{2.6}\right)} \tag{9.9}$$

Table 9.1   Peak current of a lightning strike

Peak current kA	10 kA	20 kA	30 kA	40 kA	50 kA	60 kA	70 kA	80 kA	90 kA	100 kA	200 kA
Percentage exceeding tabulated value (100 %)	95	75.6	52.1	34	22.4	15.2	11.7	7.8	5.9	4.5	0.78

where:

> $P(I)$ is the probability that the peak current in any strike will exceed I(A)
> I is the specified peak current of the strike (kA)

9.5.5 Overvoltage protection devices

In areas of high lightning density, protection is needed to limit the overvoltage during a lightning strike to the 25 kV overhead lines. The following are typical lightning arrestor devices used to control the effects of lightning surges on 25 kV conductors and in local earthed conductors:

(a) Arcing horns – protection from 25 kV transients: the space around the 25 kV insulator is highly charged during an overvoltage and the arcing horns provide a path for the discharge to bypass the surface of the insulator, preventing a flashover or breakdown of the insulator. This arc is eventually forced up the horn by thermal effects and is then eventually blown out, therefore ensuring the arcing does not continue when the voltage returns to normal (see Figure 9.9).

Figure 9.9   Dielectric breakdown of an arcing horn

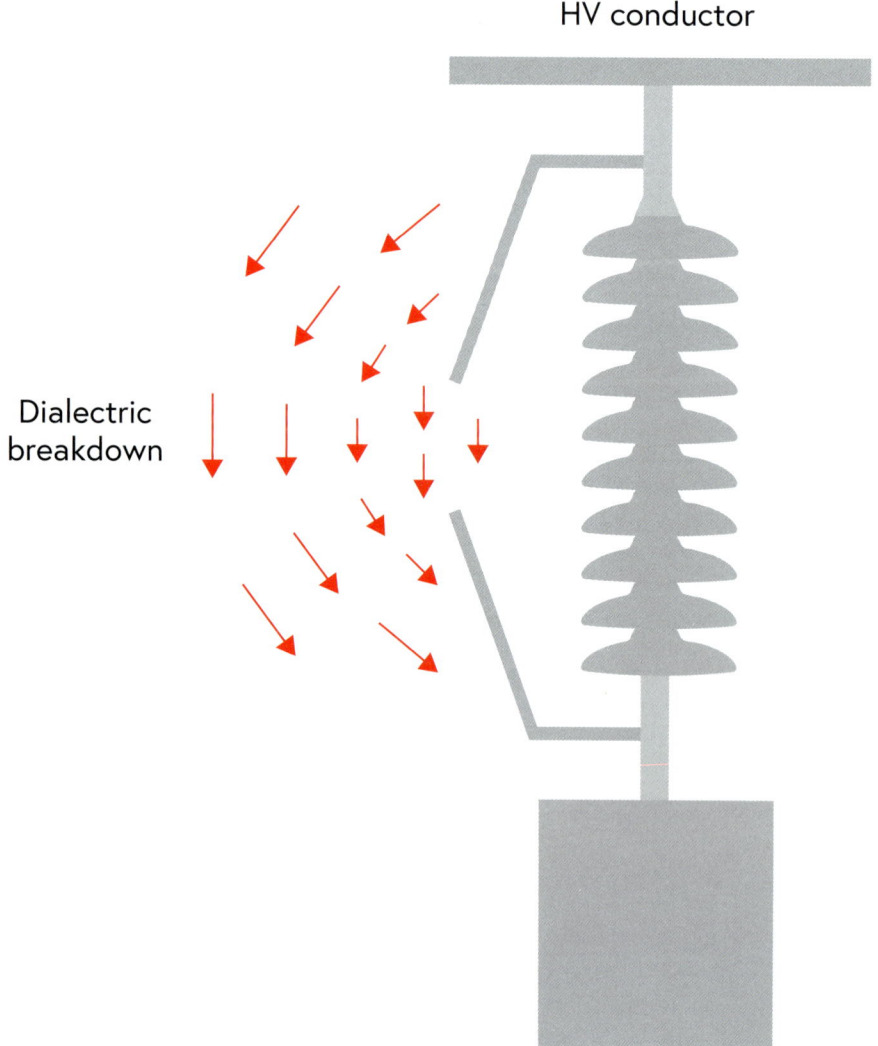

(b) Metal oxide varistor arrestors – protection from 25 kV transients: a surge arrestor is a protective device usually connected between any live conductor of an electrical system and a local earth. The arrestor limits surge voltages by diverting surge currents to earth when a given voltage is exceeded.

Surge arresters are an indispensable aid to the application of insulation coordination in 25 kV electrical overhead line systems. Valuable assets such as transformers, circuit-breakers, traction vehicles and switchgear should ideally be protected against lightning and switching overvoltage. The main task of the arrester is to protect HV equipment from the effects of overvoltage (see Figure 9.10).

Figure 9.10   HV lightning EGLA MO arrestor and spark gap

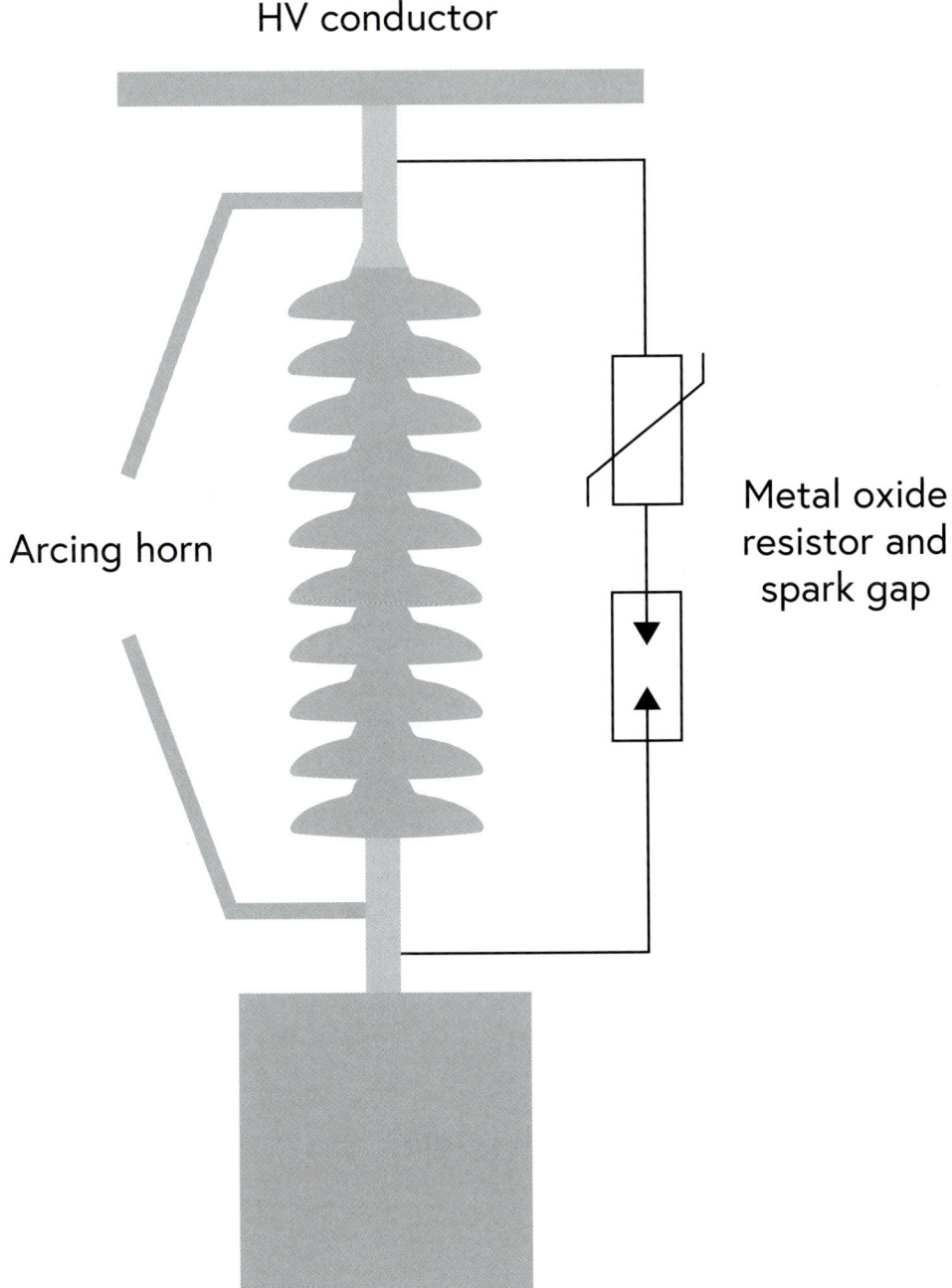

These different surge arrester technologies were and are, respectively, covered by the following standards of the IEC 60099 series:

(a) metal oxide surge arresters without gaps, all housing technologies: IEC 60099-4; IEC 60099-5;
(b) metal oxide distribution surge arresters of 52 kV rated voltage or less, with internal series gaps, porcelain or polymer-housed: IEC 60099-6;
(c) externally gapped line arresters (EGLA) for application in distribution or transmission overhead lines, porcelain or polymer-housed: IEC 60099-8.

During normal operation, the arrester should not affect the operation of the power system, and the arrester is required to withstand typical power surges without any damage (see Figure 9.8). Metal oxide varistors (MOVs) provide a high switching surge energy rating and a very low protection level, and they can absorb a high amount of energy while avoiding thermal runaways. The MOVs are characterized by their high single-impulse withstand rating. With this kind of non-linear resistor, there is only a small flow of current when the continuous operating voltage is being applied. When there are surges, however, excess energy can quickly be removed from the power system by a high discharge current.

Metal oxide non-linear resistors (MORs) have the following properties and characteristics:

(a) high resistance during normal operation of the power system;
(b) low resistance during surges to limit overvoltage; and
(c) sufficient energy absorption capability for stable operation.

For the requirement for lightning strike counters, IEC 62561-6 should be referenced.

The EGLA has two basic components: the MOR and the spark gap. The MOR is a non-linear metal oxide resistor and has roughly the same characteristics as a standard MOV arrester. The gap is generally a single air gap in series with the arrester.

All bonded structures and arresters must be connected directly through a minimum surge impedance and inductance path to a lightning earth pit. The most suitable material for this route would be with a cross-sectional area (csa) of less than 75 mm^2.

(c) **Spark gaps – for protection of segregated earths:** spark gaps are usually installed on earthed systems where there is a need to maintain segregation between two separate earths. This intentional segregation occurs at overline bridges, or segregated pipe joints and electrical services. Overvoltages occur as a result of a lightning strike and switching transients and can damage systems that are intentionally segregated, in some cases beyond repair. The spark gap is an overvoltage device and protects structures and equipment against damage caused by surge voltages. A spark gap consists of two electrodes separated by a gap that is gas-filled. When a potential difference between these electrodes exceeds the breakdown voltage of the gas, a spark forms, ionizing the gas and reducing its electrical resistance. An electric current then flows through the ionized gas until the current reduces below a minimum value called the 'holding current' (IEC 62561-3:2017 *Lightning protection system components – Part 3: Requirements for isolating spark gaps*).

9.6 Lightning protection for HV equipment

Insulated cables are used in the autotransformer AC distribution network for the 25 kV circuits of the autotransformer system. Insulated cables are used between the feeder station and the overhead lines where the connection is remote from the feeder station. Insulated cables are also used in the negative feeder circuit (-25 kV) where protection by clearance of a bare conductor to earthed structures cannot be achieved. This commonly occurs at overbridges and against station canopies.

Without surge protection, lightning strikes could damage the cable insulation and connected HV equipment (transformers). In areas of high lightning density, it is common to install surge arrestors at the cable sealing end to prevent cable insulation damage. The operating voltage level of surge protection devices must be higher than the system operating voltage and must be lower than the minimum voltage withstand level of the equipment. This minimum voltage rating is defined as the BIL of cable insulation (see Figure 9.8 Zones C, D and E).

Figure 9.11   Surge diverter (foreground) (image reproduced with permission by R. Catlow)

9.7   Lightning protection for viaducts

On a viaduct, and due to the limitation of space, it is best practice to provide a common-bonded earthing system as this protects humans from indirect contact with exposed-conductive-parts of the running rails, the overhead line masts and other exposed-conductive-parts, including the viaduct structures, LV systems, traction units and equipment cabinets.

On long viaducts (typically where the length exceeds 500 m), the viaduct requires lightning earth-termination at regular intervals (typically every 250 m). This also provides the 50/60 Hz earth reference for the common-bonded traction return system, including the LV supplies on the viaduct. Without this earth reference, the voltage of the traction return circuit, on the viaduct would float, creating the possibility of touch voltages between exposed conductive parts.

Where there is lightning earth-termination (4 Ω) on both sides of the railway and it is located at 250 m intervals, the equivalent 50/60 Hz resistance over a 1 km section is 0.5 Ω (calculation based on eight 4 Ω

Section 9 – Lightning protection for railway structures

earth rods – 4/8 Ω). This provides an adequate earth reference for LV supplies, signalling and control systems that are located on the viaduct.

9.7.1 Viaduct lightning protection systems

The viaduct substructures generally comprise reinforced concrete piers, reinforced concrete abutments and flared pier heads to support the deck. The piers and abutments are both founded on large diameter bored piles.

The viaduct main superstructures are constructed typically from deck segment cross-sections that are U-shaped. These segments are constructed as a post-tensioned segmental construction. Where a bridge is integral with the viaduct traversing a road or railway, this often includes a steel superstructure that is 40 m off the ground and may require its own specific lightning and earth arrangement.

Overhead line masts are mounted on the viaduct sections and straddle the whole viaduct.

Air-termination (overhead line): The AC overhead lines of the railway, provide a meshed network of steel structures and conductors, and this provides a protected zone for the equipment that is located under the overhead conductors and between the masts. When lightning strikes this protected zone, the current passes down the overhead line mast (or the bonding conductor 'A' in Figure 9.12), at typically 300 m intervals, to the traction return circuit and the viaduct auxiliary earth conductor. Equipotential bonding is necessary to prevent raising the potential of the viaduct with respect to adjacent metallic surfaces, creating the risk of a flashover and danger to humans and to other systems on the viaduct.

Down conductors (pier): The lightning down conductor path continues from the viaduct deck, down the pier ('B' in Figure 9.12) to the earth-termination ('C' in Figure 9.12). To achieve the most direct lightning path, the pier down conductors should align with the overhead line down conductors and the track-to-track cross-bonding.

The steel reinforcements in the viaduct pier can be designed to act as lightning down conductors; this does require choosing a number of the pier reinforcements and ensuring that they are bonded during construction. These shall be provided in accordance with the following standards: BS 7430:2011+A1:2015 and EN 50522:2010 Section 7.1 and Annex K; EN 50122-1 Section 6.3.1.1; and EN 62305-3 Annex E.4.3.7. Civil engineers often struggle to combine the requirements of a lightning conduction path with structural reinforcement; therefore, it may be necessary to install an external lightning copper down conductor and an independent lightning earth-termination as in EN 62305-3 Annex E.

Where there is a problem with metal theft, the designer should consider installing the copper down conductors within the viaduct pier. Access to the copper down conductor terminals and earth-termination is required for periodic inspection and testing.

Where pier steel reinforcements act as down conductors, they are typically located at 30 to 50 m intervals. A risk-based approach is needed on the frequency of lightning strikes, and a decision is then required for the number and position of down conductors and earth-terminations.

Earth-termination (base of the pier): Where viaduct pier steel reinforcements are used as down conductors, they need to be bonded to either the pile reinforcements or a separate earth rod that acts as the earth-termination (EN 62305-3: 2011 Annex E.5.4.3.2).

The effect of lightning strikes on the overhead lines on a railway viaduct is shown diagrammatically in Figure 9.12.

Section 9 – Lightning protection for railway structures

Figure 9.12 Viaduct lightning protection system

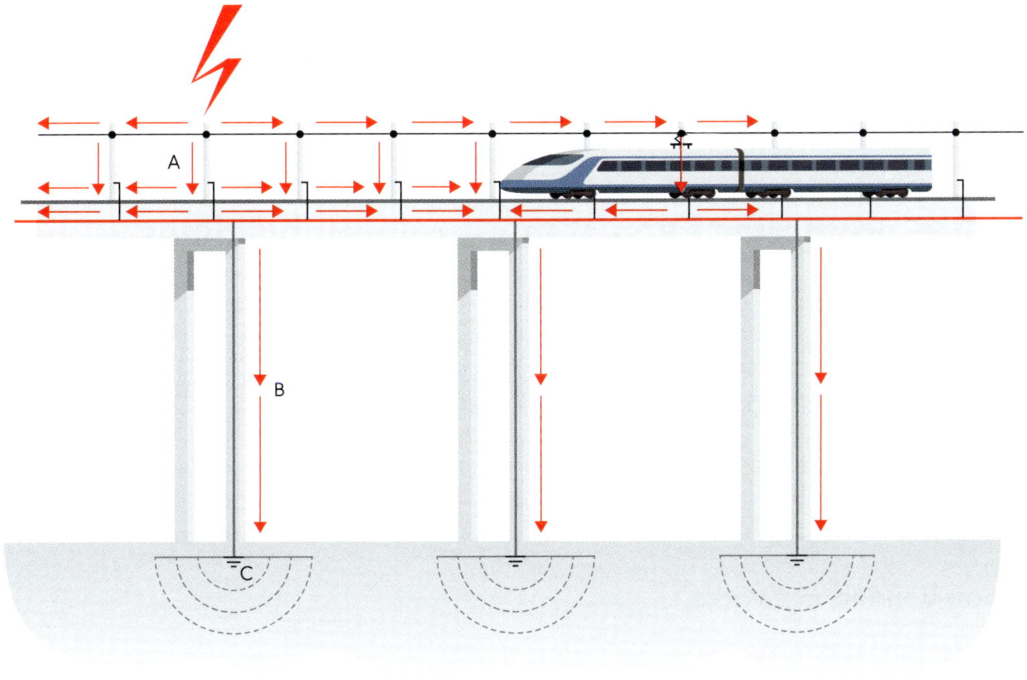

9.7.2 Lightning protection for LV systems

On viaducts, and because of close proximity, the LV systems are normally common-bonded to the traction return circuit via an auxiliary earth conductor that is also connected to the rails and AEC and down conductors. To prevent unnecessary disturbances from lightning impulses, there is a requirement to electrically segregate the LV earth from the lightning down conductors. It is preferable to provide additional protection for the LV systems, and this can be achieved through segregation with one of the following electrical arrangements:

(a) bonding LV systems to the steel running rails, allowing separation of at least 25 m from any connection to the lightning down conductor. The steel rails have a high impedance at 1 MHz and are therefore able to provide the necessary electrical segregation from the lightning down the path; or

(b) providing separate bonding locations for the LPS and the auxiliary earth conductor to the traction return system.

9.8 Lightning protection LV equipment – open route

LV non-traction supplies are required for electrical systems located in the open route, on bridges, on viaducts and in stations. The LV systems provide many operational functions of the railway, including signalling, communications, lighting, points heating, level crossings, GSM-R masts and CCTV. In exposed locations without the protection of these systems, there is an additional risk from lightning strikes.

Signalling and control systems are susceptible to conducted voltage surge, induced voltage surge and radiated frequencies due to lightning striking the railway or the ground. The overhead lines of the AC electrified railway provide protection from a direct strike and offer a protection zone to equipment that is located between the overhead line masts. To prevent indirect lightning impulse current from disturbing the LV systems inside this protection zone, it is best practice to provide a common-bonded network, including the traction return system.

Section 9 – Lightning protection for railway structures

It should be noted that, historically, signalling designers have preferred segregated earths; however, this has led to potential difference between exposed-conductive-parts and transient disturbance caused by the rise of earth potentials on LV earths, particularly during lightning strikes and following line to earth faults.

The protected zone extends to approximately 3 m outside the overhead line masts. This is based on the principle of rolling spheres (EN 62305:2011 Part 3 Section 5.2.2).

9.9 Lightning protection LV systems – railway buildings

Lightning is likely to strike the roof of stations and railway buildings, and will conduct through the structural earth and conductive parts of the building. Electrical systems are likely to be disturbed where there is no bonding network or structure LPS. Any exposed structure or building should have the necessary lightning protection systems as described in Section 9.5.1 and EN 62305:2011 Parts 3 and 4.

Two specific design requirements are required to be addressed in protecting LV equipment located in railway buildings: a common-bonded network (CBN), and zoning and surge protection.

9.9.1 Common-bonded networks[17]

The primary purpose of a CBN is to help shield people and equipment from the adverse effects of conducted and electromagnetic energy. The CBN will reduce the magnitude of the transfer function between the LPS of the building structure and the susceptors (LV systems and humans) to an acceptable level.

Principle of a mesh common-bonding network (CBN)[18]: the CBN is the principal means of effecting bonding and earthing inside a telecommunications building. The CBN bonds metallic components that are intentionally or incidentally interconnected to form the principal bonding network in a building. These components include structural steel or reinforcing rods, metallic plumbing, AC power conduit, protective conductors, cable racks and bonding conductors. The CBN design has a mesh topology and is connected to the earthing network (see Figure 9.13).

---

[17]ITU-T K.27 Section 6 and Annex B Figure 2
[18]ITU-T K.27 Section 6.2.1 and Annex 2 Figure B.2

Section 9 – Lightning protection for railway structures

Figure 9.13   Principle of a common-bonded network (CBN)

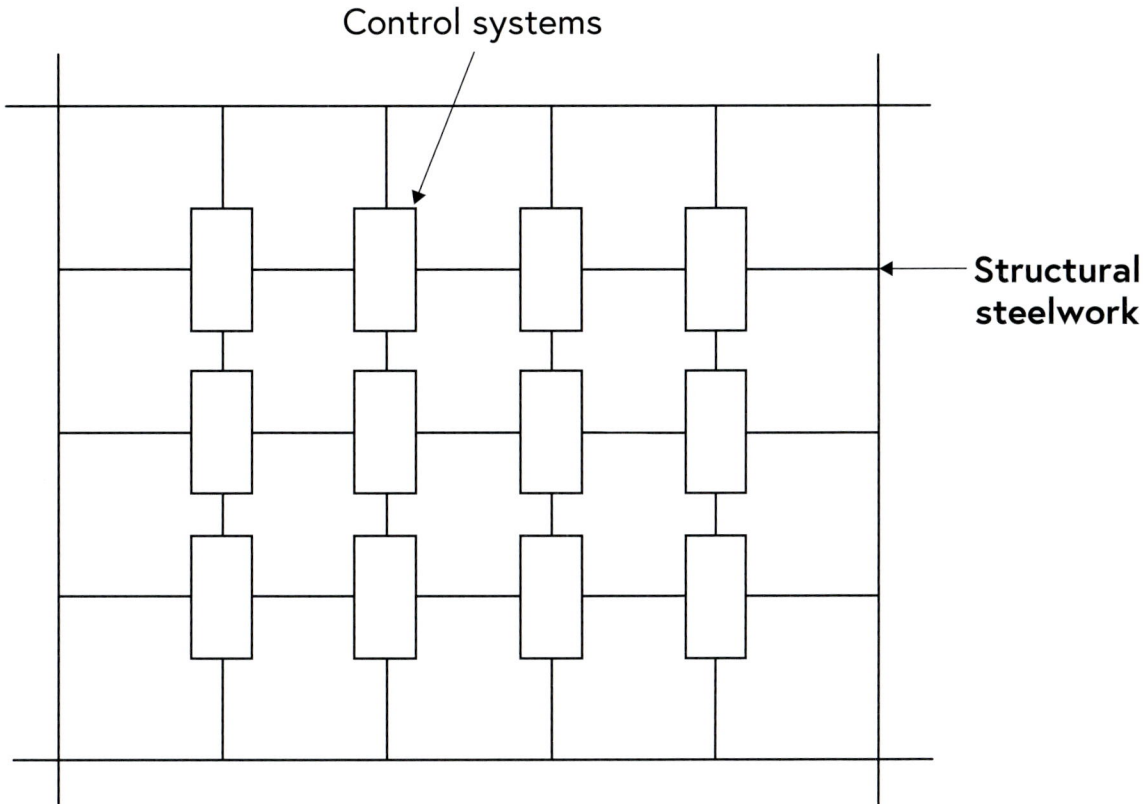

Principle of a mesh-isolated bonding network (IBN)[19]

Single-point connected structures are referred to as isolated bonding networks (IBNs) (see Figure 9.14). The main feature of an IBN is that it is isolated from the surrounding CBN, except for a single-point connection where conductors enter the system block via the transition region between the IBN and CBN.

---

[19]ITU-T K.27 Section 6.2.3 and Annex B Figure B.3

Figure 9.14 Principle of a mesh-isolated bonding network (IBN)

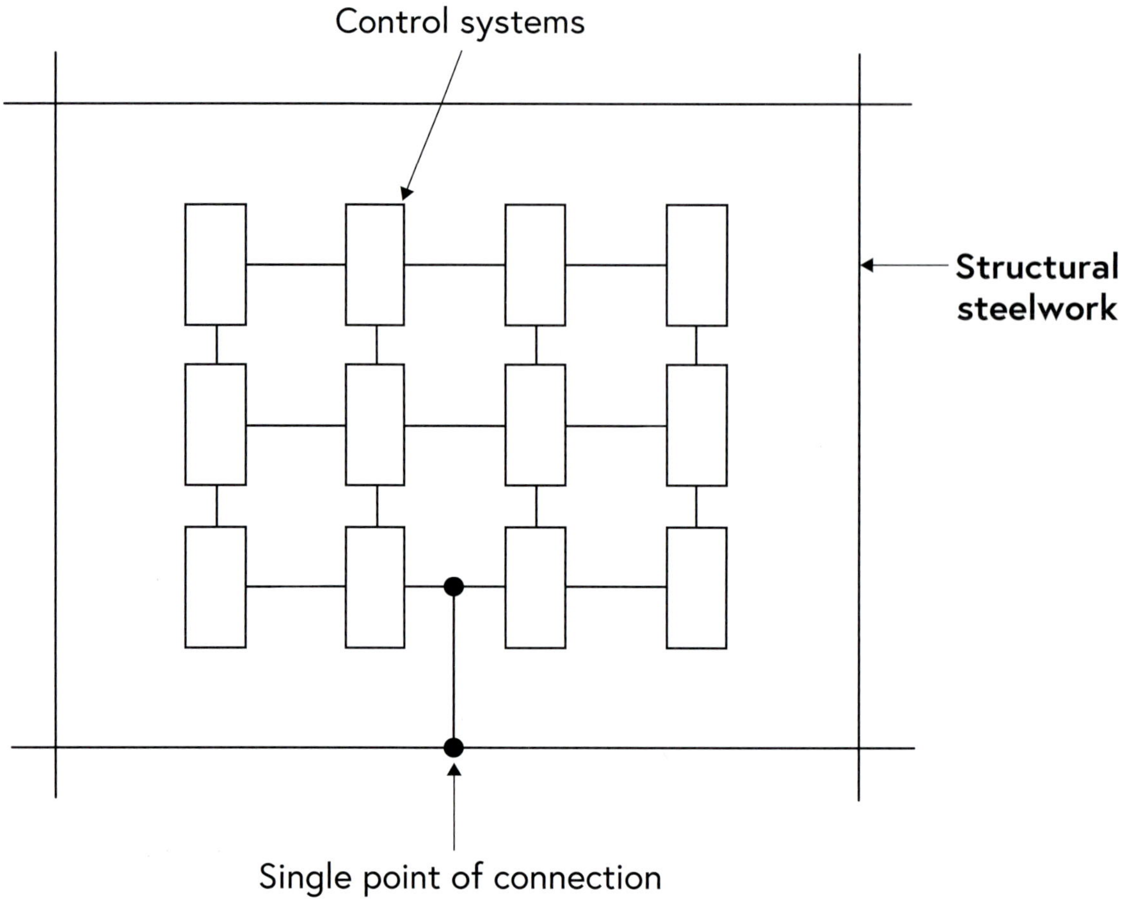

9.9.2 Zoning and surge protection measures[20]

Electrical and electronic systems are subject to damage from a lightning electromagnetic impulse (LEMP). Therefore, zoning and surge protection measures (SPMs) need to be provided to avoid failure and damage of the internal electronic and control systems.

Protection against LEMP is based on the lightning protection zone (LPZ) concept: the zone containing the systems to be protected should be divided into LPZs. These zones are theoretically assigned part of a space (or of an internal system) where the LEMP severity is compatible with the withstand level of the internal systems. Successive zones are characterized by significant changes in the impact of the lightning impulse. The boundary of an LPZ is defined by the environment and the protection measures employed.

Figure 9.15 depicts diagrammatically the zones as detailed in EN 62305 Part 4 Figure A.1, where 'H' references the radiated and magnetic field, 'I' the conducted current and 'U' the conducted voltage and equipment that is well protected against conducted surges ($U2 << U0$ and $I2 << I0$) and against radiated magnetic fields ($H2 << H0$).

---

[20]EN 62305- 4:2011 Figure A.1 – LEMP situation due to lightning strike; AS 1768 Section 5.6.

Section 9 – Lightning protection for railway structures

Figure 9.15 Lightning protection zones (LPZs)

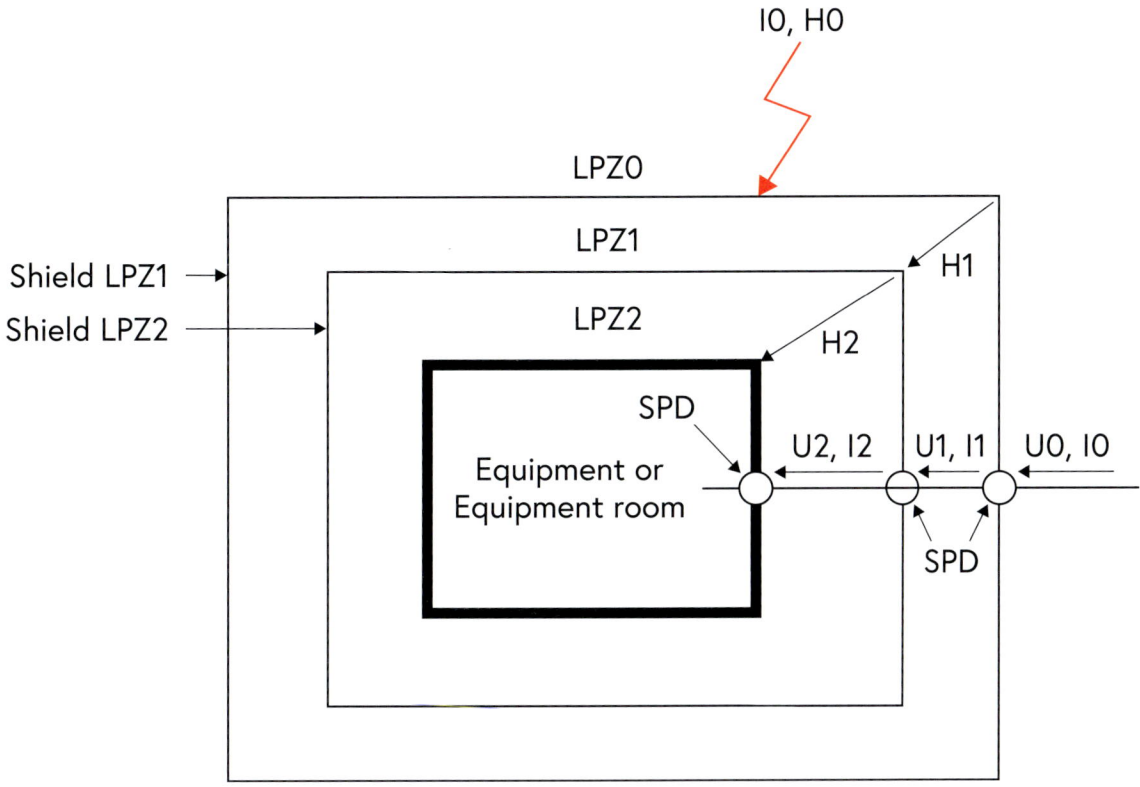

9.9.3   Principles of station bonding

Modern railway stations are made largely from a steel structure and non-conductive cladding. The overhead line masts are usually supported on the steel structure of the building, or the cantilevers can be mounted within the platform canopies.

The incoming electrical supplies are normally bonded to the station main earth terminal (MET), which is then bonded to the traction return circuit at the rails. This bonding arrangement provides two main functions: a protective conductor for a flashover to the station canopy, and equipotential bonding to ensure that passengers do not experience any voltage differences between station structures, canopies and the train body side.

Where a station is bonded to the public electricity network and also to the railway AC electrification return system, lightning has several pathways to enter a station building and cause danger, damage or disturbance, including:

(a) a direct strike to the station civil structure;
(b) direct or indirect strike to electrical services (for example, direct strike to pole-mounted supplies);
(c) indirect lightning strike to adjacent railway overhead line – with entry via the rail to the MET; and
(d) lightning-induced voltage into control cables – cables could be railway or public.

Section 9 – Lightning protection for railway structures

Each station needs a lightning protection plan to mitigate lightning impulse (direct and indirect) and prevent danger to humans and damage to station equipment rooms and systems. The following lightning protection design requirements are required in passenger stations:

(a) LPS for the structure of the building, including air-termination, down conductor and earth-termination;
(b) structure CBN;
(c) lightning protection zones;
(d) surge current protection on the incoming earth (railway, electrical services and utilities);
(e) overvoltage protection between LPZs; and
(f) surge protection device (SPD) for signal data cables.

Figure 9.16   Lightning protection for a common-bonded station

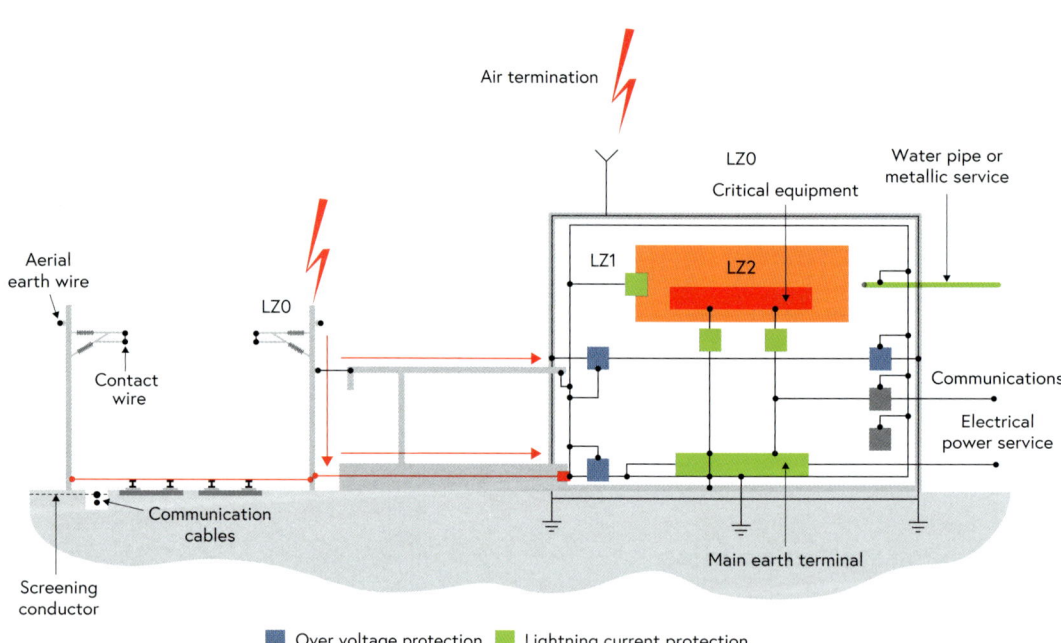

9.10   Lightning protection for traction units

Traction power is supplied through a pantograph and a roof-mounted circuit-breaker. The AC feed passes through a roof bushing and is cabled to the primary winding of the traction transformer. The main circuit-breaker and roof bushing are exposed to the lightning overvoltage of typically 1 MV for 1 μs. When lightning strikes the overhead line, this lightning surge will be transmitted along the catenary and into the pantograph of the traction unit.

In areas of high lightning intensity, the bushing, cable and traction transformer are required to be protected. Arcing horns (overvoltage protection) have historically been provided on the HV side of the locomotives and multiple units, but the usefulness of this device is now doubtful as only a few instances of the horns showing signs of flashover have been recorded. More recently, it is common for the protection to be provided by surge arrestors.

The insulation coordination requirement for vehicles now includes a clause stating that the complete vehicle requirements as detailed in EN50124-1:2017 Section 8.3 Table A.2 – OV3 are:

(a) 17.25 kV: OV3 UNi 75 kV peak; OV4 UNi 95 V peak; and
(b) 27.5 kV: OV3 UNi 125 kV peak; OV4 UNi 170 V peak.

Between 2010 and 2013, the UK rail infrastructure reported that 58 trains a year were cancelled due to damage by lightning[21]. The surge protection for a traction unit is shown in Figure 9.17.

Figure 9.17　Train 25 kV surge arrestor

9.11　Lightning hazard safety assessment

As part of the engineering management plan and safety management system, it is necessary to undertake a hazard identification (HAZID) for lightning protection. The HAZID should identify the possibility of danger to humans, damage to infrastructure and disturbance to electrical control systems. The HAZID should consider the consequence of direct and indirect lightning on the following assets (list not exclusive):

(a) civil conductive structure (for example, viaduct, bridges, canopies and stations);
(b) LPS (for example, lightning down conductor and earth-termination);
(c) overhead line masts and AEC (air-termination);
(d) permanent way (running rails);
(e) LV non-traction supplies 400/230 V supply;
(f) signalling and communication systems;
(g) signal heads and gantries;
(h) station local LV/HV supplies and equipment rooms;
(i) depot LV/HV supplies and equipment rooms; and
(j) GSM-R; lighting and CCTV masts.

---

[21]National Rail Enquiries – Train Delays - http://www.nationalrail.co.uk/service_disruptions/81158.aspx

Section 9 – Lightning protection for railway structures

Where the electrified railway is an integrated earth, the earth of each asset is common-bonded to the traction return system. As part of a design process, it is necessary to undertake an interdisciplinary review of the interconnections of earths and the consequence of lightning surges in those paths.

Equipment close to the overhead line masts or bonded to the traction return system should be identified as they can be affected by direct lightning strikes to the overhead lines and the interconnection with the aerial earth bonding.

The most likely disturbances to signalling equipment from lightning are detailed in EN 50121-4:

(a) surge immunity input/output/power/earth ports (1.2/50 µs);
(b) fast transients immunity input/output/power/earth ports (5/50 nS); and
(c) radio-frequency electromagnetic field enclosure ports (800 MHz – 6 GHz).

It should be noted that induced voltages by 50/60 Hz traction currents are not addressed in EN50121-4, and any requirement should be addressed by the project functional requirements.

9.11.1   Factors to be considered for lightning risk assessment

A lightning risk assessment, as described in EN 62305, applies to structures that may or may not contain electrical and electronic systems, including stations, control centres and depots. The standard does not address the requirements for the protection of railway bridges, viaducts and overhead electrified systems, which are an integral part of the electrified railway.

A lightning flash may cause damage depending on the construction of the structure to be protected. Some of the most important characteristics to be assessed during a risk assessment are the type of construction (conductive or non-conductive), contents and application, the type of service and the protection measures to be provided.

Railways should undertake a lightning hazard safety analysis (human safety) and risk process (electromagnetic compatibility). The following text has been adapted from EN 62305 as a guide to undertaking this analysis for exposed railway overhead lines and railways systems.

Damage: For practical applications, it is useful to distinguish between three basic types of damage that may appear as the consequence of lightning strikes:

1. D1: injury to humans by electric shock;
2. D2: physical damage to civil structures; the damage may be limited to a part of the structure or can include the entire structure; and
3. D3: operational failure of electrical and electronic systems.

The loss may be alone or in combination with others, and produce a different consequential loss in the object to be protected. The type of loss that appears does depend on the characteristics of the structure.

Loss: This is a legal term that describes the value placed on injury or damages due to an accident caused by another's negligence, a breach of contract or other wrongdoing. The amount of monetary damage relates to the effects of lightning and what has happened. The following types of loss are considered as necessary:

1. L1: loss of human life;
2. L2: loss of railway service to the public (delay in opening the railway, the downtime required for maintenance);

Section 9 – Lightning protection for railway structures

3. L3: loss of railway infrastructure (requirement to redesign);
4. L4: loss of economic value; and
5. L5: loss of reputation.

Risk: A hazard (for example, electricity, lightning) can cause harm , while the risk is the chance (high or low) that any hazard will cause somebody harm. As part of the safety management system's hazard and operability analysis (HAZOP), a risk assessment may be needed to be undertaken for the following:

1. R1: risk of loss of human life (including permanent injury);
2. R2: risk of loss of railway operational service to the public;
3. R3: risk of loss of railway infrastructure; and
4. R4: risk of loss of economic value.

The likely consequences of lightning strike to the railway are:

1. Danger to public, operation and maintenance staff;
2. Damage to 25 kV overhead line equipment (for example, insulator flashover);
3. Damage to earth and LV equipment;
4. Disturbance to axle counters and signalling track circuits;
5. Disturbance to track-connected equipment (for example, LV supplies);
6. Disruption to 25 kV operational service; and
7. Reputational loss.

▬ Section 10

Earthing and bonding at AC/DC interfaces

10.1   AC/DC interfaces

Electrification systems operate with AC or DC distribution systems and use either overhead catenary (AC or DC electrification) or a conductor rail (DC electrification). Railway companies are looking to integrate the operation of their railway systems, making it easier to provide interchanges for passengers. The close running of these railway alignments has increased the possibility of electrical and physical disturbances to both AC and DC railways.

The electric traction return circuit for both the AC and DC railways is provided by the running rails. AC electrification running rails are earthed, whereas DC electrification running rails are nominally floating with respect to the mass of the Earth. When these systems operate independently, each system is required to comply with the relevant earthing and bonding requirements as detailed in standards such as IEC 62128-1 and the railway's code of practice. However, when they coexist, and are interconnected or adjacent to each other, the technical challenges of combining the different requirements become significantly more complicated (these requirements are detailed in IEC 62128-3).

10.2   Interfaces between different electrification systems

Since the introduction of AC electrification, there are many places where AC and DC lines intersect. These intersections can take a variety of arrangements and are based on the operational requirements of the railways, including:

(a) AC overhead alignment is parallel to DC electrified lines;
(b) AC overhead lines are provided over tracks where there is also the third rail; and
(c) AC overhead contact system interconnect with DC overhead contact systems.

10.2.1   Adjacent AC/DC railways – line of route

Adjacent operation is common where a railway company repurposes an old railway alignment and is often the case with city DC metros being installed parallel to AC railways. The challenge here is that the two systems are operating on adjacent lines and the overhead contact line zone alignment can conflict with the rails, fences and infrastructure of the adjacent railway.

Figure 10.1 shows a typical alignment for the adjacent operation of AC railways parallel with DC railway overhead lines (750 V or 1,500 V) and also with the third rail (750 V).

Figure 10.1   AC/DC line of route with adjacent operation

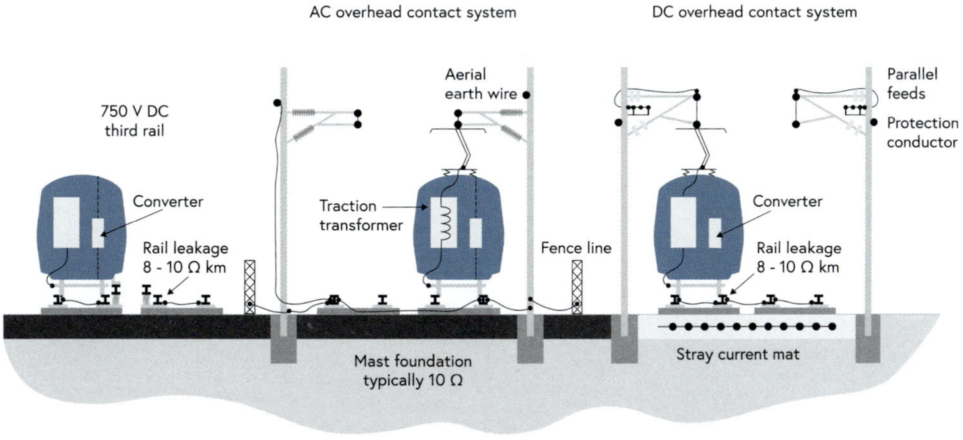

Section 10 – Earthing and bonding at AC/DC interfaces

10.2.2 Station and overline interfaces

Where a city metro transport system needs to integrate with high speed and intercity services, the operational requirement brings about more interfaces at stations and bridges. Where the station's structural steelwork, or the over or underline bridge deck are within either the overhead contact line zone (OCLZ) or the current collector zone (CCZ) of either the AC or DC railway, there is a necessity to integrate the earthing design for the two railways (see Figure 10.2).

The following are some current examples where DC and AC railways operate at dual voltage, run parallel and share stations:

UK (station and parallel interfaces):

(a) 630 V DC London Underground – fourth rail and 25 kV Network Rail;
(b) 750 V DC Docklands Light Rail – third rail and 25 kV Network Rail;
(c) London Euston and Watford Junction – third rail and 25 kV Network Rail;
(d) 750 V DC OCLS Edinburgh Tram, Manchester Metro, Birmingham Metro and 25 kV Network Rail; and
(e) 1,500 V DC Tyne and Wear Metro and 25 kV Network Rail.

UK (station and dual voltage interfaces):

(f) 25 kV High Speed 1 route to its interface with 750 V DC Network Rail third rail at Ashford International;
(g) City Thameslink route through to its interface with the 750 V DC third-rail network at Blackfriars and 25 kV Network Rail; and
(h) London overground route, Richmond to Stratford, with interface with 25 kV at Acton Central.

Hong Kong MTR (station interfaces):

(i) 1,500V DC Mei Foo, Nam Cheong, Ho Man Tin and 25 kV North-East Line;
(j) 1,500V DC Hung Hom, Kowloon Tong, Tai Wai, Diamond Hill, Ho Man Tin and 25 kV East Rail; and
(k) 1,500V DC Admiralty and 25 kV Shatin to Central Link.

USA (station, parallel and dual voltage operation):

(l) California high speed line (25 kV-0-25 kV) and Valley Transport Authority Metro (750 V DC OCS) at Mountain View (planned); and
(m) New York City, the New Haven Line of Metro–North Railroad is dual voltage –third rail and 25 kV.

Figure 10.2 shows the typical adjacent operation of a DC third rail and a 25 kV AC railway with the overline running of a 1,500 V DC metro system.

Section 10 – Earthing and bonding at AC/DC interfaces

Figure 10.2 AC/DC adjacent and overline interfaces

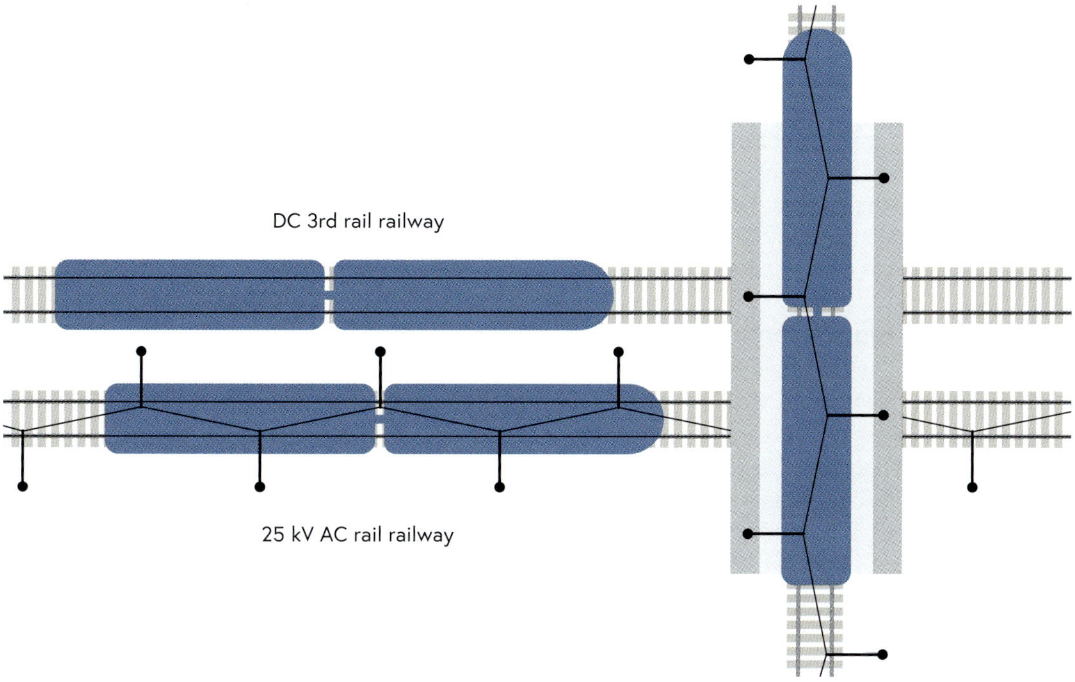

10.2.3 Dual voltage train operation

Operational and electrical interfaces are created where new AC schemes integrate with legacy DC traction systems, such as 750 V third rail, 1,500 V and 3,000 V. As networks have developed, there has been an increased operational requirement for passenger and freight services to operate dual voltage trains or have sections where trains operate on dual voltage.

Dual voltage areas

Examples of operational railways with dual voltage 25 kV and 750 V DC, exists on Chicago's Yellow Line (the 'Skokie Swift'), Network Rail (Thameslink), Network Rail (Euston to Watford), Transport for London's North London Lines, and UK High Speed 1 (Dollands Moor and Ashford International Station).

Interconnecting AC and DC infrastructures

25 kV/1,500 V DC and 25 kV/3,000 V DC operational interfaces have been created where new 25 kV high speed lines integrate with legacy traction systems, such as 1,500 V DC and 3,000 V DC overhead lines. This interface occurs predominantly in France, Italy, Belgium, Spain and the Netherlands.

10.3 Interfaces and project programme risks

Every project faces the risks of design failures. These technical failures do create operational and commercial risks and these add significantly to the project programme and budget:

(a) technical risk – the individual systems (signalling, telecommunications, traction power or electrification) may not perform as intended and may have the effect of delaying the introduction of the revenue earning service;

(b) operational risk – the railway is unable to deliver the required service performance in terms of reliability, availability, maintainability and safety (RAMS), resulting in a delay to the introduction of the revenue earning service; and

(c) commercial risk – the railway suffers delays, contractual disputes and cost overruns.

10.3.1 Engineering management

To ensure the project programme is not exposed, project managers are required who have an engineering mind capable of understanding the interfaces and coupling mechanisms and able to identify the hazards and risks to the programme. During the project development phase, engineering interfaces, hazards and project requirements need to be identified through interface and safety management activities. These hazards need to be closed out through the design, safety management, and validation and verification process:

Interface management requirements

The railway infrastructure managers are required to ensure that physical and electrical interfaces are identified:

(a) the operational and engineering interfaces between traction, non-traction power systems, signalling, track circuits and civil structures – interface matrix; and

(b) the coupling of the AC/DC interactions between railway and non-railway assets.

Safety management requirements

The railway infrastructure managers are required to ensure the HAZID process identifies potential failure modes of the earthing and bonding, return circuits, the signalling track circuits, track-based control and telecommunications systems at each specific AC/DC interface. Failure to identify these requirements at the design stage can create significant challenges in the later operational stages of the project:

(a) assess the impact of traction system on non-traction systems and other engineering disciplines, including human safety, equipment functionality and corrosion;

(b) produce engineering design solutions to reduce safety hazards to ALARP (as low as reasonably practical); and

(c) produce engineering design solutions that are not 'performance affecting'.

10.4 Interfaces and interaction zones

Interaction zones between AC and DC electrified lines have been identified within IEC 62128-3 (EN 50122-3) *Railway applications. Fixed installations. Electrical safety, earthing and the return circuit – Part 3: Mutual Interaction of AC and DC traction systems.*

The electrification interaction zone is specified in Section 6 of IEC 62128-3 and EN 50122-3. The interaction zones identify areas where there is an increased likelihood of interference due to the proximity of the AC and DC railways and due to physical separation (OCLZ and CCZ), electrical galvanic and mutual coupling. Throughout the design process, the designers and infrastructure managers of both AC and DC railways need to have identified the requirements at these interfaces.

10.4.1 Physical interfaces and clearances

Where AC and DC alignment is close, it is necessary to undertake a study of the alignment and OCLZ/CCZ for both AC and DC railways. AC and DC OCLZ may extend beyond the alignment of the railway, and it is not uncommon that OCLZ interfaces can include the following structures and electrical systems:

Section 10 – Earthing and bonding at AC/DC interfaces

(a) conductive fencing between railway alignments;
(b) overhead line masts between railway alignments;
(c) signal posts and gantries between railway alignments;
(d) equipment cabinets between railway alignments;
(e) station conductive canopy and platform structures;
(f) railway overline and underline bridges; and
(g) station overbridges (passenger).

Clearances between live overhead conductors and earthed structures (AC and DC railways) are required to be controlled in line with EN 50124 Parts 1 and 2 *Railway applications. Insulation coordination.* Insulation coordination is concerned with the selection, dimensioning and correlation of insulation both within and between items of equipment. In dimensioning the insulation, the electrical stresses and environmental conditions are required to be taken into account.

10.4.2 Electrical zone of galvanic coupling

Mutual interaction between the AC/DC return circuits is possible due to the return currents conducting through the earth; other adjacent conductive return paths include return rails, earth conductors, earthing installations of traction power supply substations and cable screens. A zone of mutual interaction galvanic coupling; IEC 62128-3 defines this as the separation distance where these interactions are likely to be active.

Interaction between DC and AC traction return systems: Galvanic coupling of DC return current may directly affect rails of any third-party asset that is less than 50 m from the DC railway (EN 50122-3 and IEC 62128-3). Galvanic coupling dominates at low frequencies when circuit impedances are low.

Dual voltage operation: The return circuit of the AC railway and the DC railway is common, and both return circuits are at the same potential. A short circuit within the AC system can cause a peak voltage on conductive structures connected to the return circuit of the DC railway. The same effects apply for conductive structures connected directly or via a voltage limiting device (VLD). The voltage across the VLD can trip without a fault on the DC side. The connection of the return circuit of the DC railway to the earthed return circuit of the AC railway increases the danger of stray current corrosion.

10.4.3 Electrical zone of mutual interaction

The mutual interaction between AC and DC systems is likely to occur in the following locations:

(a) adjacent operation of AC and DC railways;
(b) crossing of railways at overline and underline bridges;
(c) shared use of buildings or other structures; and
(d) shared station platforms.

A zone of mutual interaction is defined for inductive and capacitive coupling; IEC 62128-3 defines this as the separation distance where these interactions are likely to be active.

Induction coupling (magnetic field): AC overhead line and return conductors create a changing magnetic field (50/60 Hz). This zone of mutual interaction is based on the magnitude of the voltage that is coupled into low voltage (LV) systems. EN 50121-3 states, with preconditions, that the limit of the zone between AC and DC railway is 1,000 m (EN 50122-3 Section 6.2 and IEC 62128-3) This limit is given based on an exemplar system and depends on the conductor arrangement, length of parallelism, inducing current and frequency.

Section 10 – Earthing and bonding at AC/DC interfaces

Induction coupling (electric field): AC railway creates an electric field that can influence unearthed DC conductive structures. The preconditions in IEC 62128-3 and EN 50122-3 apply, and the distance between the AC and DC railway is less than 50 m. Voltages influence the capacitive coupling into galvanically separated parts or conductors.

10.4.4   AC/DC interface hazards

Safety hazards and risks should be considered from the start of the design and during installation. Suitable measures need to be specified to limit the voltage for human protection to the levels given in IEC 62128-3 and EN 50122-3 or the national code of practice, while also limiting the damaging effects of corrosion to conductive structures in accordance with EN 50122-2 and IEC 62128-2.

At the interface of the traction return systems, currents can flow in unintended paths. It is necessary to consider the current paths for all operational and failure modes, and the possibility of additional adverse effects on the electrical, signalling and control systems.

The following are the hazards and adverse effects listed in IEC 62128-3 and EN 50122-3 Section 4.1:

(a) touch voltages to the levels given in EN 50122-3;
(b) damaging effects of stray currents in accordance with EN 50122-2;
(c) thermal overload of conductors, screens and sheaths;
(d) thermal overload of transformers due to magnetic saturation of the cores;
(e) restriction of operation because of possible effects on the safety and correct functioning of signalling systems; and
(f) restriction of operation because of malfunction of the communication system.

Other potential hazards that should be assessed:

(g) increased risks during degraded modes or bypass of the insulated rail joints (IRJs);
(h) induced voltages on DC electrification infrastructure.
(i) damage to trains traversing IRJs.

10.5   Protective provision at AC/DC interfaces

The requirements concerning connections to the return circuit of the AC and DC railway are listed in EN 50122-1:2011+A4:2017 Section 6 (AC and DC electrical safety) and EN 50122-2: 2010 Sections 5 and 6 (DC stray current).

EN 50122-3 and IEC 62128-3 standards address specific protective provisions for fixed installations, including electrical safety, earthing and the return circuit. These requirements have been harmonized to avoid the risks of hazardous voltages and DC stray currents resulting from the mutual interaction of AC and DC electric traction systems.

10.5.1   Protective provisions for humans

At AC/DC interfaces, the earthing and bonding design must ensure that touch voltages, including those associated with objects connected to remote earth, cannot cause electrical hazards to humans. Touch voltages relate to the combined effect of both AC and DC electrification systems. In Section 7 of IEC 62128-3 and EN 50122-3, the combined requirements have been redefined with an overall reduction in the short-term and long-term limits. However, the standard does not provide any additional methodology on how this reduction can be achieved.

Section 10 – Earthing and bonding at AC/DC interfaces

The EN 50122 standards state that touch voltages and immunization of signalling track circuits have a higher priority than the protection of assets for DC stray current.

10.5.2 Bonding of rails, overhead line and conductive structures

The impedance of rails at 50/60 Hz is much greater than the DC resistance, so the return current from AC trains and short circuits on the AC overhead lines would cause significant voltage drops in the rails, which would bring a risk of electric shock if suitable means were not employed to limit the voltage on the rails.

On AC electrification systems, the scheme for earthing and bonding connects the rails to the overhead line masts and the connection from the masts to the ground by leakage through the structure foundations, which effectively earths the rails. The bonds between the masts and the rails also provide a current path that flows if a 25 kV insulator fails.

On DC electrified lines, the requirements are quite different as structural items, the earth of electrical systems, etc. must not be bonded to the rails, as the direct current can then flow through such items into the ground and cause damage by electrolytic corrosion. These requirements are detailed in EN 50122-2.

Therefore, a conflict arises in the requirements for bonding, wherever DC electrified lines are so close to AC electrified lines that it is necessary to consider bonding the two systems in order to avoid the risk of electric shock.

10.6 DC electrification – interference to AC railways

DC traction units inject the DC traction return current into the rails via the wheel set axle brushes. DC return currents of overhead and third-rail systems should only flow in the running rails of the DC traction system. The running rails are insulated to prevent any DC current from flowing through the railway structures and rails of the AC railway.

The DC return circuit requires a reliable, adequately rated path for the return of DC traction load current and short-circuit current to the DC substations' negative return bar (see Figure 10.3). The return running rails of DC railways usually float with respect to remote earth, to prevent corrosion of the railway assets and DC overhead mast foundations. DC overhead line masts and structural assets should not be permanently bonded to the rail.

10.6.1 Conductive assets within the DC overhead contact line zone

Where conductive assets are located within the OCLZ (for a DC railway) protective provisions for humans are achieved with either a non-permanent bond or a VLD at the DC substation:

(a) an overhead line fault can cause a high impedance earth fault or flashover. This creates a large negative rail voltage which is detected at the substation by a VLD. If this exceeds a defined limit, the track feeder breakers of the substation are opened clearing the overhead line fault.
(b) VLDs may be installed on separate structures and bonded to the return circuit. When there is an earth fault to a protected structure, the VLD conducts and allows adequate fault current to trip the substation track feeder breaker. Where there are many structures, this solution is a costly option and also becomes an additional requirement for maintenance.

10.6.2 Unintended DC return paths

Stray current may be best described as DC traction return current that has strayed out of the desired traction return circuit and is freely flowing in unintended conductive paths or assets. At the interface, the

Section 10 – Earthing and bonding at AC/DC interfaces

DC stray current may flow in any earth path and then return to the rails somewhere near the DC supply source.

At the AC/DC interface, it is necessary to limit any direct current flowing in the AC tracks and earth paths to levels which do not cause corrosive damage to conductive structures and utilities. There may be periods of time when the DC infrastructure or the interface is required to operate in degraded mode (or bypass for dual voltage operation) and, in this case, there may be a significant increase of stray current, so the effects should be kept to a minimum.

The conductive assets at risk include AC earthed traction return circuits, AC and DC mast foundations, steel-reinforced track-beds, steel-reinforced tunnel linings, conductive station structures, conductive overline and underline bridges, pipe bridges, parallel cable screens and armours, metallic utilities entering or crossing the railway alignment, conductive fences and neighbouring metallic structures, and DC track circuits in the AC railway.

Figure 10.3   Possible DC earth return paths in adjacent overhead line railways

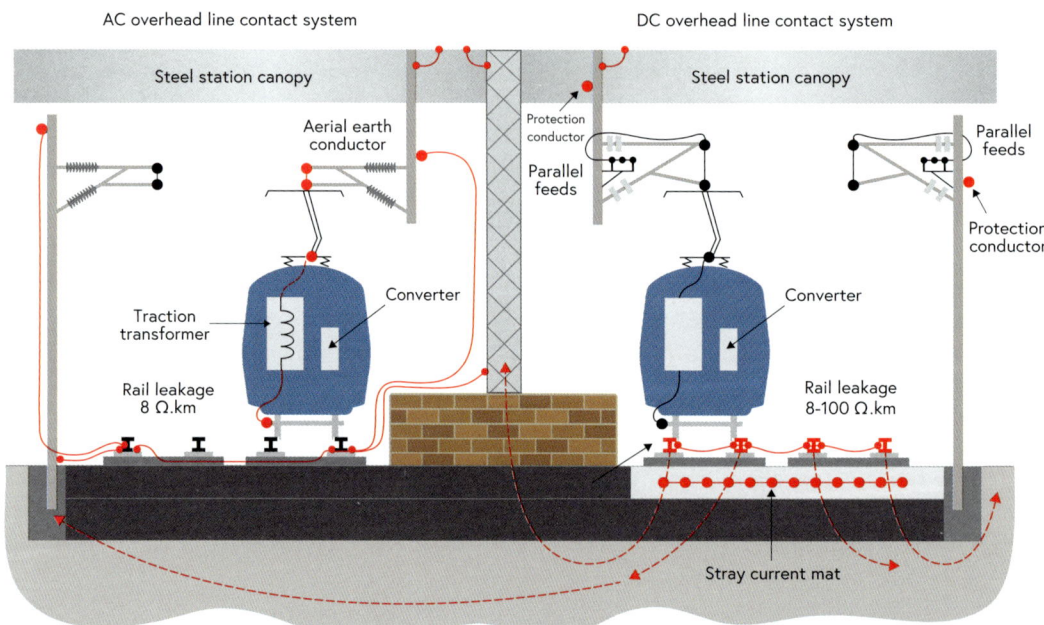

10.6.3   DC stray current protective provisions - corrosion

Direct current leakage from DC rails is an inevitable consequence of using the running rails as a mechanical support and the return circuit for the traction supply. Each track has a longitudinal resistance of typically 15 mΩ/km and insulation from earth typically 1–10 Ω.km (embedded rail), 8–100 Ω.km (ballasted), and 100–200 Ω.km (slab track). Therefore, a proportion of the traction current returning along the rails leaks through the rail insulation to the mass of Earth. This current can then flow along any parallel conductor, any utilities, or in the soil itself before returning into the running rail and the negative return busbar of the DC rectifier. Where current leaves the rail to the mass of Earth, there will be oxidation, or electrons producing a chemical reaction (see equation 10.1), visible after time as corrosion damage (see Figure 10.4):

$$Fe \rightarrow Fe^{2+} + 2e^{-}$$

(10.1)

Section 10 – Earthing and bonding at AC/DC interfaces

Corrosion of metallic objects occurs at each point where current transfers from a metallic conductor (such as a reinforcement bar in concrete) to the electrolyte (i.e. the concrete). The first signs of corrosion are generally visible on the rails and around the rail fasteners. Stray current causes corrosion damage to both the rails and any other surrounding metallic elements. In a few extreme cases, severe structural damage can occur as a result of stray current leakage.

The affected structure potential will rise when the oxidation reaction occurs. The metal is oxidized and produces metal ions into the electrolyte (soil). The stray current accelerates the corrosion rate of the track fasteners and the running rail. According to Faraday's Law of electrolysis, for every 1 A of stray current that passes through a conductive asset, this will cause 9.1 kg of corrosion in a year. Figure 10.4 shows corrosion of probably 10 mA over 10 years.

Figure 10.4   Track fasteners in a DC depot (image reproduced with permission by R. Catlow)

10.6.4   Stray current protective provisions for AC railways

In the presence of high voltage (HV) AC overhead lines, the bonds between the rails and other metal structures are necessary for electrical safety. This conductive path can become a natural path for stray current and is likely to bring about electrolytic corrosion of overhead mast structure foundations, and any protective earth that is bonded to the rails, including station earth mats, LV protective earths, buried pipes and the armour of cables.

The AC overhead line mast foundations are made of either steel-reinforced concrete or steel piles. Corrosion of the structure foundation can cause cracking of the concrete and move the mast out of alignment. To prevent the DC current from flowing in the structure foundations, it is common practice to provide mitigations, including:

(a) increasing the concrete cover for the mast foundation rebars (typically 100 mm);
(b) providing secondary insulation of the mast between the mast and the foundation mounting bolts, allowing the mast to be bonded while protecting the mast foundations and
(c) the inclusion of additional traction return rails or DC bonding cables.

10.7   AC electrification – interference to DC railways

On AC overhead electrification systems, the traction unit injects the AC traction return current into the rails via the axle brushes and the wheel set. The return path for the traction load current is via the rails, earth conductors and the earth itself (typically 30 %). The depth of the earth return path is determined by the ground's resistivity and can be 100 m or more. The greater the resistivity, the more significant the depth and magnetic loop.

Section 10 – Earthing and bonding at AC/DC interfaces

A reliable, adequately rated path must be provided for the return of AC traction load current and short-circuit current to the return current busbar of the AC feeder stations, while limiting the alternating current flowing in the DC tracks or earth paths to a level that does not cause electrical hazards to persons or potential damage to civil structures and electrical assets.

The current in the rails disperses based on Ohm's Law. The AC rails and aerial earth conductors (AECs) are typically cross-bonded every 300 to 500 m. There are parallel tracks at AC/DC interfaces, and the current also attempts to disperse between all the parallel rails and parallel conductors of the railway.

10.7.1 Conducted interference

The magnitude of the AC rail return current in the DC rails is determined by the rails' proximity to each other, the length of parallelism and the leakage of both DC and AC running rails. The AC railway's mutual coupling causes AC return current to flow in any earthed parallel path of the DC railway.

The internal impedance of the return rail for the AC railway (50 Hz) is based on two components, with one representing the internal AC resistance and the internal reactance. This is typically $0.227 + j*0.181$ Ω/km. The DC internal resistance per rail is typically 30 mΩ/km; therefore, the AC rail current and short-circuit current of the AC traction supply can produce significant voltage drops in the DC rails.

10.7.2 AC electric fields

The electric fields emanating from live AC conductors are produced whenever the overhead line conductor is energized. The electric field creates standing voltages on any unearthed conductive surfaces mounted between the ground and the overhead line, including fences, metal gutters, and unearthed overhead line conductors of adjacent DC railways. This phenomenon is addressed in Section 3.

For human protection, there is a requirement for any exposed conducting surfaces to be bonded when they are in the electric field of the AC railway. An unearthed conductive structure may give an electric shock to persons who are also touching exposed surfaces of an extraneously earthed structure.

Figure 10.5   Possible AC earth return paths in AC/DC adjacent railways

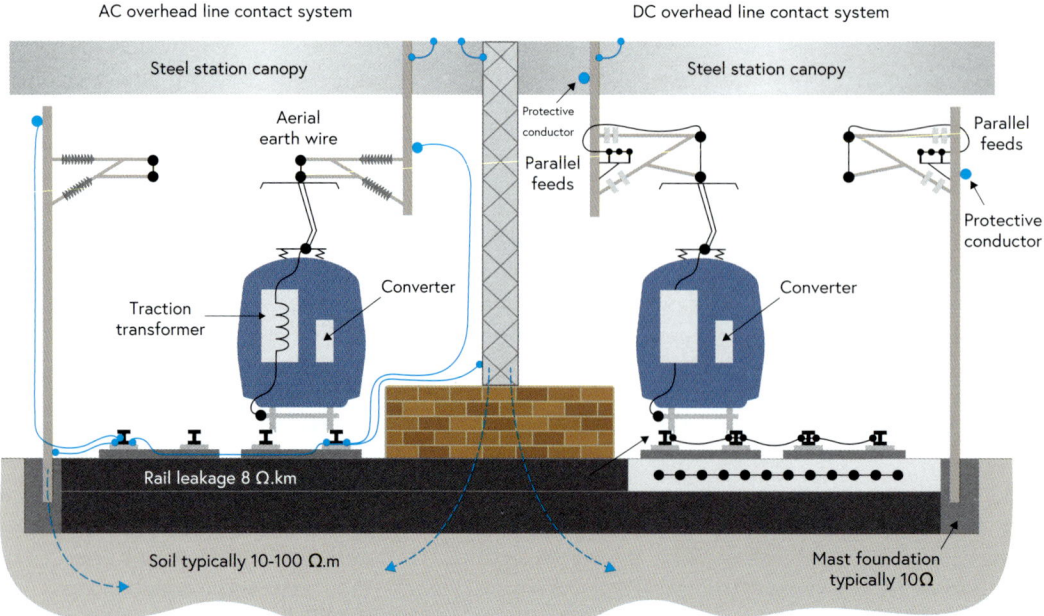

Section 10 – Earthing and bonding at AC/DC interfaces

10.7.3   AC electromagnetic induction

The AC electrification current flows in the overhead contact system and returns through the rails and the distributed earth return system. This current flow is responsible for producing a changing magnetic field (50/60 Hz) and an induction voltage into parallel lineside conductors, sheaths and armours, protective conductors and isolated overhead lines. This induction characteristic behaviour is explained in Section 3.8.1.

AC-induced voltage into the DC railway system can create a hazard for staff undertaking operation and maintenance activities, and so the associated procedures for the DC railway should therefore include mitigating these induced voltages (electric and magnetic) due to the proximity of the AC railway overhead line conductors.

Where DC overhead contact lines are de-energized and isolated to undertake work, a bond is applied between the DC contact lines and the DC rails. This is effectively applying a short circuit such that if the contact line becomes accidentally live, the protective devices will operate. The rails of DC electrified railways are intentionally insulated (50–200 Ω.km) and have a higher leakage than AC railways (10–50 Ω.km). These DC return conductors may become a legitimate AC path due to the mutual coupling during both normal train services and a short-circuit condition. The train operator should establish if the induced voltages present are a hazard, and if this is the case, then the rails of the DC railway may require additional VLDs installed to limit rail voltages or the direct earthing of the overhead contact line. Where the rails are diode earthed (not a preferred arrangement in EN 50122-2), then special procedures will be required.

10.8   Adjacent AC/DC railways – specific requirements

Railway alignments are often close, therefore specific design requirements are necessary for the bonding by both AC and DC electrification systems. This Section addresses the specific requirements for protective bonding, stray current protection and immunization where AC/DC overhead railways intersect, cross, or are parallel.

10.8.1   Protective bonding for AC and DC OCLS masts

DC and AC overhead line masts can form the outer limit for the OCLZ, which can help restrict the extent of this zone.

10.8.1.1   Protective bonding of AC Masts

For the protective provision for humans, and to clear an electrical AC earth fault, the masts are required to be bonded to the rails or the AEC. If the AC masts are in the DC OCLZ, an additional non-permanent bond is required between the DC running rails and the AC masts (see Figure 10.9).

10.8.1.2   Protective bonding of DC OCLS Masts

In compliance with Section 6.3 of EN 61140 protective measures can be provided by double or reinforced insulation for LV up to and including 1500 V DC. Therefore where the DC masts are double-insulated they do not require protective measures and can be left unbonded; the explanation for this is given in Section 6.2.3.2 of EN 50122-1 (2011 +A4 2017) & IEC 62128-1. However, where masts are in public areas it is common practice to provide lightning surge protection. If the DC masts are in the AC OCLZ, additional bonding is required to the AC traction return circuit, to clear AC earth faults (see Figure 10.9).

Section 10 – Earthing and bonding at AC/DC interfaces

Where the DC masts have a single insulator, they require protective measures as the masts cannot be bonded to rails. They can however be bonded together with a fault return conductor. During an insulator flashover or dewirement, there is low resistance to the earth and therefore the earth fault detection device can clear the DC fault.

10.8.2 Non-permanent bonding of conductive structures

AC bonding requires conductive parts to be bonded directly to either the running rails or AECs of the AC railway. Where conductive structures and electrical services are mounted inside the DC OCLZ they are not permanently bonded and rely on a non-permanent bond as in Figure 10.9.

Devices such as spark gaps are used to bond items including bridges to the rails, as in Figure 10.6. These devices prevent direct current from flowing in the bridge under normal conditions, but conduct current if there is a short circuit from the overhead lines to the bridge, so preventing a dangerous voltage on the bridge. It has been found that spark gaps, even those which are designed to reset themselves after the short circuit has been cleared, are difficult to apply in mixed AC/DC situations because, even when an AC short circuit has been cleared, direct current may continue to flow in the spark gap, therefore maintaining the arc so that the electrodes weld together.

Non-linear resistors are often used, which are much more robust. However, non-linear resistors are problematic if the short-circuit current is high, as a non-linear resistor may conduct a large current, creating an unacceptable voltage drop, or be likely to conduct too much direct current under normal conditions. The alternative approach includes a spark gap in series with a non-linear resistor. The spark gap blocks the direct current under normal conditions, while the resistor ensures that the arc is extinguished with the clearance of a short circuit.

Figure 10.6   VLD overline bridge

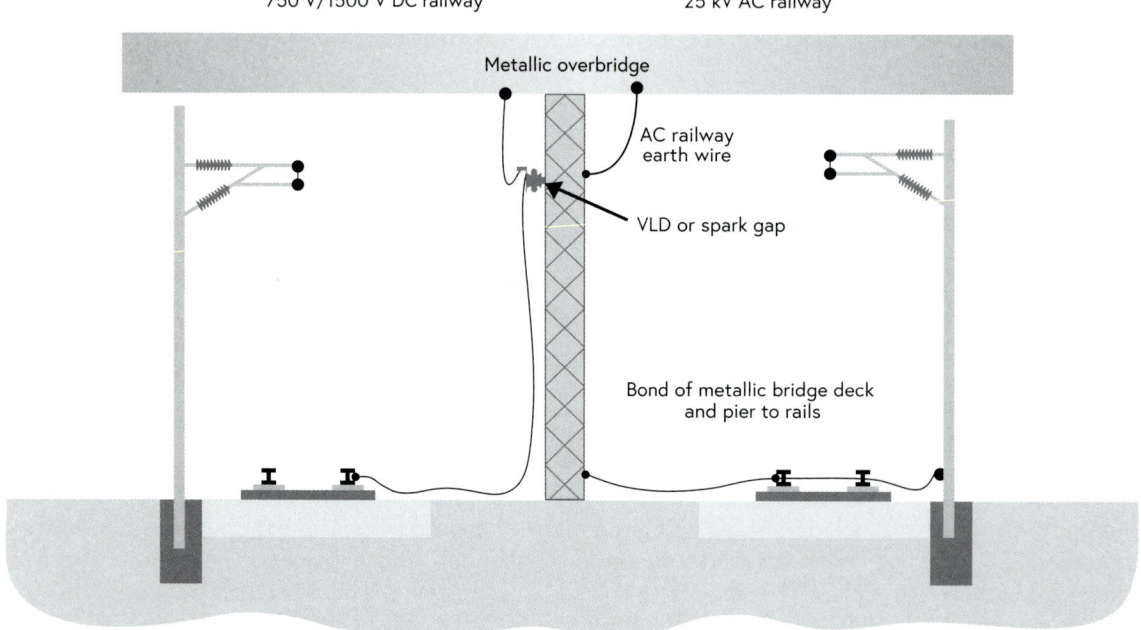

Section 10 – Earthing and bonding at AC/DC interfaces

10.8.2.1 Protective bonding for overline and underline structures

Where the overline or underline structure is within the AC OCLZ or CCZ, AC traction bonding requires exposed-conductive-parts to be bonded directly to either the AC running rails or AEC. Where the overline or underline structure is within the DC OCLZ or CCZ, DC traction bonding requires exposed-conductive-parts to have a non-permanent bond directly to either of the DC running rails (see Figure 10.6)

10.8.3 Physical interface of adjacent AC and DC lines

AC railways are operated adjacent to DC third rails and DC overhead lines (see Figures 10.1, 10.7 and 10.8).

10.8.3.1 Adjacent AC overhead and DC third-rail lines

At locations where the two different electric traction systems are adjacent, AC electrification masts, civil structures and equipment cases may accidentally come into contact with a live broken overhead contact line, or live parts of a broken or dewired current collector, thereby becoming live. This physical interface is defined by the zones, which are specified as:

(a) the AC CCZ is defined for protective provisions and is the area whose limits are in general not exceeded by an energized, dewired or broken current collector and its fragments; and

(b) the OCLZ of the AC railway is likely to impinge on the running rails, fences and metallic structures of the DC railway.

NOTE: This subject is also covered generally for AC OCLZ in Section 5.

The rails on a DC system are floating from the earth, and therefore bonding them to the AC running rails would bring about corrosion of the foundations of the AC overhead line masts. The DC running rails should therefore be connected to the AC running rails with a non-permanent bond (see Figure 10.7).

Figure 10.7   OCLZ and CCZ – adjacent AC railway and third rail

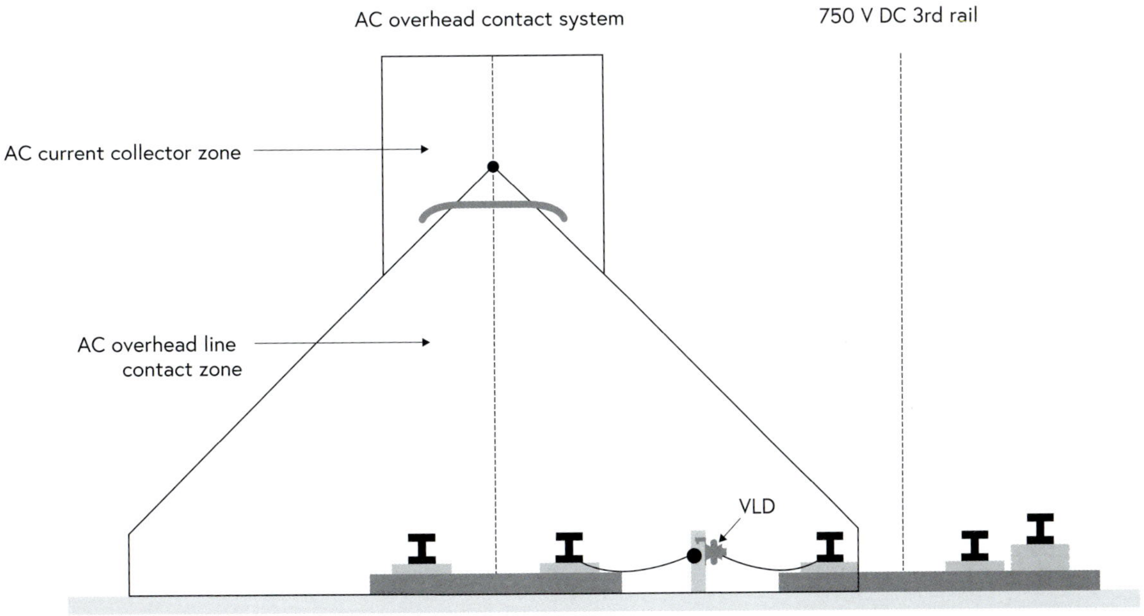

Section 10 – Earthing and bonding at AC/DC interfaces

10.8.3.2 Interface of adjacent AC overhead and DC overhead lines

At locations where the AC and DC OCLS are adjacent, electrification masts, civil structures and equipment cases may accidentally come into contact with a live broken overhead contact line or live parts of a broken or dewired current collector and become live. This physical interface is defined by the zones which are specified as:

(a) the AC and DC CCZ are defined for protective provisions and are the areas whose limits are in general not exceeded by an energized dewired or broken current collector and its fragments; and
(b) the DC and AC OCLZ is likely to impinge on the running rails, fences, and overhead line masts and the metallic structures of the adjacent railway.

The running rails on a DC system are floating from the earth and therefore bonding them to the AC running rails would bring about corrosion of the foundations of the AC overhead line masts. Therefore, the DC running rails should be connected to the AC running rails with a non-permanent bond (see Figures 10.1 and 10.9).

Figure 10.8   OCLZ and CCZ – adjacent AC and DC overhead contact system

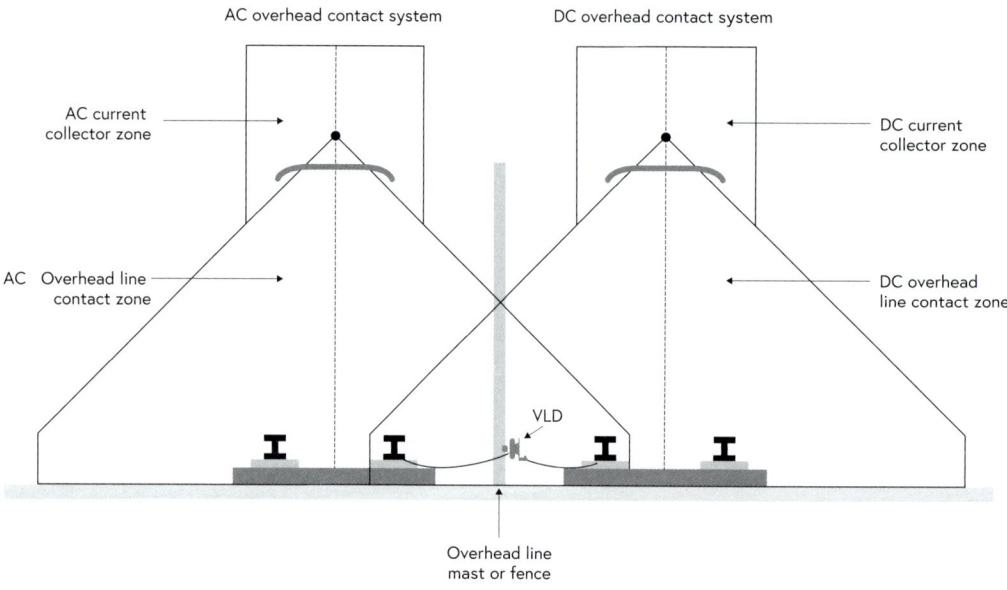

Figure 10.9   Open route OLCZ and conductive interfaces

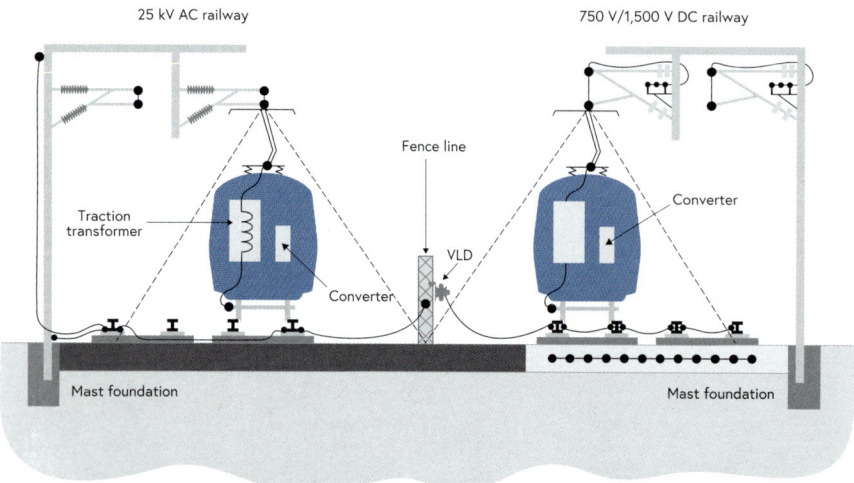

Section 10 – Earthing and bonding at AC/DC interfaces

10.8.4 Protective bonding for stations

Station structures such as overbridges and platform canopies are commonly constructed of steel, and these exposed-conductive-parts can be located within both the AC and DC railway OCLZ and CCZ.

LV bonding requires all exposed-conductive-parts of the station canopy and platform in the OCLZ and CCZ to directly bond to the station main earth terminal (MET) (see Figure 10.11). To clear AC earth faults to station metallic structures, the station MET must also be bonded to the AC traction return with a permanent bond.

The DC traction return current should not be allowed to flow through the station structure or earthing system. The AC bonds can provide a low resistance conductive path for any DC stray current flowing in the station earth system. When DC stray current flows in AC earths, the station civil structures, rebars, earth mats and grids, third-party metallic utilities, and screens and armours of electrical supply cables can be exposed to corrosion.

To clear DC earth faults to station metallic structures, the station MET must be bonded to the DC traction return with a non-permanent bond or a short-circuiting device. Following a DC earth fault to the station structure, the station short-circuit device trips and the subsequent current can clear the DC fault. However, during the fault, the current flows in both the AC and the DC traction return circuits. After the DC earth fault is cleared, the non-permanent bond (or short-circuiting device) must revert to open-circuit condition (see Figures 10.10 and 10.11).

Figure 10.10 DC platform bonding

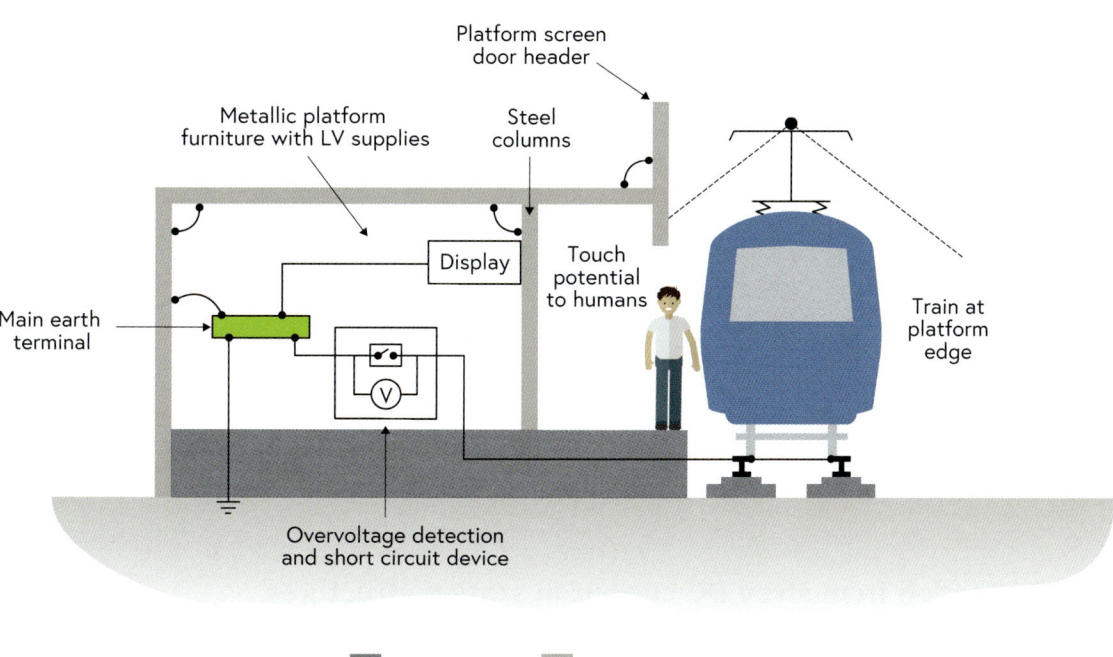

Rail insulation   Steel civil structure

Section 10 – Earthing and bonding at AC/DC interfaces

Figure 10.11   AC and DC railway island platform bonding

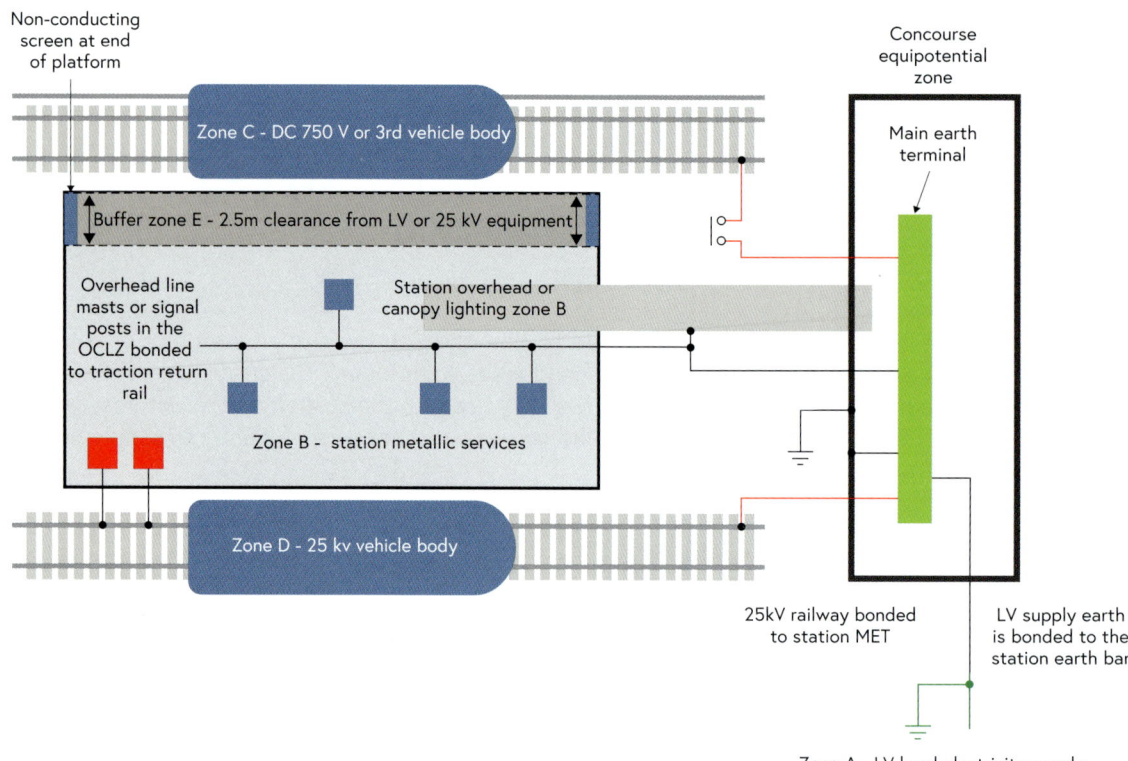

10.8.5   Protective bonding of cantilevers in station canopies

The mounting of both AC and DC cantilevers in the canopies of a shared platform creates a conflict for bonding the canopy, the station, and the electrification return path. The following bonding arrangements may be required in any dual voltage island canopy:

(a) where AC cantilevers are mounted in station canopies, they are required to be bonded to the station canopy. This structure shall then be bonded directly to the station MET and the running rails;
(b) where DC double-insulated cantilevers are mounted in station canopies, no additional bonding is required for an insulator flashover. However, a DC non-permanent bond to the DC rails may also be required if the station canopy is in the CCZ of the DC overhead lines; and
(c) where DC single-insulated cantilevers are mounted in station canopies, a fault return wire (earth wire) is required, with a DC non-permanent bond to the DC rails.

Since the station structures are bonded to the AC railway running rails, the track circuits are likely to be affected by the DC electrification fault current. Therefore, where the DC fault current can flow in the station earth and the AC traction return circuit, the track circuits should be specified as AC and DC immune.

10.8.6   Protection from stray DC current in station structures

The AC traction return circuit must be bonded to the station MET to control touch voltages within the station structure. DC traction return circuits (running rails) are required to be floating and only connected via a non-permanent bond; however, if this non-permanent bond fails, there can be corrosion damage to the structure, utilities, and station earth mat or earth rods.

The non-permanent bond, as shown in Figure 10.10 and Figure 10.11, typically includes the use of a reliable fast-acting, resettable short-circuiting device that is connected between the station MET and the DC traction return of type VLD-FI. For more details, see the example in EN 50122-1 Annex F, Chapters 6 and 7 (2011 +A4 2017), and IEC 62128-1.

10.8.7 Train detection and immunization

The close proximity of AC and DC electrification systems can create the possibility for disturbance in either infrastructure. In most cases, where both AC and DC systems share a common station platform, it is necessary to undertake a risk assessment.

The risk of interference to signalling and telecommunications systems, arising from the combined presence of AC and DC electrification, must be sufficiently low to avoid the risk of maloperation of the train detection systems. The following scenarios should therefore be assessed as part of the HAZID process:

(a) AC traction return current (including harmonics of the electrification and traction load) flowing in the track circuits of the DC-only area (long-term);
(b) AC fault current flowing in the track circuits of the DC-only area (short-term);
(c) DC traction return current (including harmonics of the electrification and traction load) flowing in the track circuits of the AC-only area (long-term); and
(d) DC fault current flowing in the track circuits of the AC-only area (short-term).

To avoid disturbance to track circuits at AC/DC interfaces, it is always advisable to use either axle counters or AC and DC immune track circuits.

10.8.8 Station control and telecommunications systems

Station control systems located within station canopies may be affected by DC or AC fault current that can flow in the station structures, protective earths and bonds. Without adequate bonding, the fault currents can damage the protective earths and bonds. It is common to provide an adequately rated common-bonding conductor in a canopy to avoid disturbance or damage to LV systems.

10.9 Dual voltage operation of trains

Electrical changeover between AC and DC electrification schemes has become more common where railway companies have interconnected their existing DC overhead (1,500 V/3,000 V) or DC third-rail networks with an AC electrification scheme. The challenge for the infrastructure operator is to operate the traction package from a high AC voltage and a lower DC voltage. In most cases, it is necessary to have a separate current collector system for either voltage.

This Section describes different layouts for 750 V, 1,500 V and 3,000 V DC, which interface with AC 50/60 Hz schemes. Examples and the principles that have been implemented in the UK and in French railway networks are described.

Section 10 – Earthing and bonding at AC/DC interfaces

10.9.1 Insulated rail joints and resistors

The insertion of insulated rail joints (IRJs) is the simplest arrangement and enables a changeover with the train on the move. A single set of IRJs are usually inserted between the AC and DC traction return circuits, and the overhead line runs continuously over the IRJs. The IRJs limit the flow of direct traction return current into the AC-only area and limit the flow of alternating traction return current into the DC-only area (see Figure 10.12).

This method does not give good suppression of stray current, but in some cases, it is sufficient if the voltage on the rails of the DC lines is kept to a low level. Adding a second pair of IRJs suppresses and prevents the current from passing across the interface.

Resistors can be added that bypass a pair of IRJs: the resistors are chosen so that the voltage reduces across the rail joints due to traction load and short-circuit current, but does not risk electric shock or damaging the trains (see Figure 10.13).

When the IRJs are bridged by a train, some DC stray current can flow through the train body. The voltage difference across the two sides of the insulated joints can appear between the vehicles of the train. This voltage can create a hazard to passengers and staff, and damage the train if a significant current can flow through the vehicle, including through bonding cables, electrical and mechanical equipment, bearings and draw gear (the coupling between train vehicles). Additionally, care must be taken in the design of the on-train electrical systems to prevent on-train length control and supply wires from providing a preferential path for traction return currents under both normal and fault conditions. This requirement is challenging to apply on lines with high traction current and or high short-circuit current. More complex arrangements of infrastructure switching around IRJs are, therefore, usually implemented.

Figure 10.12   IRJs with isolation section in AC lines

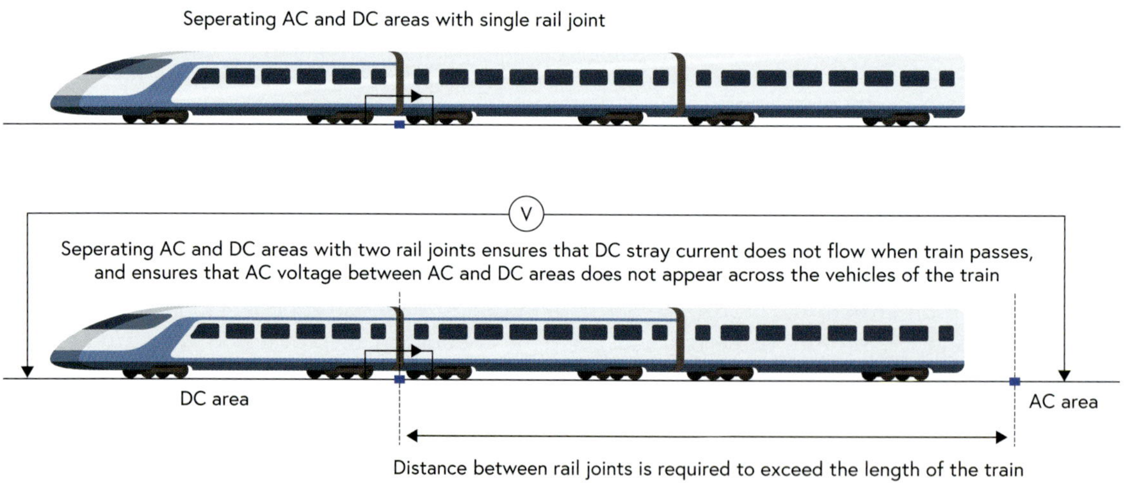

Seperating AC and DC areas with single rail joint

Seperating AC and DC areas with two rail joints ensures that DC stray current does not flow when train passes, and ensures that AC voltage between AC and DC areas does not appear across the vehicles of the train

DC area   AC area

Distance between rail joints is required to exceed the length of the train

Figure 10.13   IRJs and resistors

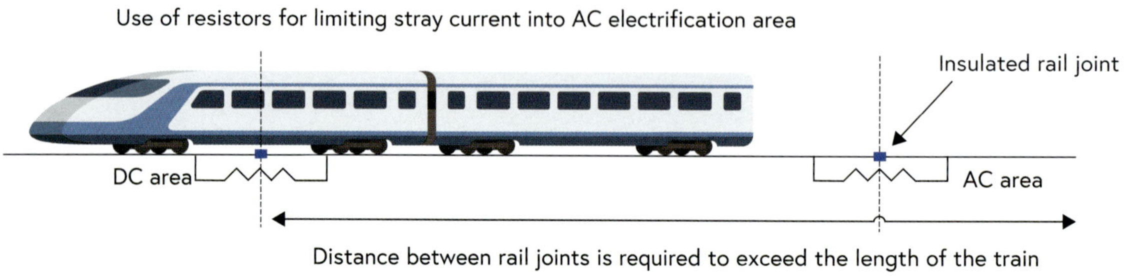

Use of resistors for limiting stray current into AC electrification area

Insulated rail joint

DC area   AC area

Distance between rail joints is required to exceed the length of the train

Section 10 – Earthing and bonding at AC/DC interfaces

10.9.2   Operational changeover – AC 50/60 Hz and 1,500 V/3,000 V DC

The mainline railways in Europe are electrified at 1,500 V DC or 3 kV DC (France and the Netherlands are 1,500 V DC, Belgium, Spain and Italy are 3 kV) and require changeover to 25 kV for entrance lines to the 25 kV high speed areas. The changeover between electrification systems usually involves the raising and lowering of pantographs on the move.

The SNCF's Sud Est region was traditionally electrified at 1,500 V DC, and most of the traditional lines used by Train à Grande Vitesse (TGV) in that region are 1,500 V DC. There were some later electrifications in the Savoie area and east of Dole on the lines towards Switzerland electrified at 25 kV AC. More recently, all of the Lignes a Grande Vitesse (LGV) lines are 25 kV AC.

When trains pass across the interface between electrification areas of 25 kV and 1,500 V DC, marker boards provide instructions for the driver to open the circuit-breaker – electrically isolating the train and lowering the pantographs. The driver is then required to adjust a switch to select the appropriate system and raise the pantographs. Pantographs and pantograph height control are automatically selected based on the voltage system selected by the driver. Once the train detects the correct supply, a dashboard indicator illuminates, and the driver can close the circuit-breaker and retake power.

The TGV operational changeover between the two electrification systems is typically carried out on the move, and the train coasts across the boundary between electrification sections.

10.9.2.1   Operational changeover using DC resonant filters

In France, series resonant filters are positioned across the IRJs where trains pass across the interface between electrification areas of 25 kV and 1,500 V DC. Each filter consists of a bank of capacitors and an inductor so that direct current is blocked, but a low impedance is presented to 50 Hz current (see Figure 10.14)

Such an installation needs to be specified very carefully, paying particular attention to the possibility that the capacitors' characteristics may change due to heating or ageing, resulting in the device going 'off-tune'.

The capacitors are rated for voltage. Therefore, it is necessary to protect the capacitors against overvoltage, which can occur following a short circuit of the 25 kV system, and transient overvoltages and where harmonics within the traction current can encounter a high impedance and create a large voltage drop. There is also a risk that equipment can be damaged when the traction load current is higher than expected.

The introduction of a series resonant filter can introduce the added complexity of electromagnetic compatibility, including high levels of stray magnetic fields from the inductors.

Figure 10.14   Use of resonant filters to limit DC stray current

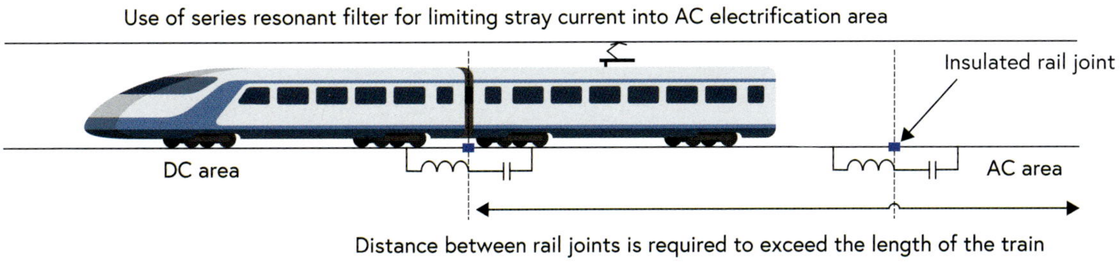

Use of series resonant filter for limiting stray current into AC electrification area

DC area · Insulated rail joint · AC area

Distance between rail joints is required to exceed the length of the train

Section 10 – Earthing and bonding at AC/DC interfaces

10.9.3 Dual voltage operation – 25 kV and 750 V DC third rail

Overhead 25 kV AC and DC third rail does enable dual voltage operation of trains. This electrical changeover enables the movement of a train from the AC-only area (25 kV operation) into a section that is a dual voltage area (25 kV and 750 V DC) and eventually into the DC-only area (750 V DC).

The design of these interfaces is bespoke since land and civil infrastructures are different in each location. So far as is practicable, standard equipment is used at the interface. However, since the designs are bespoke, appropriate safety cases for the interface's operation must be developed and accepted.

10.9.3.1 Operational requirements

The separation of the electrification systems (AC and DC) is necessary to control the DC and AC return traction current and protect the AC earthed infrastructure from stray DC current.

With an AC/DC interface, the following operational requirements are generally required. The AC/DC changeover must be designed and systems positioned to ensure safe operation during both normal and degraded modes of operation. The duration and magnitude of stray current shall be minimized and shall be ideally no more significant than the historical baseline, and its path should be limited.

10.9.3.2 Technical requirements[22]

The traction return circuit in dual voltage areas comprises both the rails and AECs. AC and DC can, therefore, usually flow in these conductors. The technical requirements of the electrification systems include:

(a) controlling the maximum permissible accessible and touch voltages as specified in EN 50122-1 and IEC 62128-1 and in IEC 62128-3 and EN 50122-3;
(b) restricting DC currents flowing in AC structures and maintaining the rail potentials within the limits of EN 50122-1 and IEC 62128-2;
(c) controlling AC rail potentials – this can be achieved by restricting the AC return currents within the AC return conductors and booster transformers;
(d) confirming that all infrastructures within the zone of influence are compatible with the traction currents present within the rail return and earth system; and
(e) controlling the DC return currents from flowing in the AC return conductors – it is therefore customary to provide additional DC return conductors which are parallel to the rail.

The requirements of 25 kV bonding within the OCLZ have been addressed in detail in Section 10.5. As part of the AC 25 kV design, an additional requirement includes segregating conductive paths parallel to the interface section. This segregation is required to ensure human safety and is shown diagrammatically in Figure 10.17.

The segregation of electrical conductors and assets typically includes:

(a) traction supply and return conductors;
(b) AECs;
(c) parallel conductors (cable screens and armours);

---

[22]N.Jordan *AC and DC Electric Railway Interfaces* IET Rail Electrification Infrastructure Systems 2009 paper

(d) parallel structures, including lineside fencing; and
(e) metallic services.

The return system's segregation is designed in conjunction with the signalling system so neither will compromise the signalling system changeover or the traction return system.

10.9.3.3 Earthing and bonding of dual-electrified lines[23]

The earthing and bonding design must ensure that touch voltages, including those associated with extraneous earth, cannot cause hazards to persons. For touch voltages due to short-circuit current, compliance is within limits such as stated in Section 9.2 of EN50122-1 for AC systems and Section 9.3 for DC systems. The protective provisions for direct and indirect contact are specified in EN 50122-1.

Bonding in dual-electrified areas

The worst situation arises if the trains can run on both systems and pass between AC and DC electrified areas. In some locations, lengths of track have AC overhead lines and DC conductor rails and both AC and DC traction currents flow in the same rails.

Tracks having both AC and DC electrification, commonly called dual-electrified tracks, present a particular problem as conductive structures should be bonded to the rails for reasons of safety in the presence of AC overhead lines.

The railway's own assets (overhead line masts, station canopies, conductive overline and underline bridges, the earth of LV electrical distribution systems) are bonded to the rails as in an AC system. The station overline and underline bridges, and signal gantries are bonded via earth wires, having their connections to the rails located close to the substation. These bonds can also provide a path for direct current with the possibility of damage. It should be noted that isolation sections and IRJs do not eliminate this problem as dual voltage trains need a length of dual-electrified track to change over between AC and DC traction. This requirement is further complicated if the dual voltage area includes a passenger station used for a changeover.

The following bonding configurations can reduce the direct current within bonded structures to an acceptably low level:

(a) additional conductors laid between the running rails and connected in parallel with the running rails reduce the resistance of the DC return circuit.
(b) bonding conductors and earth wires are reinforced to allow for direct current circulating in them.
(c) the DC substation has duplicated transformers and rectifiers provided at the AC/DC changeover. This feeding arrangement maintains the DC voltage on the rails as close to zero (or negative) as possible to protect the railway's assets against corrosion; and assets belonging to outside parties, for example, the earth of the incoming electricity supplies, are segregated from the rails as recommended in EN 50122-2.

Non-permanent bonding of conductive structures

AC bonding requires conductive parts to be bonded directly to either the running rails or AECs of the AC railway. DC bonding requires a non-permanent bond from the affected equipment to the DC traction return circuit. In AC/DC dual voltage operation, permanent bonding is not permitted because of direct

---

[23]Dr R.D. White, F. Waterland, D.C. Knights WS Atkins Rail Limited A.I.C. on electromagnetic compatibility: London, 21-09-1999 *The design of railway electrification at interfaces between AC and DC electrified lines.*

Section 10 – Earthing and bonding at AC/DC interfaces

current within conductive structures. The non-permanent bond takes the form of either spark gaps, VLDs or a short-circuiting switch (see Figure 10.15).

Figure 10.15   VLD and metallic overbridge

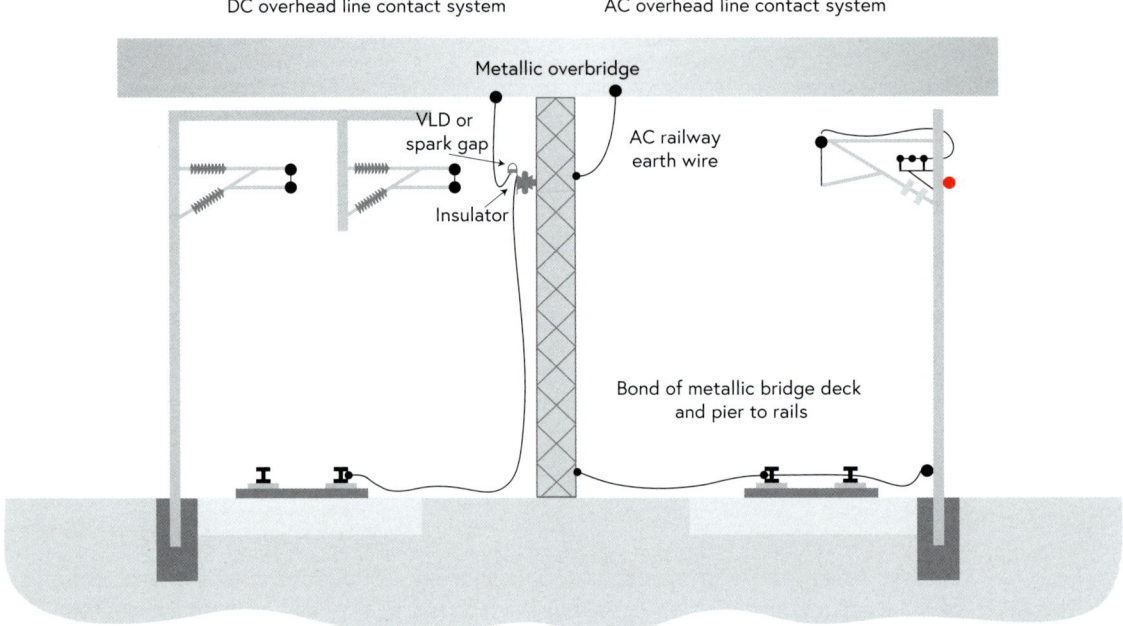

10.9.3.4   Electrical isolation sections

The isolation section utilizing IRJs can limit the flow of direct traction return current into the AC-only area and limit the flow of alternating traction return current into the DC-only area.

Two IRJs in each rail are used to create an isolation section. They are separated by a distance no less than the length of the longest train. This arrangement ensures that a train passing over them does not connect the AC-only area to the DC-only area and ensures that the full voltage difference between the AC-only and DC-only areas cannot appear between a train's vehicles.

Where two sets of insulated joints are arranged in this way, they are referred to as an 'isolating section'. When electric trains traverse this, it is also necessary to bypass the insulated joints for the trains' own return current. Therefore the following different configurations have been described in more detail.

10.9.3.5   Isolation section in AC-only lines (isolation transformers)

The isolating sections in the AC-only lines require 25 kV section insulators to be placed in the overhead lines and in-line insulators in the AECs, directly in line with the IRJs. Isolating transformers are used to transmit power past the insulated joints and section insulators (see Figures 10.18 and 10.19).

This arrangement has been applied on UK HS1, and in the Network Rail infrastructure. The arrangement has no moving parts and does not require a special control system.

Section 10 – Earthing and bonding at AC/DC interfaces

The application of the 25 kV isolating transformers is not straightforward, and careful attention must be paid to the design of the overhead line layouts and the structural bonding details to ensure that the bonds to the overhead line masts and other structures, essential for electrical safety, do not bypass the IRJs as shown in Figure 10.17.

It is best to locate such interfaces well away from extraneous items, such as stations and overline or underline bridges, that are required to be bonded to the rails.

A typical configuration that is used in the UK is shown below in Figure 10.17. This utilizes isolation transformers and IRJs (AC isolation area). This type of interface has been installed within the UK at Dalston Junction to Canonbury (North London Line), Acton, Dollands Moor (see Figure 10.16), North Poole maintenance depot and Ashford Station.

Figure 10.16   Isolating transformers at Dollands Moor (Image reproduced with permission by R. Catlow)

10.9.3.6   Protection for AC isolation and traction transformers

Ferromagnetic materials cannot support an infinite magnetic flux density as they can saturate at a certain level dictated by the material and core dimensions. This means that further increases in magnetic field force (MMF) do not result in proportional increases in magnetic field flux (Φ).

DC current in the transformer primary winding is a cause of saturation of the core. A small amount of DC current produces a voltage drop across the primary winding and consequently causes an additional magnetic flux in the core. This additional flux density bias, or offset, can push the alternating flux waveform closer to saturation in one half-cycle more than the other.

In AC isolation areas and dual voltage areas, the 25 kV traction supply transformers and circuit-breakers and 25 kV transformers on the trains must be protected against damage and mal-operation caused by the direct current circulating through the primary or secondary winding. Predetermination of DC stray currents is challenging as these values depend on several parameters, including rail conductance, leakage and ground resistivity. The isolation transformers at the AC/DC interface must

Section 10 – Earthing and bonding at AC/DC interfaces

be designed to withstand direct current flow in their windings and the possibility of the transformer iron core saturation.

The following is a typical requirement specification for transformers at AC/DC interfaces exposed to DC traction current:

'The transformer shall withstand, at full load, a DC current of 50 A continuously and be able to accept a peak exceeding 130 A for a duration of 10 minutes without damage.'

Figure 10.17   Isolation section with AC transformers

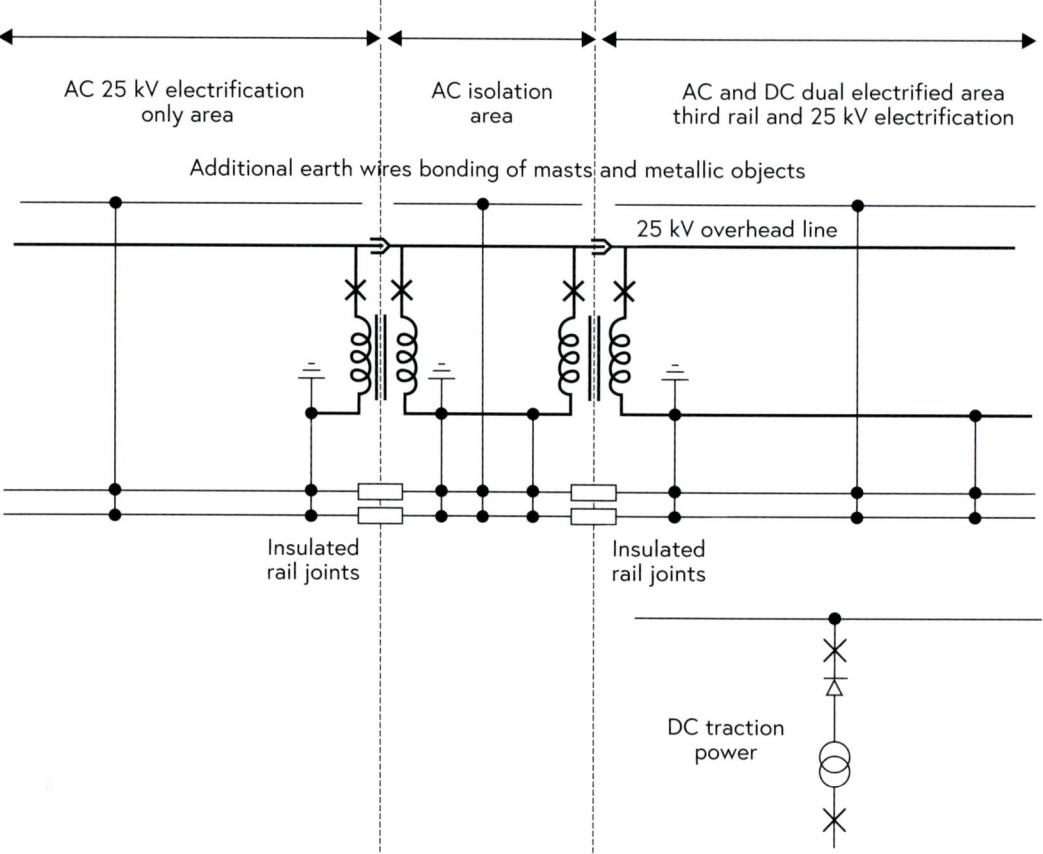

10.9.3.7   Isolation section in DC-only lines (DC contactors)

The isolating sections that are in the DC-only lines require that the conductor rails are gapped alongside the IRJs, and switches (contactors) are used to bypass the IRJs and the gaps in the conductor rails to allow trains to pass.

The operation of the DC contactors is initiated by track circuits and is arranged to ensure that the contactors at both ends of the isolating section are not closed simultaneously. This arrangement is used in the UK on the Thameslink line in central London, between City Thameslink and Blackfriars stations.

Reliable operation has been achieved, although the contactors' maintenance and ancillary equipment requires significant resources. Monitoring equipment is provided to detect excessive flows of direct current into or out of the AC lines and send an alarm to the ECR.

Section 10 – Earthing and bonding at AC/DC interfaces

A typical configuration used in the UK is shown in Figure 10.18. This utilizes DC contactors and IRJs (DC isolation area).

Figure 10.18   Isolation section with DC contactors

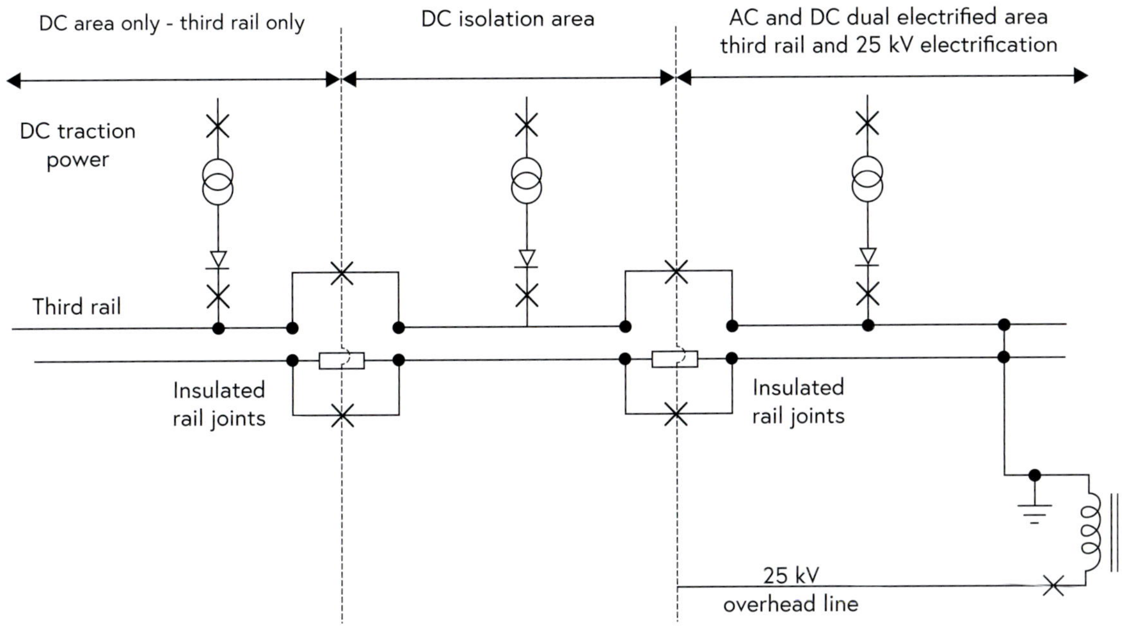

10.9.3.8   Train movements - requirements

The length of each isolation section is required to be greater than the length of the operational trains to ensure that the train is unable to bridge the running rails directly between:

(a) AC-only area and the dual voltage area; and
(b) DC-only area and the dual voltage area (see Figure 10.18).

Trains are required to pass through these isolation sections uninterrupted and without loss of electrical power. The length of the dual voltage area is normally determined by the railway alignment and includes limitations imposed by tunnels, bridges and stations.

10.9.3.9   Rating of protective bonding requirements

As the return circuits can include both AC and DC traction return current, bonding conductors must then be adequately rated so that they do not become overheated due to the combined effect of both AC and DC current flowing in them.

10.9.3.10   Train detection and immunization

Close to the AC/DC changeover area, the track circuits need to be AC and DC immune. The likelihood now is that the train detection system would be axle counters, which are inherently immune from both AC and DC traction current.

Section 10 – Earthing and bonding at AC/DC interfaces

The proximity of the track circuits in the AC-only and DC-only areas may be at risk due to the following possible degraded modes:

Degraded modes in DC-only area (when operating in bypass):

(a) AC traction return current (including harmonics of the electrification and traction load) flowing in the track circuits of the DC-only area; and

(b) AC 25 kV fault current flowing in the DC-only area.

Degraded modes in AC-only area (when operating in bypass):

(c) DC traction return current (including harmonics of the electrification and traction load) flowing in the track circuits of the AC-only area; and

(d) DC 750 V third-rail fault current flowing in the AC-only area.

It is necessary to identify that track circuits local to the AC/DC changeover check their immunity and undertake a safety case.

10.9.3.11 Degraded mode of operation

When the AC/DC changeover area is required to operate under a degraded mode, the isolation transformer or DC contactors may be bypassed causing an increased level of DC return current flowing in the AC-only area, or 50 Hz current into the DC-only area.

Bypass operation may be necessary where a transformer or DC switchgear is required to be maintained. Bypass operation may also be required following equipment failure; the following criteria should then be assessed:

(a) the level of direct current flowing from the DC-only area into the AC-only area should be controlled (time and magnitude). The main concern would be corrosion and the immunity of DC track circuits in the AC-only area.

(b) the level of alternating current flowing from the AC-only area into the DC-only area should be controlled (time and magnitude). The main concern is 50 Hz or harmonics related to 50 Hz and AC track circuits' immunity in the DC-only area.

To support any requirement for bypass of the AC/DC interface, modelling studies should be undertaken to consider the worst-case stray current penetration against any non-immunized track circuits, especially in complicated rail layouts. The immunity of track circuits is based on the magnitude of the current flowing in the AC-only and the DC-only area.

Corrosion risk to conductive structures is based on both the level of DC current and the period of exposure to that DC stray current. Typically, 'tens of amps' flowing for weeks is unacceptable.

10.9.4 Use of active compensation of stray direct current

It is possible to use thyristor-controlled rectifiers at the DC substations to control the sharing of current between substations so that the rail to earth voltage is held close to zero. Another possibility is to use reversible thyristor-controlled rectifiers as LV current sources, which, in conjunction with cables alongside the track, can act as negative boosters to prevent direct current from escaping into the AC lines.

Section 10 – Earthing and bonding at AC/DC interfaces

This configuration of AC/DC interfaces has significant benefits as it suppresses the stray current without using IRJs, enables an AC/DC interface to be fitted easily into a complicated track layout and allows a simplification of the earthing and bonding arrangements, giving the arrangements greater integrity than arrangements requiring the segregation of earths.

Figure 10.19   AC/DC active compensation

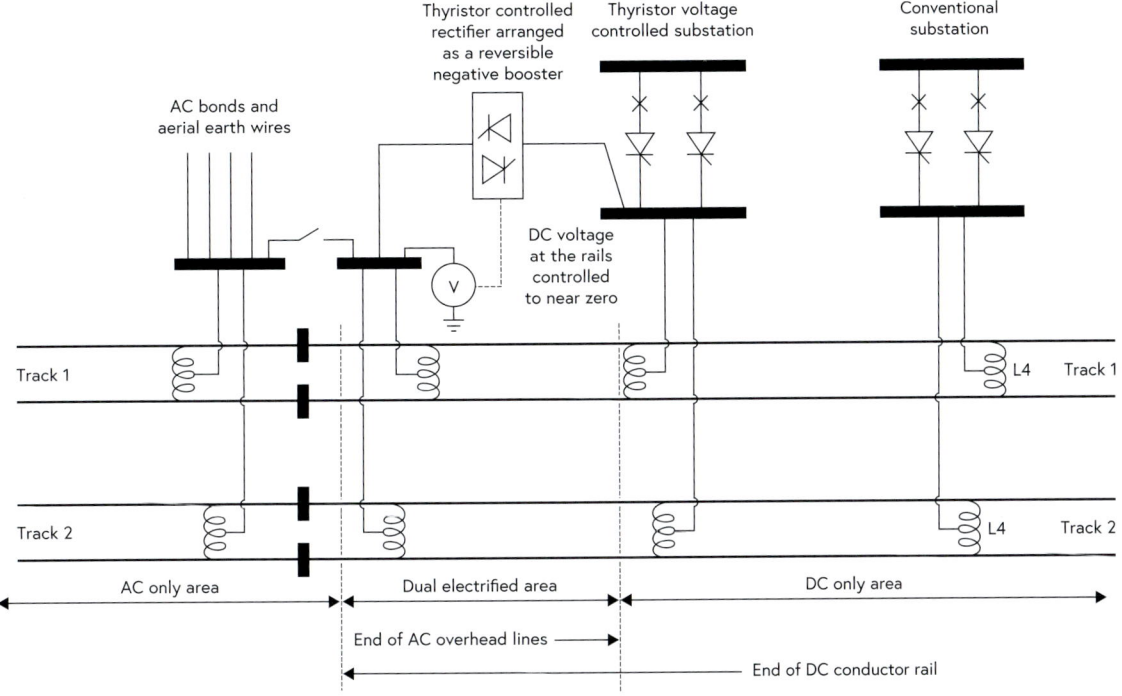

 **Section 11**

Safe working during maintenance, renewal and decommissioning

11.1 Electrical safety requirements

This Section addresses the electrical safety requirements utilized to prevent death or injury to those working on or close to live conductors. The essential safety requirements in accordance with national safety regulations have therefore been referenced.

Generally, the country/state legislation sets a safety objective that must be complied with and general technical standards are used to support compliance with the law. In some circumstances, technical standards can be mandated by a legal instrument, and in other cases, they can be carried out voluntarily.

The Section also specifically details the operational requirements for working safely during the maintenance and/or renewal of the AC overhead electrification systems.

11.2 Prevention of death or injury

The purpose of the national regulations is to ensure that precautions be taken against the risk of death or personal injury from electric shock. Electrical regulations are usually required as part of country/state health and safety legislation and duties are imposed on dutyholders in respect of systems, electrical equipment and conductors, and in respect of work activities on or near electrical equipment.

The human body responds in several ways to electrical current flowing through it. The sensation of electric shock is only one such effect and this can be extremely painful, and fatal. When a shock is received, the electric current may take multiple paths through the body and its intensity at any one point is difficult or impossible to predict. The passage of electric current may cause muscular contractions, respiratory failure, seizures, fibrillation of the heart, cardiac arrest or injury from internal burns.

Electric shock (fatal and non-fatal) to maintainers normally occurs during the period when the AC cables and overhead lines are being maintained or renewed. During this period the existing designed protective measures are essentially bypassed, therefore alternative protective measures are required to achieve safe working practices. They include:

(a) establishing clear limits within which work can be undertaken;
(b) mitigating risks from adjacent hazardous-live-parts;
(c) disconnecting and isolating of all points of supply;
(d) earthing the isolated equipment at the point of disconnection, and
(e) identifying where inadvertent re-energization can occur.

Staff shall be made aware of and understand these arrangements, and the requirements for procedural arrangements, such as permit-to-work.

11.3 Earthing and bonding – life cycle

The earthing system, its conductors, earth mats, and bonding conductors throughout their life, must be capable of distributing load current and discharging fault current without exceeding thermal and mechanical design limits based on backup protection. The earthing system performance must avoid equipment damage due to excessive earth potential rise (EPR), potential differences within the earthing system, high fault currents and high traction load currents, as these can create a heat cycle, creating annealing such that fatigue can occur.

The earthing system must be installed, operated and maintained according to national regulations, international standards and company codes of practice.

11.4 Earthing the traction return circuit

The traction return circuit must maintain step, touch and transferred potentials within the voltage limits based on the normal operating time of protection relays and circuit-breakers. combined with any additional appropriate measures (see EN 50122-1 (UK/EU), IEC 62128-1, D/MTRC/NW/DSM/ST/702/A4 (Hong Kong); General Order No176 State of California - *Rules for Overhead 25kV AC Railroad Electrification Systems for a High-Speed Rail System.*; Ref. 8880-900-677 Rev 1.00 (Western Australia)). The earthing system must also maintain its integrity for the expected lifetime of the installation with due allowance for corrosion and mechanical constraints.

The running rails are the return circuit and also act as the protective earth of the AC electrification system. Therefore, the running rails must have permanent connections and have a high degree of integrity and redundancy. During maintenance, the integrity of the traction return path is required to be maintained, even if a return bond becomes intentionally disconnected during track maintenance.

The maintenance of the overhead contact lines requires the application of additional earths to be applied to equipment that could become live or charged due to conduction, induced electric or magnetic fields and inadvertent re-energization. Typically, this is necessary during routine maintenance of switchgear, cable screens and following the dewirement of the overhead contact system.

11.5 Switching off supplies and isolation[24]

Switching off and isolation procedures are essential to ensure and prevent dangerous situations. In effect, this means cutting off the supply and ensuring that the means of disconnection are electrically robust and secured against inadvertent reclosure to protect persons working on the equipment from electric shock.

There are two steps involved in making electrical equipment safe: first, switching the electrical supplies off, and second, isolating and earthing the affected circuit. Isolation comprises the additional measures required to ensure the sources of supply remain disconnected, and separated, so that there can be no inadvertent reconnection or connection to any electrical supply.

UK Electricity at Work Regulations (EAWR) 1989 Regulation 12(2): Isolation of an electrical supply means of cutting off the supply and for isolation "the disconnection and separation of the electrical equipment from every source of electrical energy so that disconnection and separation are secure".

---

[24]EN 50488 *Railway Application – Fixed installations – Electrical protective measures for working on or near an overhead contact line system and/or its associated return circuit* Section 5.2 Dead Working

Section 11 – Safe working during maintenance, renewal and decommissioning

11.5.1   Functional requirement of switching off supplies

The following are national regulation requirements for switching off supplies and ensuring that the means of disconnection is secure:

(a) UK EAWR Regulation 12(1)(a): "where necessary to prevent danger, suitable means…shall be available for cutting off the supply of electrical energy to any electrical equipment".
(b) Queensland Australian Safety Regulations 2013: Part 3 Electrical Work; Division 1 Electrical Work on energised electrical equipment Regulation 15 'Duty to determine whether equipment is energised'.
(c) Australian Federal WHS Legislation: Regulation 155 provides that a PCBU must ensure that before electrical work is carried out on electrical equipment, the equipment is tested by a competent person to determine if it is energised.
(d) New Zealand Electrical (Safety) Regulation 2010 Regulation 104: Work on isolated high voltage fittings: "This regulation applies while a person is working on high voltage fittings that are (a) isolated from a supply of electricity."
(e) USA OSHA Standard 1910.333(a)(1) Deenergized parts: "Live parts to which an employee may be exposed shall be de-energized before the employee works on or near them, unless the employer can demonstrate that de-energizing introduces additional or increased hazards or is infeasible due to equipment design or operational limitations. Live parts that operate at less than 50 volts to ground need not be de-energized if there will be no increased exposure to electrical burns or to an explosion due to electric arcs."
(f) Hong Kong Code of Practice for the Electricity (Wiring) Regulations Code 4H Safety Precautions for Work on High Voltage Installation: "No person should carry out maintenance, repair, cleaning and testing on any part of high voltage electrical equipment unless such parts of the electrical equipment are: (i) dead; (ii) isolated from live conductors and all practical steps taken to lock off from live sources."

There may be a need to switch off electrical supplies for operational reasons or to prevent electrical danger. Switching off the circuit or feed can be by a direct manual operation or controlled by control circuits. The means of switching off are required to have the following functional requirements the EAWR Regulation 12(1)(a) (this text has been summarized):

- be capable of cutting off the supply under all likely conditions having regard to the equipment, its normal operation conditions, any abnormal operating or fault conditions, and the characteristics of the source(s) of electrical energy;
- be in a suitable location regarding the nature of the risks. The availability of people to operate the means and the speed at which operation may be necessary;
- be clearly marked to show its relationship to the equipment which it controls, unless there could be no doubt that this would be obvious to any person who may need to operate it; and
- only be common to several items of electrical equipment where it is appropriate for these to be energized and de-energized as a group.

11.5.2   Functional requirements of points of isolation

The following are the national regulations that are required to ensure that points of isolation remain disconnected and separated from electrical supplies:

(a) UK EAWR Regulation 12(1)(b), as noted in the Health and Safety Executive's (HSE) HSR25 *Memorandum of Guidance on the EAWR 1989*, requires that there must be available suitable means of ensuring that the supply remains switched off and to prevent inadvertent reconnection. HSR25 states: "This is isolation. This provision, in conjunction with safe working practices, will enable work to be carried out on electrical equipment without risk of it becoming live during the course of that work".

(b) Queensland Australian Safety Regulations 2013: Part 3 Electrical Work; Division 1 Electrical Work on energised electrical equipment Regulation 16: "De-energised equipment must not be inadvertently re-energised".

(c) Australian Federal WHS Legislation: Regulation 156 provides that a PCBU must ensure that electrical equipment that has been de-energised so that work can be carried out on it is not inadvertently re-energised while the work is being carried out.

(d) New South Wales Code of Practice Australia Section 5.1: Securing The Isolation, Locking Off, Tagging Systems; WHS Regulation Clause 155: 'Duty to determine whether equipment is energised' and WHS Regulation Clause 156: 'De-energised equipment must not be inadvertently re-energised'.

(e) New Zealand Electrical (Safety) Regulation 2010 Regulation 106: Notices when working on works and installations: "A person carrying out work on works or installations that are isolated from a power supply must, if there is a risk of unintentional enlivening of the works or installations, ensure that suitable notices warning against enlivening are fixed at a point where the power supply may be connected or restored. If works or installations have a locking facility for isolating them from the power supply, then any person isolating the works or installations must use that facility to lock the isolation".

(f) USA OSHA Standard 1910.333(b)(2): 'Lockout and Tagging': "While any employee is exposed to contact with parts of fixed electric equipment or circuits which have been de-energized, the circuits energising the parts shall be locked out or tagged or both in accordance with the requirements of this paragraph. The requirements shall be followed in the order in which they are presented (i.e. paragraph (b)(2)(i) first, then paragraph (b)(2)(ii), etc.)."

(g) USA OSHA Standard 1926.417: 'Lockout and Tagging.' 1926.417(a) Controls: "Controls that are to be deactivated during the course of work on energized or de-energized equipment or circuits shall be tagged." 1926.417(b) Equipment and circuits: "Equipment or circuits that are de-energized shall be rendered **inoperative** and shall have tags attached at all points where such equipment or circuits can be energized." 1926.417(c) Tags: "Tags shall be placed to identify plainly the equipment or circuits being worked on".

(h) Hong Kong Code of Practice for the Electricity (Wiring) Regulations Code 4H Safety Precautions for Work on High Voltage Installation: "No person should carry out maintenance, repair, cleaning and testing on any part of high voltage electrical equipment unless such parts of the electrical equipment are: (ii) isolated from live conductors and all practical steps taken to lock off from live sources."

The point of isolation is the location within an electrical system at which the electrical equipment is securely disconnected and separated from every source of electrical energy and inadvertent reconnection is prevented.

Points of isolation are required to have the following functional requirements[25]:

(a) have the capability to positively establish an air gap or other effective dielectric which, together with adequate creepage and clearance distances, must ensure that there is no likely way in which the isolation gap can fail electrically;

(b) include, where necessary, means directed at preventing unauthorized interference with or improper operation of the equipment, for example, means of locking off;

(c) be located so the accessibility and ease with which it may be employed is appropriate for the application;

(d) have clear marking to show which equipment it relates to, unless there could be no doubt that this would be obvious to any person who may need to operate it;

(e) be clearly marked to show which equipment it relates to, unless there could be no doubt that this would be obvious to any person who may need to operate it; and

(f) only be common to several items of electrical equipment where it is appropriate for these to be isolated as a group.

---

[25]UK EAWR Regulations 12(1)(b)

Devices such as isolators typically provide the means of disconnection and the means of separation sufficient to prevent the isolation gap from failing electrically. A mechanical or electrical mechanism (for example, padlock) for preventing any means of an inadvertent operation of a device used as a point of Isolation must also be provided.

In some cases, the equipment used to perform the requirement under Regulation 12(1)(a) may also serve to perform the requirement under Regulation 12(1)(b). The two functions of switching off and isolation are not the same, even though in some circumstances they are performed by the same action or by the same equipment.

11.5.3 Reliable visual position indication

IEC 62271-102:2018[26] states that a reliable visual position and reliable indication must be provided for all switchgear providing a point of disconnection. Where this cannot be achieved, the physical isolating gap must be visible, without the need for the removal of switchgear covers or housings. This includes outdoor overhead structure mounted isolators being visible at ground level and indoor isolators being visible via inspection windows.

11.5.4 Electrical hazards associated with disconnection and separation

Operational and maintenance hazards will be either generic or site-specific. Residual hazards from the system, sectioning, distribution and overhead contact line designs are required to be identified and recorded. Residual hazards that have been identified must be eliminated through design as far as reasonably practicable.

All residual operational and maintenance hazards should be recorded on AC isolation documentation and in any health and safety file. Where there is a hazard that relates to taking an isolation, then this required information must be included in the written procedures or permit-to-work.

11.6 Earthing – functional requirements

The following are national regulations that address the functional requirements for providing maintenance earths for 25 kV overhead lines:

(a) UK EAWR Regulation 13: "Adequate precautions shall be taken to prevent electrical equipment, which has been made dead in order to prevent danger while work is carried out on or near that equipment, from becoming electrically charged during that work if danger may thereby arise."

(b) Australian Federal WHS Legislation: Regulation 154 requires a PCBU to ensure that electrical work is not carried out on electrical equipment – including electrical equipment that forms part of an electrical installation – while it is energised (or 'live'), unless the requirements in relation to energised electrical work under the Division are met.

Regulation 155 provides that a PCBU must ensure that before electrical work is carried out on electrical equipment, the equipment is tested by a competent person to determine if it is energised.

(c) New Zealand Electrical (Safety) Regulation 2010 Regulation 104 Work on isolated high voltage fittings: "The person doing the work must ensure that the fittings are earthed before the work is commenced and that they remain earthed until the work is completed."

---

[26]IEC 62271-102:2018 *High-voltage switchgear and control gear – Part 102: Alternating current disconnectors and earthing switches.*

(d) USA OHSA: 1910.333(c)(3): The following are requirements that are required for overhead lines: "if work is to be performed near overhead lines, the lines shall be de-energized and grounded, or other protective measures shall be provided before work is started. If the lines are to be de-energized, arrangements shall be made with the person or organisation that operates or controls the electric circuits involved to de-energize and ground them. If protective measures, such as guarding, isolating, or insulating, are provided, these precautions shall prevent employees from contacting such lines directly with any part of their body or indirectly through conductive materials, tools, or equipment".

(e) Hong Kong Code of Practice for the Electricity (Wiring) Regulations Code 4H Safety Precautions for Work on High Voltage Installation: "No person should carry out maintenance, repair, cleaning and testing on any part of high voltage electrical equipment unless such parts of the electrical equipment are: (i) dead."

11.6.1   Railway earthing (grounding)

Generally, the country/state legislation specifies the technical standards which are used to support compliance with the law. In some circumstances, technical standards can be mandated by a legal instrument and in other cases they can be selected on a voluntary basis. This Section identifies those standards that are applicable with the functional earthing of the railway infrastructure.

Traction return circuit: The arrangement of the earthing and bonding of the traction return system should comply with the limits and bonding provisions as detailed in international standards and national code of practice. such as EN 50122-1, IEC 62128-1, D/MTRC/NW/DSM/ST/702/A4 (Hong Kong); General Order No176 Public Utilities Commission of the State of California *Rules for Overhead 25kV AC Railroad Electrification Systems for a High-Speed Rail System*, SNCF General Notice EF 4 D1 n°1 and 8880-900-677 Rev 1.00 (Western Australia).

Feeder switching stations and connected structures and LV systems: The earthing and bonding design for feeder switching stations, passenger stations and LV supplies should comply with IEEE Standard 80, BS 7430:2011+A1:2015, EN 50122, EN 61440 and EN 62305. The touch and step potentials must be designed to comply with the specific requirements of prescribed limits within the building or switching station compound without the necessity for any operational connection to the rail return circuit.

11.6.2   Circuit main earth

For work on high voltage (HV) power distribution circuits, isolation procedures should include the application of circuit main earths (CMEs) (primary earths). CME is an earth connection applied, to make apparatus safe to work on before a permit-to-work or sanction-for-test is issued, and which is nominated on the document. CMEs are required at points of isolation and additional earthing around the point of work. A minimum of one CME must be applied as part of any earthed isolation to any electrical section or subsection where access is required:

(a) CME is usually a remotely controlled isolator or load break switch. The CME must be positioned to give quick and efficient engineering access for scheduled railway infrastructure maintenance. They must be provided with high integrity bonding to mitigate the risk of a loss of continuity between the device and the traction return system.

(b) isolation safety earth (ISE) is a location at which CME can be provided by a portable earth to the overhead line equipment. An ISE is required to be provided where work is to be undertaken on equipment made dead such that there are no exposed live or charged parts.

(c) duplicate portable earths are required to be applied using portable earthing equipment to each separately isolated electrical section or subsection to be covered by the overhead line permit.

The earths are to be applied on each side of, and in proximity to or at the working party's isolation limits; and

(d) testing of the overhead line equipment (OLE): the OLE shall be tested once it has been switched off by using an approved voltage testing device.

11.6.3 Temporary or additional earths

These are particular requirements for railway electrification systems. These temporary earths are required to be applied to the overhead contact system to make conductors safe before maintenance work being undertaken on the overhead contact system and these requirements include circuit main earths (CMEs), remotely controlled earthing devices, and isolation safety earths (ISEs).

ISEs are also required to remove induced voltages from adjacent conductors. An isolated overhead line can become charged due to non-galvanic coupling (induction) by both electric and magnetic fields. This characteristic has been addressed analytically in Section 3 and is also undertaken in ITU-T Vol II *Directives concerning the protection of telecommunication lines against harmful effects from electric power and electrified railway lines*.

The application of temporary earths including ISEs is required in any operational and maintenance standard to ensure that overhead lines are earthed during installation, maintenance and decommissioning.

The following principles are based on UK best practices developed by British Rail in the 1970s:

"Where there are adjacent overhead contact lines that are energised or adjacent overhead power lines; Isolation Safety Earths must be applied at the worksite. Where a single additional earth is used, the additional earth must be visible from the worksite.

To control the induced voltage into de-energised overhead lines (< 100 m is the distance where the induced voltage is likely to create a hazard), additional Isolation Safety Earths are required to be applied; the distance is determined by the magnitude of the induced voltage and typically a maximum spacing of 400 m is required between Isolation Safety Earths.

Where there are energised adjacent or crossing power lines (in close proximity), then the maximum spacing of additional Isolation Safety Earths is required at intervals 400 m.

Where adjacent overhead contact lines are de-energised, and there are no adjacent or crossing power lines (in close proximity), the maximum spacing of additional Isolation Safety Earths is still required at intervals typically <2-3 km; this is due to the possibility of remote power lines and lightning."

11.7 Procedural arrangement for electrical isolation and limits of working

Precautions are required to be taken for work on equipment that is made dead. The following are national regulations that address the requirements of isolation procedural:

(a) UK EAWR Regulation 13: "Adequate precautions shall be taken to prevent electrical equipment, which has been made dead in order to prevent danger while work is carried out on or near that equipment, from becoming electrically charged during that work if danger may thereby arise".

(b) Australian Federal WHS Legislation: Regulation 161 also requires work to be carried out in accordance with a safe work method statement (SWMS) prepared for the work.

Regulation 166 requires a PCBU to ensure, so far as is reasonably practicable, that no person, plant or thing at the workplace comes within an unsafe distance of an overhead or underground electric line.

(c) Queensland Australian Safety Regulations 2013: Part 3 Electrical Work; Division 1 Electrical Work on energised electrical equipment Regulation 22 'How work is to be carried out'.

(d) New Zealand Electrical (Safety) Regulation 2010 Regulation 101: "The employer must take all practicable steps to provide safe working procedures for employees to follow when carrying out the work; and ensure that any associated equipment and personal protective equipment used by an employee is arranged, designed, made, tested, inspected, and maintained so that it is safe for the employee to use."

(e) Hong Kong Code of Practice for the Electricity (Wiring) Regulations Code 4H Safety Precautions for Work on High Voltage Installation: "No person should carry out maintenance, repair, cleaning and testing on any part of high voltage electrical equipment unless such parts of the electrical equipment are: (iii) effectively earthed at all points of disconnection of supply to such apparatus or between such points and the points of work."

Where maintenance requires that normal protection or clearance is removed, changed or access is required inside existing guarding, then additional measures are needed to prevent danger from the electrical hazards that may be exposed. Clear safety isolation procedures (instructions or house rules) are therefore required to isolate and maintain the overhead line.

Documentation, including permit-to-work, may form part of the written procedures, and their use is considered essential ensuring a safe system of work where this involves work on the conductors or equipment of high voltage power distribution systems. Correctly formulated and regulated permit-to-work procedures must address how the work is to be done and how the equipment has been made safe. The basic rules, however, are that there should be secure isolation from all power sources, and this is achieved with the following:

(a) isolator to be locked in position (for example, by a padlock (locked out and tagged)), and a sign should be used to indicate that maintenance work is in progress;

(b) isolation requires using specifically designed devices for this purpose;

(c) stored capacitive energy should be dissipated before the work starts;

(d) provision for preventing energization by train current collectors bridging section insulators or OCLS overlaps;

(e) where more than one maintenance worker is involved in the work, each of them should lock off the power with their own padlock (locked out and tagged); and

(f) before entering or starting work on isolated equipment, a suitably competent person must verify the effectiveness of the isolation and earthing.

11.7.1   Working on live, dead or charged OLE parts[27]

EN 50110-1 *Operation of electrical installations*, EN 50488 *Railway applications – Fixed installations – Electrical protective measures for working on or near an overhead contact line system and/or its associated return circuit*; and NFPA 70E *Standard for Electrical Safety in the Workplace* set out the requirements for the safe operation of and work activity on, with or near electrical installations. These requirements apply to operational, working and maintenance procedures. It applies to all electrical work activities as well as non-electrical work activities such as building work near overhead lines or underground cables.

---

[27]EN 50488 *Railway Application – Fixed installations – Electrical protective measures for working on or near an overhead contact line system and/or its associated return circuit* Section 5.3 Working near to hazardous live parts

Section 11 – Safe working during maintenance, renewal and decommissioning

Standing surface of 2.75 m

The railway earthing and bonding standard EN 50122-1, and as discussed in Section 6, gives the clearance for the standing surface of 2.75 m for touching in a straight line. This standing surface of 2.75 m is a restricted area and only applicable to locations where operational and maintenance staff have lawful access.

This provides clearance against direct contact from platform surfaces, raised platforms, etc., with live parts of an overhead contact line system (OCLS). These clearances are for basic protection and are the minimum values that shall be maintained at all temperatures and in the full range of electrical and mechanical loads of the conduct.

There are areas of the railway with OCLS that are less well controlled and are accessed by many different persons with different competencies (for example, substation compound, rolling stock and maintenance facilities), and therefore this figure needs careful application by each national railway. Where long tools are used (for example, scaffolding, tree cutting), special procedures are necessary.

Recommended distance in the air for work activities

When working within 2.75 m, additional control measures are usually required. The vicinity zone (IEC 60050 651-21-04) is the limited space outside the live working zone where specific precautions are taken to avoid encroaching into the live working area:

(a) Danger zone EN 50522:2010 and IEC 61936-1 Section 3.5.5 – Encroaching into the live area will create an electrical hazard; 'danger zone' area limited by the minimum clearance around live parts without complete protection against direct contact.

> **NOTE:** Infringing the danger zone is considered the same as touching live parts.

(b) Vicinity zone EN 50522:2010 and IEC 61936-1 Section 3.5.6 –The vicinity zone is "a limited space surrounding the 'danger zone'"; it combines air distances: electrical distance, ergonomic component and minimum working distance.

Railway infrastructure operators are required to specify the limits of their vicinity zone and danger zones. The following is an example shown in Figure 11.1 and is based on Table 3 in EN 50488:

- Danger zone typically extends up to 600 mm from any live or charged parts. Except for work that requires the use of voltage testing devices, live line tools and live line measuring devices, no other work should be attempted.

  This is broadly protecting against insulation breakdown and equivalent to 'reinforced insulation' following EN 50124-1 (insulation of hazardous-live-parts which provides a degree of protection against electric shock equivalent to double insulation).

- The vicinity zone for 25 kV AC electrification systems typically extends a distance of 1 m from the danger zone. This equates to a distance of 1.6 m from any live or charged part; this is the limit where hazards may exist and this, therefore, reflects a boundary where risk assessment is required.

Figure 11.1   Vicinity and danger zones

11.7.2   Permit-to-work[28]

The permit-to-work is a protective measure that authorizes certain people to work on a defined part of an electrical system within a specified time frame. It sets out the precautions required to complete the work safely, based on a risk assessment.

National regulations for permits-to-work include:

(a) UK EAWR Regulation 13, as noted in HSR25 *Memorandum of Guidance on the EAWR 1989*: "Written Procedures. – It may also be appropriate for the safety isolation procedures to be formalised in written instructions or house rules. 'Permits-to-work' may form part of the written procedures and their use is considered essential to ensuring a safe system of work where this involves work on the conductors or equipment of high voltage power distribution systems".

(b) Hong Kong Code of Practice for the Electricity (Wiring) Regulations Code 4H Safety Precautions for Work on High Voltage Installation: "No person should carry out maintenance, repair, cleaning and testing on any part of high voltage electrical equipment unless such parts of the electrical equipment are: (v) released for work by issue of a permit-to-work. All working on, or testing of, high voltage equipment connected to a system should be authorised by a permit-to-work or a sanction-for-test respectively following the procedures set out in Code 21D."

---

[28]IEC 60050-903:2013, 903-01017 & EN 50488 *Railway Application – Fixed installations – Electrical protective measures for working on or near an overhead contact line system and/or its associated return circuit* Section 3.1.1.15

Permits-to-work should also be issued and checked and signed off as being completed by someone competent to do so, and who is not involved in undertaking the work. A permit-to-work should include:

(a) a formal, written, safe system of work to control potentially hazardous activities. The permit must detail the work to be done, and the precautions to be taken (for instance, they may involve limiting the movement of overhead cranes, the precautions for working near live 25 kV systems).
(b) declarations from the people authorizing the work and carrying out the work. Where necessary, it requires a declaration from those involved in shift handover procedures or extensions to work.
(c) a declaration from the permit originator that it is ready for normal use before equipment or overhead lines are put back into service.

More information can be found at: https://www.hse.gov.uk/humanfactors/topics/ptw.htm

11.8 Maintenance and inspection of earthing and bonding

Regulations and codes of practice require maintenance and inspection of the earthing and bonding. BS 7430[29] requires there to be a maintenance regime that comprises a visual inspection carried out annually. This is part of the inspection and routine maintenance as part of planned maintenance and that should occur typically every five years.

National regulations for requirements for maintenance include:

(a) UK EAWR Regulation 4 (3): "Every work activity, including operation, use and maintenance of a system and work near a system, shall be carried out in such a manner as not to give rise, so far as is reasonably practicable, to danger."
(b) Queensland Australian Safety Regulations 2013: Part 10 Electrical Supply; Division 7 maintenance of works: "An electricity entity must ensure the integrity of the insulation of the relevant part of the electrical entity's works is inspected and maintained."
(c) South Australia Electricity Regulations Schedule 4 – Requirements for earthing and electrical protection systems Part 12 – Maintenance of Earthing Systems.

11.8.1 Visual inspection

This is the inspection of above-ground earth conductors, connections, guards, etc., for evidence of corrosion, decay, signs of burning, vandalism or theft. The inspection should visually check all earthing and bonding electrodes and connectors, particularly noting the condition of any bi-metallic connectors (for example, cable lugs), looking for corrosive damage and missing connections.

The inspection of a railway electrical infrastructure should typically include:

Lineside:

(a) bonding connection of the traction return circuit including rails, aerial earth conductors (AECs), buried earth conductors and OCLS masts;
(b) bonding between the rails and switching stations negative bar;
(c) bonding of any metal enclosures, metallic fencing, overline and underline bridges, metal walkway, cable trays, fire hydrants, etc.;
(d) bonding of lightning down conductors and to lightning earth pits; and
(e) bonding of auxiliary 25 kV transformers.

---

[29]BS 7430:2011+A1:2015 *Code of practice for protective earthing of electrical installations*

Passenger station:

(f) bonding of the main incoming HV earth normally connected either to the casing of a transformer, HV switchgear or within the LV distribution board;

(g) bonding between a station main earth terminal (MET) and electrical earth mats;

(h) bonding of LV passenger station MET to traction return circuit;

(i) bonding of platform screen doors (traction and LV bonding);

(j) bonding of station canopies, metallic fences and overbridges; and

(k) where the station has separated HV/LV earthing, verification that this separation is maintained.

AC switching station (feeder, mid-point, autotransformer location):

(l) bonding between traction return conductors and traction return bar;

(m) bonding between a substation MET and electrical earth mats; and

(n) where the switching station has separated HV/LV earthing, verification that the separation is maintained.

11.8.2   Routine maintenance/examination

Routine maintenance should include visual inspection of equipment and bonding and, where necessary, excavation to examine buried earth conductors, earth rods, and earth plates.

Records of inspection/examination and testing should be taken and made available for subsequent maintenance visits.

The examination should be in four parts.

This inspection under examination should cover the same items as listed in the visual inspection.

Part 1: Inspection – The inspection is only looking for obvious wear corrosion or defects. Under inspection as part of the examination, this should determine whether the conductor or connections can last until the next examination. Where any real doubt exists, sections of the conductor and/or connections should be replaced or connections remade. The results of the inspection and any remedial work should be recorded.

Part 2: Joints – A resistance measurement should be taken across all exposed accessible joints/ connections using a micro-ohmmeter. Any joint where the resistance value is excessive should be broken down, cleaned and remade or replaced.

Part 3: Bonding checks – The integrity of bonding of each item on site should be checked using a micro-ohmmeter.

Part 4: Earth resistance – As part of the routine maintenance, testing is required and should include: resistance tests across joints, checking the integrity of the bonding, measurement of the substation earth value (to be compared with the design specification), and earthing integrity of any LV segregation to the traction or third-party earth.

A measurement of the earth mat values (HV and LV where appropriate) is required and compared to the designed value.

Where the HV and LV earth are separate, the resistance between the two earths should be measured indicating the effective separation.

11.9 Maintenance and renewal of electrical equipment

Civil infrastructure is usually designed for a life of 120 years, the main steelwork for 80 years, contact wire for 30–40 years, and heavy electrical equipment (transformer, motors and switchgear, etc.) typically require replacement between 25 and 50 years. Therefore, decommissioning and renewals of installed equipment and structures will occur throughout the life cycle of the infrastructure.

During any infrastructure maintenance or renewal, including the maintenance of civil structures, the overhead line, and the return system including the running rails, the AC supply must be disconnected, earthed and made secure. In most cases, a method statement and permit-to-work (for AC equipment) are required.

11.9.1 Earthing – overhead line during maintenance

The overhead contact line system (OCLS) must maintain its vertical and horizontal position. If the spatial position of the pantograph is not correct, higher forces and kinetic energy will affect the pantograph, and if left uncorrected, the conductors and mechanical supports of both the OCLS and pantograph will wear and fatigue more quickly. Therefore, there is a requirement for maintenance and electrical isolation and earthing.

11.9.2 Earthing – overhead line during renewals

Renewal of overhead line equipment and electrical distribution equipment is an inevitable requirement during the assets' life cycle. The renewal will require isolation of the OCLS from all sources of supply, earthing and effective steps taken to ensure that it is dead and cannot inadvertently become re-energized or dangerously charged.

National regulations may require secure marking or otherwise suitably labelled equipment, circuits, switches, etc. to guard against inadvertent re-energization.

11.9.3 Bonding – track renewals or repairs

National regulations and railway codes of practice will require uninterrupted protection for passengers and maintainers during the removal or replacement of the traction return circuit.

Many railways operate 24/7 and therefore maintenance is undertaken alongside an energized or operational railway. This will require that hazards are identified and procedures are undertaken to ensure safe operation and maintenance:

(a) AC traction return circuit, of the operational railway, shall not be interrupted;
(b) adequately rated traction return circuit shall be maintained during the maintenance period;
(c) the integrity of the continuity bonding of any operational tracks (parallel) shall not be compromised;
(d) where rails act as protective earths and are required to be replaced, any bonding for the protection earth of structures and equipotential bonding shall not be compromised; this would require additional bonds to any affected infrastructure; and
(e) touch potentials at all bonded structures to be maintained within the required limits.

To undertake any track renewals, these requirements and hazards should be included within the permit-to-work and method statement.

11.9.4 AC electrification cable feeding

Where the railway uses AC cables to feed remote electrical sections, existing parallel tracks can be used as the traction return circuit. Where these rails are required to be replaced, a hazard exists and should be recorded in the hazard log (maintenance). Renewal of the rails would require a permit-to-work and method statement, including an isolation of the cable feed. Alternative continuity bonding of these rails should be provided.

The AC cable will require continuity of the return path to be maintained, and touch potentials of the cable screen and armour controlled.

11.10 Decommissioned equipment

When an overhead line is to be decommissioned, there is a hazard that the conductors could become re-energized or dangerously charged, and therefore maintenance procedures are required to ensure safe practice for the dismantlers.

UK EAWR Regulation 13, as noted in HSR25: "Before electrical equipment is decommissioned, dismantled or abandoned for any reason, it must be disconnected from all sources of supply and effective steps taken to ensure that it is dead and cannot inadvertently become re-energised or dangerously charged. It may be necessary to securely mark or otherwise suitably label equipment, circuits, switches etc., to guard against inadvertent re-energisation."

# Appendix A

Definitions and abbreviations

Generally, the definitions extracted from EN 50122-1(IEC 62128-1) and IEC 60050-442-01-21 as appropriate have been used.

Touch voltages

touch voltage (accessible) – That part of the rail potential under operating conditions which can be bridged by persons, the conductive path being conventionally from hand to both feet through the body or from hand-to-hand.

touch voltage (effective) (U_{te}) – The voltage between conductive parts when touched simultaneously by a person.

touch voltage (prospective) – The voltage between simultaneously accessible conductive parts when those conductive parts are not being touched by a person touch voltage – *see* (effective) touch voltage.

Potentials

potentials 'step potential' – The voltage between the feet of a person.

potentials 'transfer potential' – The voltage that is transferred galvanically between two remote points.

potentials 'mesh potential' – The voltage difference between two points on an earthed system.

potentials 'rise of earth potentials' – The voltage due to conductivity of earth at location of a fault indirect contact – electric contact of persons or animals with exposed-conductive-parts which have become live under fault conditions.

potential 'rail potential' (U_{RE}) – The voltage occurring between running rails and earth.

General terms for the electrical feeding of the overhead line

Autotransformer (2-phase) systems

autotransformer substation – Installation to supply a contact line system and at which the voltage of a primary supply system, and in certain cases the frequency, is transformed to the voltage and the frequency of the contact line (ref EN 50388, Section 3.16).

autotransformer feeder switching station – The location where electrical section of the route fed by individual track feeder circuit-breakers within the area supplied by the substation.

autotransformer switching station – Installation from which electrical energy can be distributed to different feeding sections or from which different feeding sections can be switched on and off or can be interconnected.

autotransformer site – Location site for autotransformers. The installation from which electrical energy can be distributed to different feeding sections of the route fed by individual track breakers within the area supplied by the substation.

Appendix A – Definitions and abbreviations

autotransformer feeder conductor – A conductor carrying the negative phase of the two-phase autotransformer supply.

feeder – An electrical conductor, like a cable or overhead line between the contact line and a substation or a switching station which is fed by a circuit-breaker (ref EN 50122-1).

Single-phase systems

Installation to supply a contact line system and at which the voltage of a primary supply system, and in certain cases the frequency, is transformed to the voltage and the frequency of the contact line. This is usually a 25 kV/132 kV transformer.

Single-phase feeder switching station – This is the location where electrical section of the route fed by individual track feeder circuit-breakers within the area supplied by the substation [EN 50119:2009, 3.3.2].

switching station (or traction switching station) – Installation from which electrical energy can be distributed to different feeding sections or from which different feeding sections can be switched on and off or can be interconnected.

traction switching station (classic) – An installation from which electrical energy can be distributed to different feeding sections or from which different feeding sections can be switched on and off or can be interconnected feeding section electrical section of the route fed by individual track feeder circuit-breakers within the area supplied by the substation [EN 50119:2009, 3.3.2].

General terms

aerial earth conductor (AEC) – A conductor electrically connecting together the steelwork of two or more overhead line structures or a number of overhead line small part steelwork assemblies and bonded to a traction return rail or to the centre tap of an impedance bond.

basic insulation – Insulation of hazardous-live-parts, which provides basic protection. NOTE: this concept does not apply to insulation used exclusively for functional purposes [IEV IEC 50050ref 195-06-06].

basic protection – Protection against electric shock under fault-free conditions [IEV ref 195-06-01].

bonding – The electrical connection of two or more conductive parts to ensure a continuous path for electric current, or that all non-continuous metallic conductors (overhead line structures, location cases, handrails, etc.) are held at the same potential.

carrier wire neutral section (CWNS) – The CWNS is made up of a number of consecutive insulated overlaps to form reliable separation between adjacent feeder station phases and eliminating the possibility of pantographs bridging between phases.

charged – An item that has acquired a charge either because it is live or because it has become charged by other means such as static or induction charging, or has retained or regained a charge due to capacitance effects even though it may be disconnected from the rest of the system.

circuit protective conductors (cpcs) – A protective conductor connecting exposed-conductive-parts to the main earthing terminal.

circuit main earth (CME) – An earth connection applied to make apparatus safe to work on before a permit-to-work or sanction-for-test is issued, and which is nominated on the document.

Appendix A – Definitions and abbreviations

Class I Equipment – Equipment in which protection against electric shock does not rely on basic insulation only, but which includes means for the connection of exposed-conductive-parts to a protective conductor in the fixed wiring of the installation [see EN 61140].

Class II Equipment – Equipment in which protection against electric shock does not rely on basic insulation only, but in which additional safety precautions such as supplementary insulation are provided, there being no provision for the connection of exposed metalwork of the equipment to a protective conductor, and no reliance upon precautions to be taken in the fixed wiring of the installation [*source* EN 61140].

common-bonding conductor (CBC) – A protective conductor capable of carrying full traction current under fault conditions providing a parallel path to circuit protective conductors (cpcs).

contact line system – A system that distributes the electrical energy to the trains running on the route and transmits it to the trains by means of current collectors.

continuity bonds – A bond provided across each electrical discontinuity in the traction return rails (e.g. at points and crossings, trap points, expansion joints, redundant insulated rail joints, movable bridges and rail weighbridges).

cross-bond (traction) – A bond between the traction return rails of the same track or adjacent tracks.

cross-bonded area – Areas where the OLE longitudinal earthing conductors are bonded together and to traction return rails (with or without impedance bonds) of all tracks.

danger/hazard zone [IEV IEC 50050 ref 903-01-03] – A hazard zone is any space within and/or around a product, process or service in which persons, or livestock can be exposed to a hazard. The space around live or charged parts in which the insulation level to prevent electrical danger is not assured when reaching into or entering it without protective measures.

dead – A conductor that is neither live nor charged and danger is prevented while work is carried out. Where the voltage impressed on a conductor is controlled such that it remains within the permissible limits specified in EN 50122, the conductor is considered dead.

direct contact – Electric contact of persons or animals with live parts [IEV IEC 50050 ref 195-06-03-modified] or sufficiently close that danger may arise.

disconnected – Equipment (or a part of an electrical system) that is not connected to any source of electrical energy.

disconnection – A contact opening in a pole so as to ensure the equivalent of basic insulation between live parts and those parts intended to be disconnected.

double insulation – Insulation comprising both basic insulation and supplementary insulation [IEV IEC 50050 ref 195- 06-07].

earth – Ground or earth is a reference point in an electrical circuit from which voltages are measured, a common return path for electric current, or a direct physical connection to the earth.

electric shock – A dangerous physiological effect resulting from the passing of an electric current through the human body or livestock [IEV IEC 50050 ref 195-01-04].

Appendix A – Definitions and abbreviations

Earthing (grounding) system – An earthing system or grounding system connects specific parts of an electric power system with the ground, typically the Earth's conductive surface, for safety and functional purposes. The choice of earthing system can affect the safety and electromagnetic compatibility of the installation.

earth wire – A conductor electrically connecting together the steelwork of two or more overhead line structures or a number of overhead line small part steelwork assemblies and bonded to a traction return rail or to the centre tap of an impedance bond.

earth electrode – An earth rod or tape often laid in a grid format making an earth mat.

earthing conductor – A protective conductor connecting the main earthing terminal of an installation to an earth electrode or to other means of earthing.

earthed isolation – The entire process of disconnection, separation, providing securely isolated equipment, earthing and the issue of relevant safety documentation.

earthed – A conductor that is connected to the general mass of Earth by conductors of sufficient strength and current-carrying capability to discharge electrical energy to earth.

exposed-conductive-part – A conductive part which can be touched and which is not normally live, but which may become live under fault conditions [source IEV IEC 50050: 442-01-21], such as exposed metalwork, exposed metal parts, exposed metal surfaces or exposed metal services and is within the overhead contact line zone or the pantograph zone. Excludes speed restriction signs, drain covers, metal stakes securing troughing and sections of fencing shorter than 3 m.

earth potential rise (EPR) – Occurs when a large current flows to earth through an earth grid impedance. The potential relative to a distant point on the Earth is highest at the point where current enters the ground, and declines with distance from the source. Ground potential rise is a concern in the design of electrical substations because the high potential may be a hazard to people or equipment.

electromotive force (EMF) – The characteristic of any energy source capable of driving electric charge around a circuit. It is abbreviated E in the international metric system but also, popularly, as emf.

emergency switch-off – The opening operation of a switching device intended to remove electrical power from an electrical installation to avert or alleviate a hazardous situation.

extraneous-conductive-part – A conductive part not forming part of the electrical installation and liable to introduce an electric potential, generally the electric potential of a local earth [IEV IEC 50050 ref 195-06-11].

fast transient earth (FTE) - An earthing system of low impedance to carry high frequency currents (e.g. lightning, switching surges) to earth.

feeder – Electrical conductor like a cable or overhead line between the contact line and a substation or a switching station which is fed by a circuit-breaker.

functional earth – To provide an earth to ensure the correct operation of a device, usually a common reference voltage between remote items of equipment. Earthing of a point or points in a system or in an installation or in equipment, for purposes other than electrical safety, such as for proper functioning of electrical equipment.

functional insulation – Insulation between conductive parts, necessary for the proper functioning of the equipment [IEV IEC 50050 ref 195-02-41].

Appendix A – Definitions and abbreviations

functional supply point – The point that the power supply distribution system connects with an item of supplied equipment and may be co-located with signalling equipment.

high integrity bonding – Bonding that provides an appropriate level of resilience through a combination of multiple bonding connections, diverse routing and appropriate mechanical protection measures. Each connection must be rated to carry the current for the duration of the backup protection to operate (without annealing), while maintaining voltage levels to EN 50122-1.

Global System for Mobile Communications-Railway (GSM-R) – An international wireless communications standard for railway communication and applications. A sub-system of the European Rail Traffic Management System (ERTMS), it is used for communication between train and railway regulation control centres.

indirect contact – Electric contact of persons or animals with exposed conductive parts which have become live under fault conditions [source IEV IEC 60050-826-12-04].

impedance bond – A device which, while allowing the traction return current to flow freely, so impedes the flow of track circuit current as virtually to isolate two track circuits, one from another. On tracks where both rails are traction return rails and are equipped with double-rail track circuits, all bonding to the traction return rails shall be made via impedance bonds. Duplicate bonds shall be provided between each traction return rail and the impedance bond.

isolated – Equipment (or part of an electrical system) which is disconnected and separated by a safe distance (the isolating gap) from all sources of electrical energy in such a way that the disconnection is secure (i.e. it cannot be re-energized accidentally or inadvertently.

isolation – The disconnection and separation of electrical equipment from every source of electrical energy in such a way that this disconnection and separation is secure.

isolation safety earth (ISE) – The location at which the overhead line equipment can be earthed, for the purpose of issuing safety documentation.

live – Equipment that is at a voltage by being connected to a source of electricity.

main earthing terminal (MET) – The terminal or bar provided for the connection of protective conductors, including protective bonding conductors, and conductors for functional earthing, if any, to the means of earthing.

metallic service – A service having an exposed metallic surface (e.g. a gas pipe, water pipe, conduit, or metal sheathed cable).

overhead contact line (OCL) – Contact line placed above (or beside) the upper limit of the rail vehicle gauge and supplying vehicles with electric energy through roof-mounted current collection equipment [IEV IEC 50050 ref 811-33-02].

overhead contact line system (OCLS) – Contact line system using an overhead contact line to supply current for use by traction units.

overhead contact line zone (OCLZ) and current collector zone (CCZ) - The zones whose limits are not exceeded in general, by a live, broken overhead line conductor or by a live, broken or dewired pantograph respectively.

Appendix A – Definitions and abbreviations

operation and maintenance (O&M) – The functions, duties and labour associated with the daily operations and normal repairs, replacement of parts and structural components, and other activities needed to preserve an asset so that it continues to provide acceptable services and achieves its expected life.

point of isolation – The point at which separation of the electrical equipment from every source of electrical energy is achieved in such a way that disconnection and separation is secure.

principal supply point (PSP) – The point of common connection of one or more electrical sources from which energy may be distributed.

protective bonding conductor – Protective conductor provided for protective equipotential bonding.

protective conductor – A conductor used for some measures of protection against electric shock as defined in BS 7671.

protective earth (PE) – To provide protection against electric shock in the event that a conductive case of an item of equipment may become live in respect to earth in the event of a fault.

protective earthed neutral (PEN) conductor – A single conductor that has the combined function of providing the neutral and protective earth conductor in a TN-C-S earthing arrangement. A PEN conductor is normally, but not exclusively, used with an LV PME supply service earthing system [Wiring Regulations BS 7671].

rail-to-rail cross-bonds – Rails of each track not equipped with track circuits shall be bonded together at buffer stops, toes of points and at intermediate locations.

rail-to-earth resistance, rail leakages - Electrical resistance between the running rails and the earth or structure earth.

reinforced insulation [IEV IEC 50050 ref195-06-09] – Insulation of hazardous-live-parts, which provides a degree of protection against electric shock equivalent to double insulation.

reinforcing feeder – Overhead conductor mounted adjacent to the overhead contact line, and directly connected to it at frequent intervals, in order to increase the effective cross-sectional area of the overhead contact line.

return circuit – All conductors which form the intended path for the traction return current and the current under fault conditions [EN 50122-1:2011+A1:2011].

return conductor – Conductor paralleling the track return system and connected to the running rails at periodic intervals [EN 50122-1:2011+A1:2011].

signalling right side failure – A mode of failure, which causes a piece of signalling equipment to cease functioning without compromising the safety of trains.

signalling wrong side failure – A wrong side failure occurs when signalling equipment or a system does not fail safe. In other words, a failure occurs which could lead to an accident.

section insulator – Sectioning point formed by insulators inserted in a continuous run of a contact line, with skids or similar devices to maintain continuous electrical contact with the collector.

structure bond – A bond connecting the steelwork of an overhead line equipment structure, or bridge, or other metal structure, to the traction return circuit.

Appendix A – Definitions and abbreviations

structure earth – Construction made of metallic parts or construction including interconnected metallic structural parts, which can be used as an earth electrode.

structure outdoor mounted switchgear (SMOS) – Switchgear that is mounted on overhead line structures.

traction substation (classic) – Installation to supply a contact line system and at which the voltage of a primary supply system, and in certain cases the frequency, is transformed to the voltage and the frequency of the contact line.

traction switching station (classic) – An installation from which electrical energy can be distributed to different feeding sections or from which different feeding sections can be switched on and off or can be interconnected feeding section electrical section of the route fed by individual track feeder circuit-breakers within the area supplied by the substation [EN 50119:2009, 3.3.2].

traction earth – The distributed earth system formed by the traction return rails intentionally connected to the general mass of Earth by the foundations of the overhead line structures.

transposition bond (traction) – A bond connecting two traction return rails where the traction return rail changes from one side of the track to the other.

track-to-track cross-bonds – Bonding provided to connect together all traction return rails at each of the following locations; switching stations (with and without SMOS), feeder stations (bonds are duplicated), at intermediate locations, at buffer stops and beyond the last overhead line structure on part-electrified tracks.

traction return system – A return circuit of all conductors which form the intended path for the traction return current.

traction return rail – A running rail or rails intentionally carrying traction return current.

vicinity zone [IEV IEC 50050 ref 651-21-04] – The limited space outside the live working zone where specific precautions are taken to avoid encroaching into the live working.

voltage - extra-low – Not exceeding 50 V AC whether between conductors or to earth.

voltage limiting device (VLD) – Protective device whose function is to prevent existence of an impermissible high touch voltage.

voltage - low – Exceeding extra-low voltage but not exceeding 1,000 V AC between conductors, or 600 V AC between conductors and earth.

voltage - high – Exceeding low voltage but not exceeding 66 kV AC between conductors, or 40 kV AC between conductors and earth.

Cable insulation and sheath terminology

chlorinated polyethylene (CPE)

ethylene propylene rubber (EPR)

ethylene propylene diene monomer (EPDM)

low smoke zero halogen (LSZH)

polyethylene (PE)

thermoplastic polyvinyl chloride (PVC)

thermosetting cross-linked polyethylene (XLPE)

Network Rail terminology

overhead line equipment (OLE) – The name given by railway engineers to the assembly of masts, gantries and wires found along electrified railways. The purpose of all of this steel and cable is to supply power to make electric trains move.

designated earthing point (DEP) – A location used to apply an earth to the overhead line equipment to support requirements for routine maintenance.

fast transient earth – A single stake earth rod (for the purposes of this document) of intended low impedance via short length connections.

red bonds – Each bond which, if disconnected, could result in either the bond itself or the equipment to which it is connected rising to a voltage in excess of the accessible or touch voltage values quoted in the relevant specifications.

yellow bonds – Bonds required by the signalling system design for track circuit integrity provided at the specified locations and designated yellow bonds.

signal rail – A track circuit arrangement where only one rail (the signal rail) is used with IRJs to separate the track circuits.

traction return rail – An electrically continuous rail used for traction return purposes.

 **Appendix B**

Regulations and standards

European standards

EN 50119 *Railway applications -Fixed installations -Electric traction overhead contact lines*

EN 50122-1;2017 IEC 62128-1 *Railway applications fixed installations, protective provisions relating to electrical safety and earthing*

BS EN 50122-1 *Railway applications. Fixed installations. Electrical safety, earthing and the return circuit - Part 1. Protective provisions against electric shock*

EN 50124-1+A2 (2017) *Railway applications - Insulation coordination - Part 1: Basic requirements - Clearances and creepage distances for all electrical and electronic equipment*

EN 50124-2 *Railway applications - Insulation coordination - Part 2: Overvoltages and related protection*

EN 50125-2 *Railway applications - Environmental conditions for equipment - Part 2: Fixed electrical installations*

EN 50126 (series) *Railway applications — The specification and demonstration of Reliability, Availability, Maintainability and Safety (RAMS)*

EN 50149:2012 *Railway applications — Fixed installations — Electric traction — Copper and copper alloy grooved contact wires*

EN 50152-1+A1 (2013) *Railway applications - Fixed installations - Particular requirements for alternating current switchgear Part 1: Circuit-breakers with nominal voltage above 1 kV*

EN 50152-2 *Railway applications - Fixed installations - Particular requirements for alternating current switchgear Part 2: Disconnectors, earthing switches and switches with nominal voltage above 1 kV*

EN 50152-3-1 *Railway applications - Fixed installations - Particular requirements for a.c. switchgear - Part 3-1: Measurement, control and protection devices for specific use in a.c. traction systems - Devices*

EN 50152-3-2 *Railway applications - Fixed installations - Particular requirements for a.c. switchgear - Part 3-2: Measurement, control and protection devices for specific use in a.c. traction systems - Current transformers*

EN 50152-3-3 *Railway applications - Fixed installations - Particular requirements for a.c. switchgear - Part 3-3: Measurement, control and protection devices for specific use in a.c. traction systems - Voltage transformers - Voltage Transformers*

EN 50163+A1 *Railway applications - Supply voltages of traction systems*

EN 50238 IEC 62427 *Railway Applications Fixed Installations Electronic power convertors for substations*

EN 50310 *Telecommunications bonding networks for buildings and other structures 2016*

Appendix B – Regulations and standards

EN 50329 *Railway applications – Fixed installations – Traction transformers*

EN 50388 *Railway Applications — Power supply and rolling stock — Technical criteria for the coordination between power supply (substation) and rolling stock to achieve interoperability*

EN 50488 *Railway applications. Fixed installations. Electrical protective measures for working on or near an overhead contact line system and/or its associated return circuit*

EN 50443 *Effects of electromagnetic interference on pipelines caused by high voltage a.c. electric traction systems and/or high voltage a.c. power supply systems*

EN 50522 *Earthing of power installations exceeding 1 kV a.c.*

EN 50562 *Railway applications - Fixed installations - Process, protective measures and demonstration of safety for electric traction systems*

EN 50633 *Railway applications - Fixed installations - Protection principles for AC and DC electric traction systems*

EN 60060-1 *High-voltage test techniques - Part 1: General definitions and test requirements*

EN 60060-2 *High-voltage test techniques - Part 2: Measuring systems*

EN 60060-3 *High-voltage test techniques - Part 3: Definitions and requirements for on-site testing*

EN 61000-5-1 *Electromagnetic compatibility (EMC) - Part 5: Installation and mitigation guidelines - Section 1: General considerations - Basic EMC publication*

EN 61000-5-2 *Electromagnetic compatibility (EMC) - Part 5: Installation and mitigation guidelines - Section 2: Earthing and cabling*

EN 61140 *Protection against electric shock — Common aspects for installation and equipment*

EN 62305-1 *Protection against lightning Part 1: General principles*

EN 62305-2 *Protection against lightning Part 2: Risk management*

EN 62305-3 *Protection against lightning Part 3: Physical damage to structures and life hazard*

EN 50124:2005 *Railway applications. Insulation coordination. Basic requirements. Clearances and creepage distances for all electrical and electronic equipment*

HD 60364-1 *Low-voltage electrical installations - Part 1: Fundamental principles, assessment of general characteristics, definitions*

HD 60364-41-4-41 *Low-voltage electrical installations – Part 4-41: Protection for safety – Protection against electric shock*

Appendix B – Regulations and standards

IEC Standards

IEC 60099-4 *Surge arresters - Part 4: Metal-oxide surge arresters without gaps for a.c. systems*

IEC 60099-5:2018 *Surge arresters - Part 5: Selection and application recommendations*

IEC 60479-1:2005 *Effects of current on human beings and livestock - Part 1: General aspects*

Technical Specification 41–24. *Guidelines for the design, installation, testing and maintenance of main earthing systems in substations* Energy Networks Association, 1992.

IEC 60364-1 *Low-voltage electrical installations – Part 1: Fundamental principles, assessment of general characteristics, definitions*

IEC 60364-41-4-41 *Low-voltage electrical installations – Part 4-41: Protection for safety – Protection against electric shock*

IEC 62128-1 *Railway applications – Fixed installations – Electrical safety, earthing and the return circuit – Part 1: Protective provisions against electric shock*

IEC 62128-3:2013 *Fixed installations - Electrical safety, earthing and the return circuit - Part 3: Mutual interaction of a.c. and d.c. traction systems*

IEC 61936-1 *Power installations exceeding 1 kV AC and 1,5 kV DC - Part 1: AC*

ITU-T Standards

Recommendation K-27 *Bonding configuration and earthing inside a telecommunication building* ITU-T, 1996.

ITU-T Vol VI *Directives concerning the protection of telecommunication lines against harmful effects from electric power and electrified railway lines -Volume VI Danger, damage and disturbance*

ITU-T Vol II *Directives concerning the protection of telecommunication lines against harmful effects from electric power and electrified railway lines -Volume II: Calculating induced voltages and currents in practical cases*

ITU-T Vol VI *Directive concerning the protection of telecommunication lines against harmful effects from electric power and electrified railway lines*

United Kingdom

Regulations: The UK Electricity at Work Regulations 1989;

Appendix B – Regulations and standards

UK ENA

ENA *Engineering Recommendation G12 issue 4 2015.*

Engineering Recommendation *EREC S34 Issue 2, November 2018 A guide for assessing the rise of earth potential at electrical installations*

UK Main Line Railway

GE/RT8270 *Assessment of Compatibility of Rolling Stock and Infrastructure Iss 3 2015*

GL/RT/1210 *AC Energy Subsystem and Interfaces to Rolling Stock Subsystem Iss 2 2019*

GL/GN1610 *Guidance on AC Energy Subsystem and Interfaces to Rolling Stock Subsystem*

BR 13422 *Iss 1 50Hz Single Phase AC Electrification, Immunisation of Signalling and Telecommunications Systems Against Electrical Interference*

Network Rail NR/L2/ELP/21085 *Earthing and Bonding on A.C. Electrified Railways*

RIS-1855-ENE *Low Voltage Power Supplies in Electrified Areas*

RIS-1800-ENE *Rail Industry Standard for Network and Depot Interface Management - Isolation Documentation*

RIS-0725-CCS *Electromagnetic Compatibility of Train Detection Infrastructure with Rail Vehicles*

UK Standards for earthing

BS 7671 *Requirements for Electrical Installations. IEE Wiring Regulations 18th Edition* BSI, 2018

BS 7430:2011+A1:2015 *Code of practice for protective earthing of electrical installations*

UK High Speed 1 (HS1)

000-GDS-LCEEN-00041-05 *2x25kV Earthing and Bonding Principles*

United States Standards and Regulations

USA Occupational Safety and Health Administration: Title: Selection and use of work practices: Subpart:1910 Subpart S; Subpart Title: Electrical; Standard Number: 1910

USA Occupational Safety and Health Administration: Title: Safety and Health Regulations for Construction Subpart:1926 Subpart K; Title- Electrical; Standard Number 1926

Appendix B – Regulations and standards

American National Standards Institute (ANSI):

J-STD-607 A*Commercial Building Grounding (Earthing) and Bonding Requirements for Telecommunications*

California Public Utilities Commission – General Orders (CPUC):

GO-95 – 2012b *Rules for Overhead Electric Line Construction STATE OF CALIFORNIA*

General Order No176 *Public Utilities Commission of the State of California. Rules for Overhead 25kV AC Railroad Electrification Systems for a High-Speed Rail System*

PROPOSED RESOLUTION Agenda ID# 15207 *Requirement for Caltrain 25kVAC Railroad Electrification System*

National Fire Protection Association (NFPA):

NFPA 70*National Electrical Code (NEC)*

NFPA 780 (2011) *Standard for the Installation of Lightning Protection Systems*

Underwriters Laboratories Inc. (UL):

UL 467*Grounding and Bonding Equipment*

AREMA American Railway Engineering and Maintenance-of-Way Association

Part 7 Traction Electrification System Grounding & Bonding.

Institute of Electrical and Electronic Engineers (IEEE):

IEEE 81 *Guide for Measuring Earth Resistivity, Ground Impedance, and Earth Surface Potential of a Grounding System*

IEEE 80 *IEEE Guide for Safety in AC Substation Grounding*

142[TM] *IEEE Recommended Practice for Grounding of Industrial and Commercial Power Systems*

IEEE 998 *IEEE Guide for Direct Lightning Stroke Shielding of Substations*

Electric transmission line fundamentals Edwin M Anderson 1987

Appendix B – Regulations and standards

Hong Kong Standards

Code of Practice for the Electricity (Wiring) Regulations 2015 Edition

Section 7.2A: Electrical and Mechanical Systems - Signalling System (Non-FOA) D/MTRC/NW&MARW/DSM/ST/700/A2.

Section 7.2B: Electrical and Mechanical Systems - Signalling System (FOA) D/MTRC/NW&MARW/DSM/ST/700/A2.

Section 7.5.8: Electrical and Mechanical Systems – Earthing and Stray Current Collection D/MTRC/NW&MARW/DSM/ST/700/A2.

Section 7.5: Electrical and Mechanical Systems – Power Supply Systems D/MTRC/NW&MARW/DSM/ST/700/A2.

Hong Kong Electricity (Wiring) Regulations (Cap. 406, section 59) 1 June 1992.

French Standards

EDF4D1n°1 *Ligne Electrifiees En Courant Alteratif Monophase (Single Phase Alternating Electrified Line)* SNCF

China Standards

TB10621-2014 P J1942-2014 *Code for Design of High-speed Railway*

Australian/New Zealand Standards and Regulations

ENA EG1-2006 *Substation Earthing Guide*

AS/NZS 60479.1:2010 *Effects of current on human beings and livestock*

AS/NZS 4853:2012 *Electrical hazards on metallic pipelines*

AS/NZS 3000:2018 *Standard for Wiring Rules*

AS/NZS 3835:2006 *Earth potential rise—Protection of telecommunications network users, personnel and plant*

AS/NZS 4853:2012 *Electrical hazards on metallic pipelines*

AS 1768 *Lightning protection*

AS 2067:2016 *Substations and high voltage installations exceeding 1 kV a.c.*

Appendix B – Regulations and standards

AS 3008 *Electrical installations - Selection of cables Cables for alternating voltages up to and including 0.6/1 kV - Typical Australian installation conditions*

ENA EG1 *Electrical Network Association – Substation Earthing Guide*

8880-900-677 Rev 1.00 *Western Australia Earthing and Bonding in the 25 kV AC Electrified Area*

AS 7708 *Signalling Earthing and Surge Protection*

Australia Federal Work Health and Safety Regulations 2011

Australia Queensland Legislation Electrical Safety Regulation 2013

Australia NSW Code of Practice Managing Electrical Risk in the Workplace

Southern Australia Electricity Regulations 2012

New Zealand Electrical (Safety) Regulation 2010

Index

Index